Mutation research

Mutation research

Problems, results and perspectives

CHARLOTTE AUERBACH F.R.S.

Professor Emeritus, Institute of Animal Genetics,
University of Edinburgh

LONDON
CHAPMAN AND HALL

A Halsted Press Book
John Wiley & Sons, Inc., New York

First published 1976
by Chapman and Hall Ltd
11 *New Fetter Lane, London EC4P 4EE*

Printed in Great Britain by
Redwood Burn Ltd,
Trowbridge and Esher

ISBN 0 412 11280 9

Distributed in the U.S.A. by Halsted Press,
a Division of John Wiley & Sons, Inc., New York

Library of Congress Cataloging in Publication Data

Auerbach, Charlotte.
Mutation research.

1. Mutation (Biology) I. Title. [DNLM:
1. Mutation. QH460 A917m]
QH460.A93 1976 575.2'92 75-17592
ISBN 0-470-03670-2

This book
is dedicated to the memory
of H.J. Muller

Contents

Illustrations

TABLES

Preface

This book is intended for the senior undergraduate (Honours student) in genetics, and for the postgraduate who wants a survey of the whole field or information on a special area within it. In order to cater for readers with such different requirements, I have made the list of references unusually large for a textbook. It includes classical papers as well as very recent ones (to the end of 1974); reviews as well as specialized articles; elementary expositions from Scientific American as well as highly technical papers from journals on genetics and molecular biology. In areas of active research, I have given preference to the latest references, which will lead the reader to earlier ones. In addition to the references at the end of each chapter, a bibliography at the end of the book lists relevant books and general reviews.

Apart from the first chapter, the book is not written as a history of mutation research; but throughout I have tried to emphasize the continuity of the problems, concepts and ideas. The reader will find many examples of this. Muller's once famous and then almost forgotten classification of genes by their action has now been given biochemical reality by studies of gene action *in vitro*. The problem of whether mutations can arise in non-replicating genomes is one of the oldest in mutation research; yet an unequivocal solution was obtained only recently with bacteriophage. Bateson's presence-absence theory of mutation took long to die, especially in relation to X-rays, but was finally refuted by nucleotide sequencing. While some of the old problems have been solved by new techniques, many of them are still with us, waiting for new methods of attack. I hope that some readers will be stimulated to go back to the classical papers, e.g. by Muller and Stadler. They will find not only beautiful work beautifully presented, but also discussions of fundamental problems in great depth and breadth.

A second truth that I want to drive home is that, in their fundamental responses to a natural or artificial mutagenic stimulus, the genes of all organisms from virus to man resemble each other. The early mutation workers never underestimated the importance of this truth, independent of whether – like Demerec – they experimented on a variety of organisms or – like Muller – remained committed to one (*Drosophila*). Muller, indeed, was the first to suggest that mutagenesis might be profitably studied in the 'naked' genes of bacteriophage. His review articles were of general validity for their time because they were based on all available evidence from macro- and micro-organims. In our age of specialization, this is no longer so. The field is now so large and diversified that it has become impossible for any one mutation worker to remain well informed about developments outside his special area. Unfortunately, however, these areas are defined by organism rather than by problem. Mutation workers consider themselves specialists on bacteriophage, *E.coli,* yeast, *Neurospora, Drosophila, Vicia faba* etc; not as being committed to the analysis of a special *problem*, even though their experiments may be restricted to one *organism*. This attitude results in general conclusions based on specialized evidence. Such conclusions may be, and often are, wrong; but they may have a long life span thanks to the same compartmentalization of interest that first gave birth to them. Wherever possible, I have tried to draw cross-connections between observations on different organisms. They are not difficult to trace: supersuppressors, unstable genes, mutagen specificity, photorepair and dark repair are among many phenomena that are encountered, and can be analysed, in a variety of prokaryotic and eukaryotic species. The freedom to choose the most suitable species for answering a special problem should stimulate cooperation between mutation workers. To be sure, there are differences between organisms in their mutational behaviour. Set against the background of an essential similarity at the primary level of DNA, these may become important tools for analysing the roles of chromosome structure, metabolic patterns and other cellular and organismal factors in mutagenesis.

This brings me to the final, and principal, point that I have stressed throughout this book: the need to realize that mutagenesis is much more than a physicochemical reaction of environmental agents with DNA. It is a biological process and, like all other biological processes, it is deeply enmeshed in the structural and biochemical complexities of the cell. A little more than ten years ago, after the spectacular success of the molecular analysis of mutational changes in DNA, mutation research seemed to be in danger of becoming relegated to a branch of nucleic acid chemistry. Indeed, I remember a very famous geneticist saying to me at that time that the best way to study mutagenesis was to treat DNA in the test tube with a mutagen. If this were true, mutation research would now be more or less at an end; but fortunately for the research worker it is far from true. This naive picture, which never gained the allegiance of biologically-minded geneticists, broke down under the impact of findings that implicate cellular processes such as repair and replication in mutagenesis. These findings have started a new era of mutation research, with many new problems being added to the old and still unsolved ones.

In a previous book ('Mutation Methods', see Bibliography), I have discussed the rationale and basic protocols of the most important techniques in mutation research. In the present book, I have incorporated only as much methodology as seemed indispensable for an understanding of the results and conclusions. I know from teaching experience that it is tedious for the student to go through a course on methodology before he knows the particular problems that the methods are meant to solve. I have therefore inserted sections, and one whole chapter, on methods wherever it seemed required. It may be useful to list them here. Chapter 2 brings the genetic test for translocations in *Drosophila*, and the generally applicable specific-locus test for the detection of visible mutations. The best known *Drosophila* tests – the test for sex-linked lethals and the attached-X tests for sex-linked visibles and large deletions – are described in Chapter 5; the same chapter brings also the extension of the sex-linked lethal test to the detection of delayed mutations. Chapter 10 includes methods for mutation research on flowering plants, especially maize. Chapter 11 is wholly devoted to methods of mutation research on micro-organisms and cell cultures.

I owe thanks to many who have helped me with the writing of this book. First of all to my publishers, who have been understanding enough not to hustle me beyond my customary slow speed of writing. Then to their, anonymous, referees both of whom have made very useful remarks and have rescued the book from important omissions, mis-statements and ambiguities. Several colleagues here and abroad have critically and helpfully read sections of the book, or have supplied me with information; to all of them I am very grateful. Most of all I owe thanks to my colleague, Dr. D.J. Bond, of this institute, who has read very carefully every word of the book. I am sure that without his criticism, the book would contain many more obscurities; he is not responsible for those that remain. I am very grateful to Professor D.S. Falconer for letting me use the facilities of the Genetics Department. Mr E.D. Roberts has been very helpful and ingenious in making a number of original drawings. Mrs Mardi Denell, Dr Clara Queiroz and Mrs Jacqueline Stewart have saved me much time and effort by collecting references, tables and illustrations, and Mrs Valerie Rennie has been a careful and intelligent typist. For proof-reading and the compilation of the index and glossary I have had the competent help of Dr Joan Dawes. To all these helpers I extend thanks. I also wish to thank the Leverhulme Trust for generously financing the work of these assistants through a Leverhulme Emeritus Fellowship grant.

Institute of Animal Genetics
University of Edinburgh

January 1975 C.A.

CHAPTER ONE

The history of mutation research

In this introductory chapter, I want to present a brief review of the history of mutation research. I shall emphasize, not so much individual discoveries and their originators, as the problems that have instigated and dominated research in successive periods. This will set the stage for the rest of the book. We shall follow the shifting pattern of problems and of means to attack them from the earliest beginning of mutation research to the present day. We shall see that a few problems have been solved; that others have remained with us, although often in modified guise; that new problems keep on arising from new findings. Without this last feature, mutation research would no longer be a living branch of science. We shall see that, judged by its unsolved and challenging problems, it is still very much alive.

The history of mutation research may be divided into five periods. During the *first period* – from 1900 to 1927 – the concepts of mutation and mutation rates were developed, the basic questions about the nature of mutation were formulated and techniques for measuring mutation rates were worked out. The *second period* started in 1927 with the discovery of the mutagenic action of X-rays. This provided a uniquely suitable tool for probing into the nature of mutation. By the end of this period, in the late thirties, a unified theory of mutation had been constructed, the so-called 'target theory'. It has found its expression in books by three physicists: two comprehensive and monumental treatises by Timoféeff-Ressovsky and Zimmer and by Lea, and one short brilliant book by Schrödinger, who looked at mutation from the quantum-mechanical point of view (see Bibliography at the end of the book).

The *third period* started shortly before the Second World War when the discovery of chemical mutagens raised new problems, and the introduction of micro-organisms provided new means of experimental analysis, especially for the mutagenic action of ultraviolet light. Most of the findings made in the third period demanded a chemical model of the gene for interpretation. Since no such model was available, the whole period was characterized by excitement in the experimental field coupled with frustration in the theoretical approaches. This changed in 1953, when Watson and Crick, basing themselves on the proved role of DNA as carrier of genetic information (1,2) and on the crystallographic studies by Wilkins and Rosalind Franklin (3,4), proposed the model of the double helix (5), which accounted for the fidelity of gene replication and held out the promise that mutations would be explicable in terms of nucleic acid chemistry. This discovery inaugurated the *fourth period* of mutation research, which was dominated by studies on the chemistry of deoxyribonucleic acid. It has had a highly successful run and continues to provide information on the action of mutagens on DNA. Meanwhile, however, there has been a growing realization that cellular mechanisms, such as repair and expression, play important roles in the mutation process. During the last few years, the study of these cellular steps in mutagenesis has been pushed more and more into the foreground of interest. This has started the present *fifth period* in which the physical and chemical knowledge gained during the preceding ones is used as an indispensable basis for analysing mutation as a biological process.

In this introductory chapter, I want to characterize very briefly each of these periods by its main achievements and its main problems. In later chapters, these achievements and problems will be dealt with in greater detail but no longer in historical order.

The first period (1900-1927)

One of the re-discoverers of Mendel's laws, the Dutch botanist Hugo de Vries, coined the term 'mutation' for sudden hereditary changes In *Oenothera lamarckiana,* the evening primrose (6). Although it later turned out that these changes were due not to mutation but to polyploidy, polysomy or rare recombination events in a very unusual karyotype, the term has been preserved for changes in the quality, quantity and arrangement of genes. We shall see later that, in a different sense, the *relation between recombination and mutation* has become a latter-day problem. In de Vries' time, no such conceptual relationship was possible because recombination was still undiscovered. Several other basic problems were, however, already posed in these early days. One concerned the *nature of mutation.* Bateson (7) put forward the presence-absence theory, according to which all mutations are due to losses of normal genes. Although it is obvious that evolution cannot have been due to a succession of loss mutations, refutation of the presence-absence theory by experiment was extraordinarily difficult. The two main observations that contradicted it, multiple alleles and reverse mutations, could be fitted into the theory by additional assumptions. Conclusive evidence against the theory has been obtained only recently through analysis at the molecular level. The other problem concerned the *time of origin of mutations.* While some investigators thought that mutations always are due to 'errors' of gene replication, others believed that non-replicating genes can also mutate. Although there is now good evidence that the latter point of view is correct, mutation in resting genes is still difficult to explain in molecular terms.

In 1910, Muller joined the group of brilliant young people who, under T.H. Morgan, proved the chromosome theory of inheritance by experiments on *Drosophila.* Muller became interested in the nature and origin of the mutations that provided the tools for these remarkable investigations. He realized that the nature of mutation was inextricably linked with the nature of the gene. In an address given in 1921 (8) he stressed that the gene has one unique property without which evolution could not have taken place. This is its property of self-propagation in the sense that 'when the structure of the gene becomes changed, ... the catalytic property of the gene may become correspondingly changed in such a way as to leave it still *auto*catalytic'. He inferred that there must be 'some general feature of gene construction – common to all genes – which gives each one a *general* autocatalytic power – a 'carte blanche' to build up material of whatever specific sort it itself happens to be composed of'. It is this *central problem of mutation* that Muller tried to approach by genetical means and that he lived to see solved by crystallographic and chemical ones.

From the methodological point of view, the most fruitful concept that Muller introduced into mutation research was that of the 'mutation rate' (9). He was the first to point out that mutations, although very rare, occur at measurable rates, and he set about to find means for measuring these low rates. This was long before the days of micro-organism genetics, and techniques had to be evolved for such slow-breeding organisms as *Drosophila* and maize. Those that Muller developed for *Drosophila* are essentially still the same that are used today (Chapter 5). The concept of mutation rate and the means of measuring it made it possible to exploit at once to the full the new possibilities that were opened up by the discovery of the first clearly effective mutagen: X-rays. This discovery started what I consider to be the second period of mutation research.

The fundamental *problems* raised during the first period were, in order of importance:

(1) What is the property of the genetic material that enables it to replicate identically even after mutation?

(2) Are all mutations loss of genetic material?

(3) Do mutations arise only at gene replication?

The second period: 1927 to the beginning of the Second World War

In 1927, at the third International Congress of Genetics in Berlin, H.J. Muller presented data that demonstrated unambiguously the capacity of X-rays to produce mutations in *Drosophila.* In 1928, L.J. Stadler showed the same for maize. There followed two decades in which X-ray mutagenesis dominated almost all branches of genetics. It provided a powerful tool that could be put to the most varied uses. Most important for our purpose are those findings that threw light on the nature of mutation, and the new problems that arose out of this new insight. The early results have been summarized by Timoféeff-Ressovsky in 1934 and 1939 (10,11) and in a book written jointly with K.G. Zimmer (Bibliography).

Intergenic and intragenic changes

At an early stage, it was discovered that radiation produces chromosome rearrangements as well as point mutations, and that rearrangements may resemble point mutations in phenotype and segregation ratio. This led to a revival of the presence-absence theory in a new guise. While the older theory had only considered deletions as sources of mutation, the new findings showed that apparent mutations could also be due to position effects of rearrangements.

Some geneticists, in particular Goldschmidt (12) and Stadler (13), went so far as to claim that *all* mutations may be due to rearrangements of one kind or another. From an evolutionary point of view, this theory is as untenable as the old presence-absence theory for, if all mutations are due to loss or rearrangements of existing genes, then the lost or rearranged chromosome sections must already have differed from each other qualitatively, and these primary differences must have arisen through qualitative changes. Stadler did, indeed, appreciate this difficulty and restricted his theory to mutations that turn up in the laboratory. In this form, the theory was as difficult to disprove as the presence-absence theory. Applied to *induced* mutations, it is still with us. Every new mutagen confronts us afresh with the necessity of distinguishing between intergenic and intragenic changes, and this distinction is not easily made.

The relation between crossing-over and rearrangements

While crossing-over resembles translocation by being an exchange between pieces of chromosomes, it differs from it in several essential points. Unlike translocation, crossing-over is restricted to homologous chromosomes; the points of exchange are exactly matched, and crossover chromosomes differ from non-crossover ones in nothing but the arrangement of alleles they carry. Nevertheless, the finding that X-rays modify the frequency of crossing-over in *Drosphila* females and produce crossing-over in *Drosophila* males where it does not normally occur, raised the question of a connection between the mechanism of crossing-over and that of rearrangement formation. This resulted in two opposite hypotheses, one of which considered rearrangements as a sub-class of crossovers, while the other considered crossovers – at least those produced by X-rays – as a sub-class of rearrangements. On the one hand, it was considered that rearrangements are due to illegitimate crossing-over between non-homologous chromosomes. This theory was disproved toward the end of the period; in a greatly modified form (14), it was revived more recently and still plays a role in the interpretation of certain cytological observations. (See the section 'Exchange Hypothesis' in the review by H.J. Evans; Bibliography.) On the other hand, X-ray induced crossovers were considered to be 'pseudo-crossovers', due to translocations between homologous chromosomes. Efforts to test this hypothesis have continued until the present day and now are extended also to chemically induced crossovers. Since at least *some* rearrangements mimic point mutations, the hypothesis implies also a connection between these and crossing-over. This aspect of it forms one of the most intriguing problems of modern mutation research.

Dose effect curves

By far the most important tool for analysing the action of radiation on chromosomes and genes were dose-effect curves that relate the frequencies of different types of genetic effect to the dose of ionizing radiation. The outstanding success of this approach has resulted in a somewhat uncritical reliance on kinetic data for inferring the action of mutagens. This has raised problems that were not apparent in the early X-ray work and that we shall have to consider in their place.

The target theory

About 10 years after the discovery of the mutagenic action of X-rays, the results obtained so far were shown to be well-accommodated by a theory that impresses by its wide sweep and its combination of conceptual simplicity with mathematical sophistication. According to this theory, the genes and the presumed links between them are 'targets' for 'hits' by energy packets that are delivered to them by ionizing radiation. A hit in a gene produces a point mutation; a hit in a link between genes, a chromosome break. In its simplest form, the theory equated the gene molecule with the target for point mutations; in this form, it encountered difficulties from the start and was never accepted by some mutation workers, notably Muller. In its less rigorous form, which takes account of energy transfer over short distances, the theory has retained its validity for mutagenesis by X-rays and other ionizing radiation. Today, it plays a role in the interpretation of mutagenesis by UV and chemicals.

Spontaneous mutations

Attempts to fit spontaneous mutability into the target theory were not very successful. Calculations (15a,b) showed that naturally occurring ionizing radiation was far too weak to account for the rates at which spontaneous mutation and chromosome aberrations occur. Since no effective chemical mutagens had yet been discovered, it was assumed that all spontaneous mutations arose from random fluctuations in micro-temperature, the gene serving as target for chance concentrations of heat motion in its vicinity. Although this may well be true for a fraction of spontaneous mutations, it certainly is not the whole explanation. Replication errors were completely disregarded in the theory, and some of the experimental evidence was marred by sources of inaccuracy that were not then realized. A special problem was raised by the existence of unstable or 'ever-sporting' genes which, in somatic or germinal tissues, or in both, mutate with frequencies that are quite outside the range of normal mutation rates. At first these observations threw doubt on the unit nature of the gene. When this was shown to be unjustified, the emphasis shifted to the question: why are some genes so much more mutable than others? In this form, the problem has remained with us. We now know that *the phenomenon of instability* covers a variety of very different mechanisms, some of which are not yet understood. Their analysis forms one of the most fascinating areas of present-day mutation research.

Ultraviolet light

During the thirties, the mutagenic effects of ultraviolet light were demonstrated for a variety of organisms: *Drosophila,* maize, snapdragon, a fungus, a liverwort. In maize, the analysis was carried farther by Stadler and his collaborators, but the outstanding importance of UV for the analysis of the mutation process was established only during the following periods.

In the course of the second period of mutation research, some of *the problems* raised during the first re-appeared in a new guise and gained importance through their bearing on X-ray mutagenesis.

(1) Is there an essential difference between intergenic and intragenic point mutation? In particular, do X-rays produce point mutations?

(2) What is the relation between crossing-over and chromosome rearrangements? In particular, do X-rays produce true crossovers, or are the apparent crossovers in reality translocations between homologous chromosomes? Implicit in this problem is that of a possible connection between crossing-over and point mutations.

(3) Is there an essential difference between stable and unstable genes?

The third period: beginning of the Second World War to 1953

Two developments inaugurated this period: the discovery of chemical mutagens (16), and the introduction of micro-organisms as objects of mutation studies (17). The paths that grew out from these starting points remained more or less separate during the whole of this period. It is true that chemical mutagens were applied to micro-organisms, but almost all important findings on chemical mutagenesis at that time came from experiments with *Drosophila* , while most important findings on UV-mutagenesis came from experiments on micro-organisms. One crucial conclusion emerged independently from both lines of approach: the existence of a *premutational or potential lesion* that precedes the production of a mutation. In micro-organisms, the conditions that favour or inhibit the fixation of UV-induced premutations could be analysed and led to the discovery of repair mechanisms. In *Drosophila,* the premutational state was found to be capable of replication, so that new *unstable genes* could be created by chemical treatment. Applied to chromosome breakage, the concept of *a potential or latent break* preceding the realized break had already been mooted by X-ray workers; this concept was greatly strengthened by results obtained with chemical agents.

Experiments with *Drosophila* threw new light on problems concerning chromosome structure and *rearrangement formation.* Chemicals were shown to produce very different ratios between point mutations and chromosome rearrangements, and this strengthened the concept of the chromosome as a series of genes separated by intergenic linkers. The earlier conclusion that rearrangements are formed by a two-step mechanism – breakage followed by reunion – gained support from the finding that equal numbers of chromosome breaks may yield very different numbers of rearrangements depending on the mutagen used.

Dose-effect curves for UV and chemical mutagens could not so readily be fitted to target models as those obtained for X-rays, and this raised doubts about the strict applicability of the target concept to mutagens other than ionizing radiation.

Comparisons between the types of mutation produced by radiation and different chemicals showed that the spectra of mutations were not always the same. This *mutagen specificity* was first discovered in flowering plants where the proportions of different types of chlorophyll mutations were shown to depend on the mutagen used.

It was subsequently confirmed for reverse mutations (from a specific biochemical requirement to independence) in micro-organisms, where the very large numbers screened established the reality of the phenomenon beyond doubt. In the absence of a chemical model of the gene, speculation on the nature of these specificities seemed impossible. Although we shall see that this defeatism was only partially warranted, it effectively discouraged a follow-up of these cases.

The old problem of whether *mutation* arises only *in replicating genes* could be approached in a new way by scoring mutations in bacteria during growth in continuous culture (18). This led to the wholly unexpected conclusion that both spontaneous and induced mutations are independent of the number of cell cycles per unit time. Recent repetitions of these experiments with a slightly modified technique suggest that this conclusion may have been oversimplified.

For the mutation worker, these years were exciting as well as frustrating. The excitement stemmed from the many new findings under the impact of which the old target theory started to burst at the seams. The frustration stemmed from the absence of a chemical model of the gene, which made it impossible to replace the target theory by a more comprehensive one that would be based on chemical as well as physical considerations. Attempts to infer the chemical nature of the gene from the chemical nature of mutagens were as unsuccessful as had been attempts to infer the nature of cancer from the nature of carcinogens. On the contrary, the great diversity of effective mutagens made it even more difficult than before to envisage a mechanism that would allow faithful reproduction of all the different chemical changes that may give rise to mutations.

So many *problems* had accumulated at the end of this period that it is difficult to pinpoint the most important ones.

(1) The mystery of the *fidelity of gene replication* was deepened with the arrival of a host of chemical mutagens which, presumably, produced mutations by different chemical changes in the gene.

(2) The mystery of *unstable genes* was not brought nearer to a solution when it was found that such genes could be created *de novo* by chemical treatment.

(3) The problem of whether *mutations* can arise *in non-replicating genes* got a new twist by the paradoxical results obtained from bacteria grown in continuous culture.

(4) The complexities of *dose-effect curves* for mutagens other

than ionizing radiation threw doubt on the target concept.

(5) The model of the chromosome as an array of genes separated by *intergenic linkers* gained support from work with chemical mutagens, but crucial evidence was still lacking.

(6) Attempts to explain *mutagen specificity* appeared futile in the absence of a chemical model of the gene.

(7) Even less was it possible to speculate on the nature of the *premutated state of the gene* and of the *latent chromosome lesions* that were shown to precede mutation and chromosome breakage.

An interesting comparison between the state of mutation research at the beginning and the end of this period emerges from the two Cold Spring Harbor Symposia of 1941 and 1951 (19,20).

The fourth period: 1953-about 1965

This period is less clearly defined in time than the previous ones. It is true that it started with a discovery that revolutionized not only mutation research but the whole of genetics and much of biology; but it took a few years before the double helix was generally accepted as the true structure of the genetic material. At the other end of the period, there was no single outstanding discovery that terminated the almost exclusive preoccupation of mutation research with nucleic acid chemistry, but rather a gradual accumulation of observations that could not be explained at this level.

With the acceptance of the Watson-Crick model of the gene, the *central problem of mutation research* was solved once and for all. What Muller had called the 'general autocatalytic power' of the gene that transcends all specific mutational changes was now found to be inherent in the complementary nature of DNA. Several other problems that had been inherited from the preceding periods found a full or partial solution during these years, often – but not always – by application of the new molecular insight and techniques.

The doubt whether *true intragenic changes* are found among laboratory mutations was finally laid to rest when amino acid sequencing of mutant polypeptides revealed internal defects and errors (21). The presumed *division of the chromosome into genes and intergenic linkers* has been verified only recently for prokaryotes and eukaryotes (22,23). The question whether spontaneous *mutations* can arise *in non-replicating genes* was answered in the affirmative for various organisms (Chapter 21). The action of many mutagens could be explained from their reactions with DNA, and this accounted for some cases of *mutagen specificity.* Others could not be so interpreted and remained an unsolved problem. Not only mutated genes but also some of the *premutational states* produced by UV and chemicals could be described in chemical terms. Mosaic mutations were no longer a problem but seen to be a consequence of the double-stranded nature of DNA. Instead, the *origin of non-mosaic mutations* became a question that could not be answered by considering solely the reaction between DNA and mutagens. A molecular mechanism was found to account for the delayed appearance of some mutations, but it could not explain *instabilities* that replicated as such. These have remained an intriguing problem. The application of kinetic data and of the *target concept* to UV and chemical mutagens met with partial success but was soon found to be subject to severe limitations.

In general, it may be said of this period that its great achievement was the formulation of a biochemical or 'molecular' theory of mutation which was as impressive as the biophysical target theory formulated 20 years earlier. But, like the target theory, it was too limited to cover the whole of the mutation process. Most of the observations that had dethroned the target theory as a comprehensive model of mutagenesis could not be fitted into the new chemical theory either; this applies, e.g., to photorepair, unstable genes, and many cases of mutagen specificity. In order to understand these phenomena, one has to look at mutation as a *biological* process set into the framework of the living cell with its enzymes, its membranes, its growth and metabolism. This realization started the present period of mutation research.

The fifth period: (about 1965 onwards)

Research in the present period is based firmly on the achievements of the preceding ones. It makes use of the concept of mutation rates developed in the first period, of the target concept put forward in the second, of the Watson-Crick model that inaugurated the fourth, and of the molecular interpretations that now can be given to observations collected during the third. The reactions between mutagens and DNA continue to claim interest, in particular in regard to new mutagens; but the emphasis has shifted from DNA into the cell as a whole (see *Mutation as cellular process,* Bibliography). At present the making or marring of potential mutations and chromosome breaks by repair processes is in the centre of interest; but sufficient observations have already accumulated to show that other cellular events, such as translation or membrane formation, may be of equal importance. Some phenomena, e.g. unstable genes, paramutation, or certain cases of mutagen specificity, are likely to defy molecular analysis for some time to come, but this in no way lessens their importance as objects of mutation research. Biological phenomena in general are detected *in vivo* and have to be analysed in cells and organisms before they can be studied at the molecular level. Meanwhile, analysis by purely biological methods may lead to new findings that, in turn, may become potential objects of molecular studies.

At present, the growing concern with genetic hazards from environmental mutagens leads to a channelling of much money and effort into mutagen-testing. While this work clearly is of great practical importance and is likely also to provide information of theoretical interest, it carries with it the danger that mutation research may become merely a branch of applied science. This would be a great pity. Cancer research has shown that, while routine testing accumulates much useful knowledge, sudden leaps of progress usually take off from results in some area of fundamental research not obviously connected to a practical application. Almost certainly, the same is true for attempts to control the mutation process both negatively, by minimizing danger from environmental mutagens, and positively, by the planned use of mutagens for the production of improved crops. Quite apart from its probable practical repercussions, fundamental mutation research at the cellular level is likely to reveal biological complexities and biochemical concatenations that may be missed in other kinds of study. One of the major objects of this book is to bring to the reader the realization of some of these fundamental problems and to stimulate his interest in tackling them.

References

1. Avery, O.T., MacLeod, C.M. and McCarty, M. (1944), 'Studies on the chemical nature of the substance inducing transformation of pneumococcal types', *J. Exp. Med.* **79**, 137-158.
2. Hershey, A.D. and Chase, M. (1952) 'Independent functions of viral protein and nucleic acid in growth of bacteriophage', *J. Gen. Physiol.* **36**, 39-56.

3. Wilkins, M.H.F., Stokes, A.R. and Wilson, H.R.,(1953),'Molecular structure of desoxypentose nucleic acids', *Nature,* **171**, 738-740.
4. Franklin, R.F. and Gosling R.G. (1953), 'Molecular configuration in sodium thymonucleate', *Nature,* **171**, 740-741.
5. Watson, J.D. and Crick, F.H.C., (1953), 'The structure of DNA', *Cold Spring Harbor Symp. Quant. Biol.* **18**, 123-131.
6. de Vries, H. (1909), *The Mutation Theory. Experiments and Observations on the Origin of Species in the Vegetable Kingdom.* Open Court Publ. Co., Chicago.
7. Bateson, W. (1928), 'A suggestion as to the nature of the "Walnut" comb in fowls', Proceedings of the Cambridge Philosophical Society XIII, 1905, In *Scientific Papers of William Bateson,* **II**, 135-138.
8. Muller, H.J. (1922), 'Variation Due to Change in the Individual Gene', *Am. Nat.* **56**, 32-50.
9. Muller, H.J. (1928), " 'The measurement of gene mutation rate in *Drosophila,* its high variability and its dependence upon temperature', *Genetics* **13**, 279-357.
10. Timoféeff-Ressovsky, N.W. (1934), 'The experimental production of mutations', *Biol. Rev.* **9**, 411-457.
11. Timoféeff-Ressovsky, N.W. (1939), 'Le mécanisme des mutations et la structure du gène.', *Actualités scientifiques et industrielles* **812**.
12. Goldschmidt, R. (1946), 'Position Effect and the theory of the Corpuscular Gene.', *Experientia* **2**, 250-256.
13. Stadler, L.J. (1954), 'The Gene', *Science* **120**, 811-819.
14. Revell, S.H. (1959), 'The accurate estimation of chromatid breakage, and its relevance to a new interpretation of chromatid aberrations induced by ionizing radiations', *Proc. Roy. Soc. Lond. B.* **150**, 563-589.

15a. Muller, H.J. and Mott-Smith, L.M. (1930), 'Evidence that natural radioactivity is inadequate to explain the frequency of "natural" mutations', *Proc. Nat. Acad. Sci. U.S.A.* **16**, 277-285.

15b. Sparrow, A.H. (1950), Tolerance of *Tradescantia* to continuous exposures to gamma radiation from cobalt 60, *Genetics* **35**, 135.

16. Auerbach, C. (1949), 'Chemical Mutagenesis' *Biol. Rev.* **24**, 355-391.
17. Cold Spring Harbor Symposia on Quantitative Biology (11) (1946), 'Heredity and Variation in Micro-organisms', *The Biological Laboratory, New York.*

18. Novick, A. and Szilard, L. (1950), 'Description of the Chemostat', *Science* **112**, 715-716.
19. 'Genes and chromosomes, structure and organization', (1941), *Cold Spring Harbor Symp. Quant. Biol.* **9**.
20. 'Genes and Mutations' (1951), *Cold Spring Harbor Symp. Quant. Biol.* **16**.
21. Henning, U. and Yanofsky C. (1962), 'Amino acid replacements associated with reversion and recombination within the A gene', *Proc. Nat. Acad. Sci. U.S.A.* **18**, 1497-1504.
22. Rechler, M.M. and Martin R.G. (1970), 'The intercistronic divide: Translation of an intercistronic region in the histidine operon of *Salmonella typhimurium*', *Nature* **226**, 908-911.
23. Miller, O.L.Jr. and Hamkalo, B.A. (1972), 'Visualization of RNA synthesis on chromosomes', *Int. Rev. Cytol.* **33**, 1-25.

CHAPTER TWO

Classification of mutational changes at the chromosomal level

If mutation in the widest sense is defined as a sudden heritable change, the term covers many disparate phenomena, only some of which will be considered in this book. In the first place, we can distinguish between cytoplasmic and nuclear mutations. This distinction is usually not difficult to make because at meiosis nuclear mutations segregate and recombine according to Mendel's laws, while cytoplasmic ones do not. In prokaryotes, which lack meiosis, linked recombination with known nuclear genes proves the nuclear nature of a new mutation. The same applies to eukaryotes without meiotic stages and to somatic cell cultures. In the former, mitotic recombination and linked losses of functions provide evidence of nuclear mutation. In the latter, linked losses of functions and correlations between loss of function and loss of chromosome material serves the purpose.

In this book we shall deal only with nuclear changes. These can be classified into three main types: (*a*) changes in the number of chromosomes, (*b*) changes in the number and arrangement of whole genes (intergenic or structural changes), and (*c*) changes in individual genes (intragenic changes or mutations in the narrower sense).

(*a*) Changes in chromosome number result in polyploidy, haploidy, polysomy, monosomy or nullisomy. All these phenmena have played a great role in plant evolution and are used by plant breeders for the development of new strains and the analysis of existing ones. In nature and origin, they are essentially different from the remaining mutational changes and will not be discussed further, except briefly in Chapter 23 as possible sources of genetic hazards from environmental agents.

(*b*) Intergenic changes may take many forms: deletion or duplication of individual genes or strings of genes; insertion of a chromosome segment into a new position; exchange of pieces between chromosomes. In general, they are produced by the same agents that produce intragenic mutations. In many cases, it is difficult or impossible to distinguish between an intergenic and an intragenic change.

(*c*) Intragenic changes or gene mutations are, by definition, changes in individual genes. As already mentioned, it is often impossible to distinguish them from structural changes that are small enough to segregate and recombine like single genes. In these cases, it is better to speak of point mutations, a term that covers both true gene-mutations and rearrangements that mimic them. In some prokaryotes and fungi, the molecular approach has made it possible to detect, describe and classify true gene mutations. With the gradual replacement of the term 'gene' by the more clearly defined term 'cistron', it has become customary to speak of intracistronic rather than of gene mutations.

Intergenic and intragenic changes will form the main material for this book. 'Paramutational' changes, which occur in controlling elements rather than in structural genes, will be discussed briefly in Chapter 21.

Intergenic changes

The basic event that, in eukaryotes, results in intergenic changes is chromosome breakage. Indirect evidence shows that most breaks restitute again in the old order and remain undetected (see Lea, 1946: Bibliography). Those that do not restitute may remain open or they may rejoin in a new way to form a rearrangement. In prokaryotes, chromosome breakage leads to cell death, and there is no conclusive evidence that rearrangements can be formed by breakage and reunion. The discussion of intergenic changes in this chapter will therefore be limited to eukaryotes. In some of the later chapters we shall meet with cases of intergenic rearrangements in prokaryotes.

Unrejoined (open) breaks. The breakage-fusion-bridge cycle

A broken chromosome yields two fragments, one of which lacks the centromere and is called acentric in contrast to the other, centric, one. The acentric fragment fails to congregate on the spindle and is lost at mitosis. The fate of the centric fragment depends on whether the broken end can 'heal', that is, carry out the normal function of a chromosome end at mitosis. Where this is the case e.g., in the maize plant, chromosome breakage leads to cells that contain the centric fragment without the complementary acentric one. These cells have a *terminal deletion*,* or *terminal deficiency*. In diploid cells, terminal deletions may survive if they are not too large. Genetically, they can be recognized by the uncovering of recessive markers on the homologous normal chromosome. In the endosperm of maize, in *Drosophila* and probably in most animals, broken chromosome ends do not heal but retain their tendency to stick to other broken ends. If, at the time of breakage, none such are available in the same cell, an opportunity for rejoining arises after separation of the broken chromosome into chromatids, each of which is broken in the same position. Rejoining between them may occur in two different ways: (*a*) The centromere-proximal fragment of one chromatid rejoins with the centromere-distal one of the other and vice versa; this is equivalent to sister-strand crossing-over and without genetical consequences. (*b*) The two centromere-proximal fragments join, yielding a dicentric, and the two centromere-distal fragments join, yielding an acentric fragment. The latter gets lost, while the former is drawn out into a bridge between opposite spindle poles at anaphase. If the bridge persists, it prevents separation of the chromosomes at telophase and so stops development. If it breaks under tension, it results in two daughter cells with broken chromosomes and these, in turn, carry out a 'breakage-fusion-bridge' cycle. Since the bridge does not usually break exactly in the centre, the complementary fragments are unequal, one lacking a piece which the other carries in duplicate. As deficiencies accumulate in successive cycles, the cell line becomes less and less viable and eventually dies out. In the triploid maize endosperm (Chapter 10), the course of the breakage-fusion-bridge cycle can be inferred from the pattern it forms through the sequential appearance of recessive markers

* While originally the term deletion was applied to losses of intercalary chromosome pieces, and the term deficiency to losses of terminal pieces, the two terms are now used interchangeably and will be so used in this book. In general, the term deletion is preferred by students of micro-organisms and plants, while students of *Drosophila* frequently use the term deficiency.

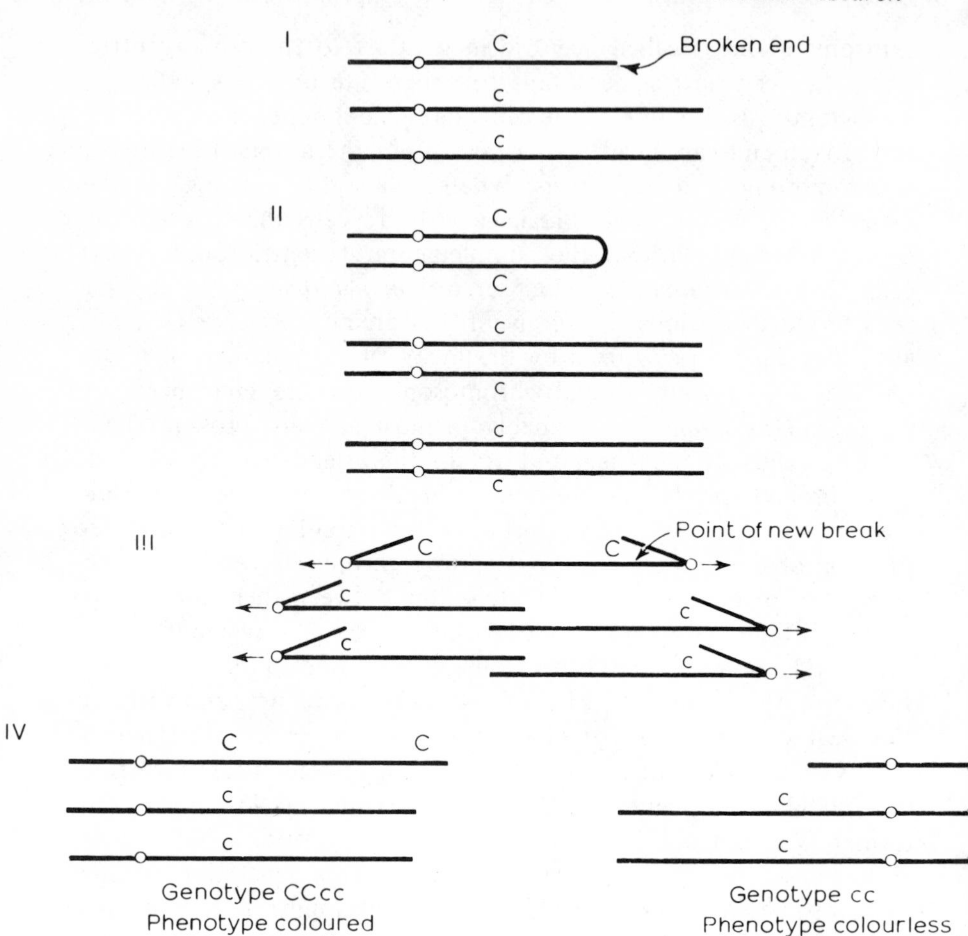

Fig. 2.1. The origin of variegated aleurone colour in maize kernels through the operation of the breakage-fusion-bridge cycle. C = gene for purple aleurone colour, c = its recessive allele. The aleurone is the outer layer of the triploid endosperm. (Source: Fig. 7.4, Srb, Owen and Edgar 1965 (bibl.). Courtesy authors)

on the normal chromosomes (Fig. 2.1).

Thus, by one means or another, open chromosome breaks cause death of the cell line in which they have occurred. When an open break is introduced into a zygote by either the male or the female gamete, this zygote dies at an early stage of development through what is called a 'dominant lethal'. Since the frequency of zygotic deaths is an easily measured parameter, dominant lethals in mice have become a favourite criterion in tests for potential mutagens in the human environment.

Chromosome rearrangements*

These require the presence of two or more chromosome breaks in the same cell and sufficiently close to each other for the broken ends to establish contact. The outstanding feature of rearrangement formation, and one that distinguishes it essentially from crossing-over, is its indifference to the polarity of the chromosome in relation to the centromere. When homologous chromosomes exchange pieces by crossing-over, polarity is always preserved. The centromere-proximal part of one chromosome links up with the centromere-distal part of the homologous one, so that each crossover chromosome has one centromere ('eucentric') and carries its genes in their original order. In rearrangement formation, this is different. In addition to 'symmetric' rearrangements which, like crossovers, preserve the polarity with regard to the centromere, there are 'asymmetric' ones in which the two centromere-proximal parts of the broken chromosomes join to form a dicentric, while the two centromere-distal ones join to form an acentric fragment. Symmetric exchanges yield eucentric chromosomes; asymmetric exchanges yield aneucentric chromosomes. (Fig. 2.2). Similarly, while in double crossing-over polarity

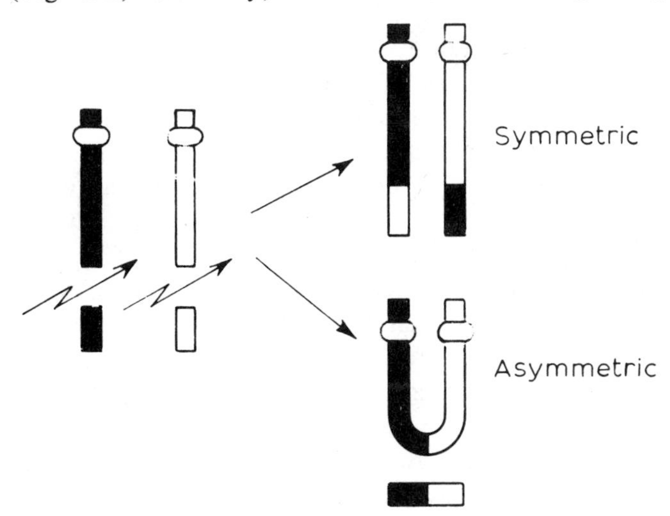

Fig. 2.2. Schematic representation of symmetric and asymmetric two-hit aberrations. (Fig. 1, Wolff (23). Courtesy Academic Press)

* Also called 'aberrations' or 'structural changes'.

in respect to the centromere and the order of the genes on the chromosome are always conserved, a piece of a broken chromosome may be re-inserted in the inverse order, changing its polarity and the order of its genes. We shall now look in more detail at the most important types of rearrangement.

Deletions, deficiencies

Terminal deletions require only one break and have been discussed above. Intercalary (or interstitial) deletions arise from chromosomes that have been broken into three fragments by two breaks. Loss of the middle fragment, accompanied by rejoining of the distal and proximal one, yields an interstitial deletion. Like terminal deletions, interstitial ones are usually but not always (1) lethal in haploid cells but – if not very large – may be tolerated in diploid ones (2). They can then be detected by the uncovering of marker genes on the normal homologue. Cytologically, interstitial deletions are difficult to detect except in chromosomes with distinct patterns of chromomeres or bands. We shall come back to the detection of deletions at the end of this chapter when we deal with criteria for distinguishing them from intragenic changes (p. 24).

Inversions

Like an interstitial deletion, an inversion requires two breaks in the same chromosome. These result in an inversion when the middle piece is re-inserted with reversed polarity. Except in organisms and chromosome regions that are subject to a position effect, inversions do not produce phenotypic effects and do not impair viability. This leads to important conclusions on the transcription process by messenger RNA. First, the normal functioning of genes inside an inversion shows that the direction in which a gene is transcribed is determined by the gene itself and not by the chromosome of which it forms a part. Second, since the sugar-phosphate backbones of the two complementary strands of DNA have opposite polarity, an inverted piece can be fitted into its original double helix only after having rotated round its own axis. Since only one strand of DNA (the reading-strand) serves as template for messenger-RNA, it follows that the reading strand need not be continuous throughout the chromosome but may switch sides. For prokaryotes, there is indeed evidence that direction of reading and position of reading strand in the double helix may change from one gene to another (3,4).

Inversions are detected most easily through their inhibitory action on crossing-over and were known as crossover inhibitors before their true nature was understood, The *C* in the famous *ClB* strain of *Drosophila* (Chapter 5) stands for a crossover-inhibitor which is now known to consist of a large inversion in the X-chromosome. The shortage of crossover progeny from heterozygotes has two essentially different causes. (*a*) The inversion interferes with pairing. If it is short, it does not pair at all with its homologous region and crossing-over cannot take place. (*b*) If it is long it may pair by loop formation. Single crossovers within a loop prevent the formation of viable zygotes either by forming an anaphase bridge or by yielding duplication-deficiency gametes (for details see textbooks of cytogenetics).

Reciprocal translocations

When two breaks have occurred in different chromosomes, the broken ends of one chromosome may rejoin with those of the other. Asymmetrical rejoining leads to loss of the cell line and, after mutagenic treatment of germ cells, is a cause of dominant lethality. Symmetrical rejoining yields new chromosomes in which the distal part of the original ones have been exchanged (Fig. 2.2). This is called a reciprocal translocation or, by cytologists, a segmental interchange. Non-reciprocal translocations, in which a piece of one chromosome is inserted into another chromosome, require three breaks and will be mentioned later as a source of duplications. Reciprocal translocations, like inversions, do not usually affect the phenotype except by position effect. Unlike inversions, they interfere with the orderly process of meiosis. In a heterozygote for a translocation, the two translocated chromosomes and their normal homologues form a quadrivalent which, because of its pairing configuration, is often called a translocation cross (Fig. 2.3). Depending on the way the four chromosomes segregate at meiosis, six types of gamete are formed. One carries both normal homologues and yields normal progeny. Four carry one translocated and one normal chromosome: these lack one chromosome segment and have another one in duplicate. They yield zygotes that die of chromosome imbalance and contribute to the pool of dominant lethals. They form somewhat more than half the zygotes and lead to semi-sterility of a translocation heterozygote. The last type of gamete carries both translocated chromosomes. On fertilization, it produces a translocation

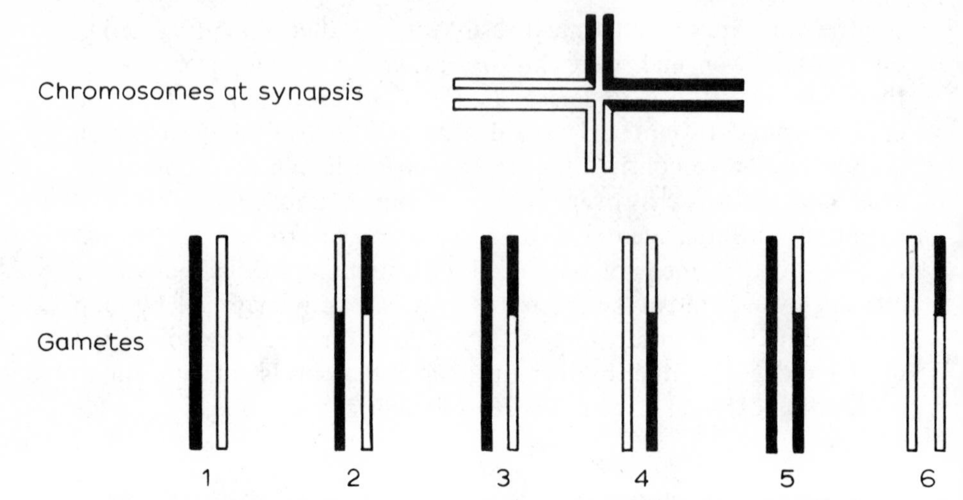

Fig. 2.3. The translocation cross in meiosis of a translocation heterozygote, and the gametes formed. In matings to a normal individual, only types 1 and 6 yield viable progeny. (Fig. 2, Snell (24). Courtesy Genetics Business Office, Austin, Texas)

heterozygote that behaves like the original one in being semi-sterile and producing about 50% semi-sterile progeny. Thus, semi-sterility behaves as a dominant character in translocation heterozygotes and can be used as diagnostic criterion preparatory to or together with cytological examinations. In *Drosophila,* translocations are detected by the creation of linkage between previously unlinked markers. In organisms with crossing-over in both sexes, this linkage is soon destroyed by crossing-over except for markers very close to the points of translocations. In *Drosophila,* however, the fortunate circumstance that there is no crossing-over in the male makes it possible to detect and preserve translocations by carrying them always through males. In addition, this permits identification of a translocation not only by true, physical linkage between markers but also by pseudo-linkage of markers outside the translocation itself because only the original combinations of chromosomes – both normal or both translocated – yield viable progeny. Without crossing-over, therefore, *all* markers on these chromosome pairs will remain together and will appear linked (Fig. 2.4). Translocations may also act as crossing-over inhibitors; in a recent investigation on *Drosophila* (5), selection for suppression of crossing-over turned out to be an unexpectedly efficient means for the screening of X-ray induced translocations.

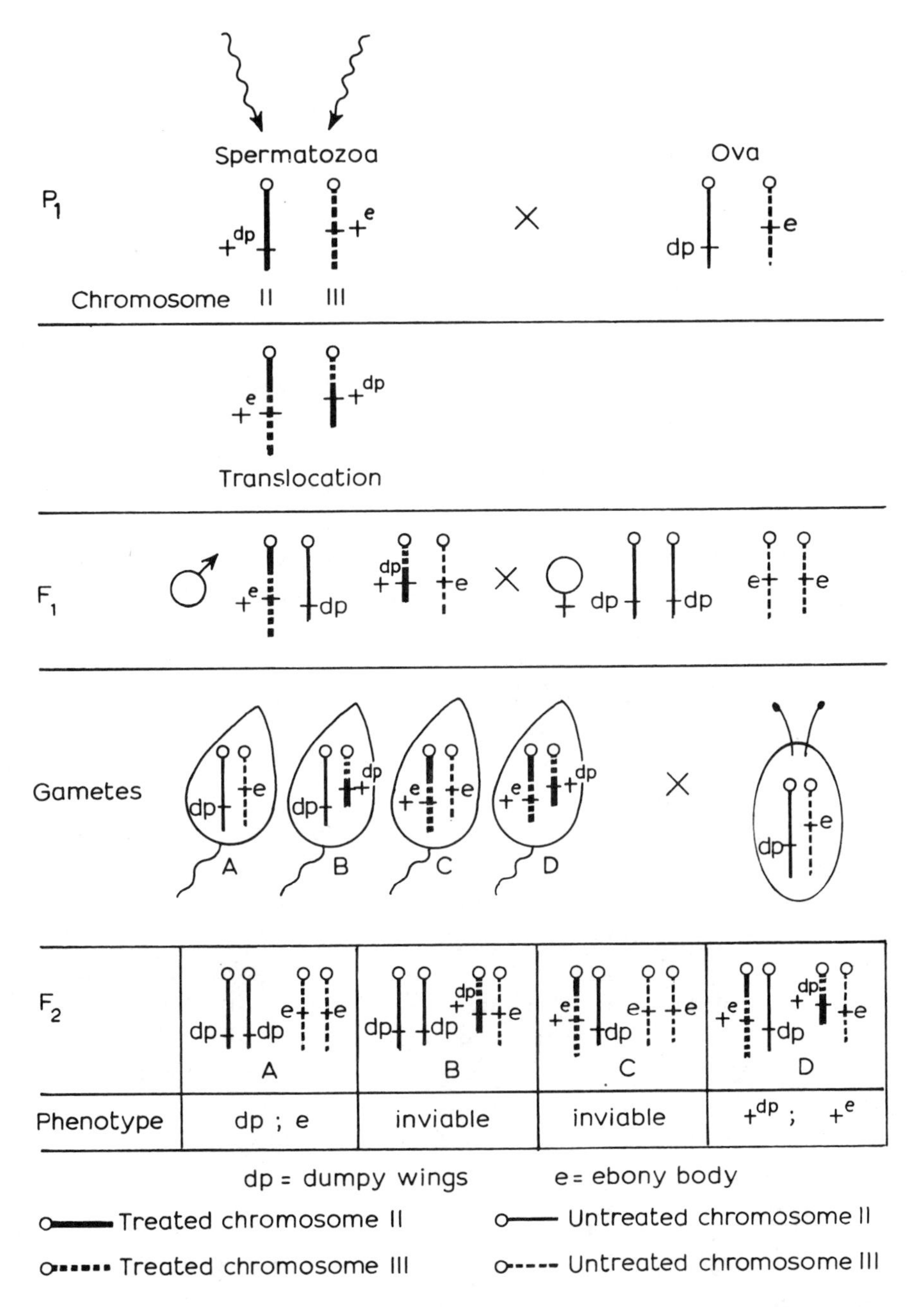

Fig. 2.4. The detection of translocations between autosomes II and III in *Drosophila melanogaster*. *dp* = dumpy wings. *e* = ebony body. (Fig. 7, Auerbach, 1962 (bibl.). Courtesy author)

Insertional translocations. Duplications

These require three chromosome breaks. Two of these – in the same chromosome – cut out a piece, which is then inserted in normal or inverted order into the third break, which may be in the same or in a different chromosome. The resulting chromosome set is still complete, for the piece that is lacking from one position is now present in another. At meiosis, however, the deleted segment and the insertion may become separated from each other by segregation or crossing-over, and zygotes are produced that carry either a deletion or a duplication by itself. While the former usually are harmful (see above), the latter are not, and duplications are not easily discovered.

From the early days of genetics, duplications have been presumed to play a major role in evolution by freeing redundant genes for new functions (6,7). Newer findings support the hypothesis. Thus, proteins carrying out related functions but controlled by different genes often have a common basic structure, suggesting that these genes arose by mutations in duplicates of a common ancestral gene (8,9). Recent research has added the realization that duplications frequently act as means for amplifying the action of certain genes (10,11). Of special interest in this context are 'repeats', in which the duplicate segments are immediately adjoining each other as 'tandem repeat' (*abcdbcde* ...) or 'reverse repeat' (*abcddcbe*...). Once a repeat has arisen, it can, by oblique crossing-over between the first segment in one chromosome and the second in the other, give rise to triplications and still further amplification of the same gene material (12,13). It should be stressed, however, that a first repeat is not likely to originate in this way because the extreme accuracy of the crossing-over mechanism does not allow the pairing of adjoining genes unless these have previously been made identical or similar by duplication. This stricture does not apply to intracistronic changes, where repeats of small sequences of nucleotides may be created by oblique crossing-over within a repetitive base sequence (14).

Intragenic changes. Criteria for distinguishing them from deletions of whole genes

Their physiological and molecular aspects will be discussed in the next chapter. Here we shall only discuss means of distinguishing them from intergenic changes. One source of misclassification arises from position effects (15). Where these are due to large rearrangements, they can usually be detected by genetical or cytological means. Position effects of minute inversions or translocations may be missed but are probably rare. A special type of position effect, which has been found in maize and *Drosophila* and may well occur in most or all eukaryotes, is that due to insertion or removal of a controlling element next to a gene or even within a gene. While the event creating this situation is characterized by an unusual kind of variability, the result, once established, may be indistinguishable from mutation in a structural gene. We shall come back to this in Chapter 22.

The main difficulty – and one which often cannot be resolved – lies in the distinction between deletion of a gene and a mutation that renders this gene inactive. A variety of criteria have been used to make this distinction. Most of them can at best establish the *presence* of very small deficiencies; the *absence* of even a minute deficiency and, by implication, the presence of a gene mutation is established only *per exclusionem*: a principle that always leaves room for doubt.

In *bacteriophage,* single-gene deletions have been discovered by loss of buoyant density or by electron microscopy of heteroduplices between a deleted and a normal strand of DNA (16,17): the normal strand forms a loop opposite the deletion (Fig. 2.5). In the salivary gland chromosomes of *Drosophila,* loop formation between chromosomes rather than between strands of DNA can be used to detect sizeable deficiencies. Smaller deficiencies are detected by the absence of a few bands or a single band – sometimes even part of a band – from the pattern. Mutations not connected with visible changes in the salivary band pattern are attributed to intragenic changes. Although there is a very good correlation between bands and genes (18), some doubt remains in many cases of presumed gene mutations. This is even more so when one accepts Crick's recent suggestion that it is the interbands and not the bands that represent the genes (19).

In *haploid* cells, a deletion for several genes can sometimes be discovered by the concomitant loss of function of several adjoining genes. In *E. coli,* mutants that have lost tryptophan prototrophy together with resistance to coliphage are recognized as deletions spanning the loci responsible for these functions (20). Other tests are based on the fact that, in the haploid state, a very large proportion

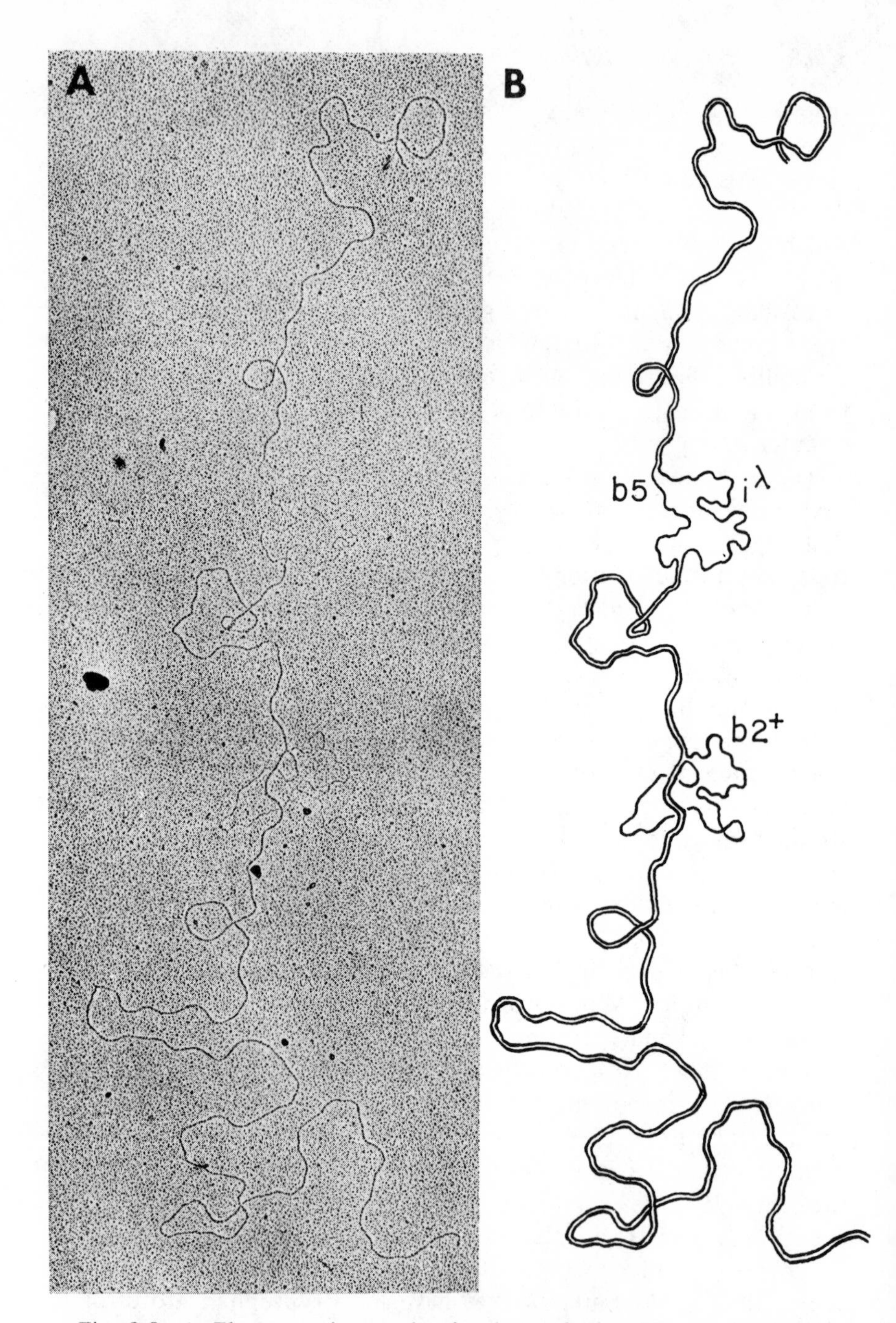

Fig. 2.5. *A*. Electron micrograph of a heteroduplex DNA molecule of phage λ. *B*. Interpretive drawing; the letters and numbers refer to specific deletions. (Part of Fig. 3, Westmoreland (17). Courtesy Am. Ass. Advancement of Science)

of all genes is necessary for survival. Thus, even if this should not apply to the gene in question, a deletion covering only a few of its neighbours is likely to act as lethal. This criterion is especially useful in flowering plants, where development of the male gamete includes several cell cycles in the haplophase (Fig. 10.1). Very few deficiencies can survive this stage, and plants that are heterozygous for even a small deficiency form only half the amount of fully fertile pollen grains. In *Neurospora,* the viability criterion has been used for detecting deficiencies in the *ad* 3 region (21). This region contains two genes concerned with adenine biosynthesis, 3A and 3B. They are separated by a small region which is required for survival of the haploid mycelium. Mutations that inactivate either the A or the B gene yield colonies on medium that has been supplemented with adenine. Deficiencies that reach into the intervening region do not survive even on supplemented medium and can be rescued only by heterokaryon formation with a non-deleted nucleus. Deficiencies of either the A or the B gene alone cannot be detected by this method.

In *diploid* organisms, a deletion may be spotted by its effect on recombination. Closely linked markers fail to recombine when the homologous chromosome carries a deletion that spans several of them. Moreover, the distance between outside markers will be shortened (22). Other tests on diploids utilize the fact that a deletion fails to cover recessive genes on the homologous chromosome. Uncovering of more than one recessive marker in the same region is taken as evidence for a deletion. Thus, in mice, mutations to dilute coat colour (*d*) that, in heterozygotes, expose also the effect of the closely linked gene for short ears (*se*) are attributed to deficiencies. Deletions for single genes may sometimes be recognized by the fact that the heterozygote for a recessive gene is more extreme in phenotype than the homozygote for the same gene. We shall come back to this 'exaggeration effect' in the next chapter.

The *specific locus test* is a well-known method for scoring visible mutations, including small deletions, in diploid organisms. It consists of crossing treated wild-type individuals with untreated ones that are homozygous for a number of recessive visible mutations. A deletion or mutation at one of these loci yields an F_1 individual in which this particular gene is uncovered.

Scheme of a 'specific loci' test

$$P_1 \quad \overset{\text{treated}}{\frac{+^a}{+^a}\ \frac{+^b}{+^b}\ \frac{+^c}{+^c}} \quad \times \quad \overset{\text{treated}}{\frac{a}{a}\ \frac{b}{b}\ \frac{c}{c}}$$

$$\text{gametes} \quad +^a\ +^b\ +^c \qquad\qquad a\ b\ c$$

$$F_1 \quad \frac{+^a}{a}\ \frac{+^b}{b}\ \frac{+^c}{c} \quad \text{phenotypically wild-type.}$$

A treated gamete that carries, e.g., a mutation from $+^b$ to b or a deficiency for b yields a mutant offspring $\frac{+^a}{a}\ \frac{b}{b}\ \frac{+^c}{c}$ of phenotype b.

In order to decide whether one is dealing with a mutation or a deficiency, the mutated locus is tested for viability in homozygotes, hemizygotes (*Drosophila* males in the case of sex-linked genes), or haploids (maize pollen grains). If it acts as a lethal, it is supposed to be a deletion. If it is viable, it is supposed to be a gene mutation. Since, however, some recessive gene mutations may have lethal as well as visible effects, while some small deficiencies may be viable even in the haploid state (1), the distinction is again not absolute.

The only means of identifying a gene mutation positively by a genetical test – as opposed to molecular ones (Chapter 3) – is a test for revertibility. An inactive gene that can revert to activity cannot be due to a deficiency. The snag here lies in the necessity to exclude suppressors as sources of spurious reverse mutations; we shall deal with this in Chapter 4.

References

1. Muller, H.J. (1935), 'A viable two-gene deficiency phenotypically resembling the corresponding hypomorphic mutations', *Journal of Heredity,* **26**, 469-478.
2. Lindsley, D.L., Sandler, L., Baker, B.S., Carpenter, A.T.C., Denell, R.E., Hall, J.C., Jacobs, P.A., Miklos, G.L.G., Davis, B.K., Gethmann, R.C., Hardy, R.W., Hessler, A., Miller, S.M. Nozawa, H., Parry, D.M. and Gould-Somero, M., (1972), 'Segmental aneuploidy and the genetic gross structure of the *Drosophila* genome', *Genetics,* **71**, 157-184.

3. Spiegelman, W.G., Reichardt, L.F., Yaniv, M., Heinemann, S.F., Kaiser, A.D. and Eisen, H. (1972), 'Bidirectional transcription and the regulation of phage λ repressor synthesis', *Proc. Nat. Acad. Sci. U.S.A.* **69**, 3156-3160.
4. Guha, A., Saturen, Y. and Szybalski, W. (1971), 'Divergent orientation of transcription from the biotin locus of *Escherichia coli*', *J. Mol. Biol.*, **56**, 53-62.
5. Roberts, F.A. (1970), 'Screening for X-ray induced crossover in *Drosophila melanogaster*: prevalence and effectiveness of translocations', *Genetics,* **65**, 429-448.
6. Muller, H.J. (1936), 'Bar duplication', *Science,* **83**, 528-530.
7. Lewis, E.B. (1951), 'Pseudoallelism and gene evolution', *Cold Spring Harbor Symp. Quant. Biol.,* **16**, 159-174.
8. Ingram, V.M. (1961), 'Gene evolution and the haemoglobins', *Nature,* **189**, 704-708.
9. Khan, N.A. and Hayes, R.H. (1972), 'Genetic redundancy in yeast: non-identical products in a polymeric gene system', *Molec. Gen. Genetics,* **118**, 279-285.
10. Ritossa, F.M., Atwood, K.C. and Spiegelman, S. (1966), 'On the redundancy of DNA complementary to amino acid transfer RNA and its absence from the nucleolar organizer region of *Drosophila melanogaster*', *Genetics,* **54**, 663-676.
11. Weinberg, E.S., Birnstiel, M.L., Purdom, I.F. and Williamson, R. (1972), 'Genes coding for polysomal 9S RNA of sea urchins: conservation and divergence', *Nature,* **240**, 225-228.
12. Slizynska, H. (1968), 'Triplications and the problem of non-homologous crossing-over', *Genet. Res. Camb.* **11**, 206-208.
13. Parma, D.H., Ingraham, L.J. and Snyder, M. (1972), 'Tandem duplications of the rII region of the bacteriophage T4D', *Genetics,* **71**, 319-355.
14. Smithies, O., Connell, G.E. and Dixon, G.H. (1962), 'Chromosomal rearrangements and the evolution of haptoglobin genes', *Nature,* **196**, 232-236.
15. Lewis, E.B. (1950), 'The Phenomenon of Position Effect', *Advances in Genetics III* (ed. M. Demerec), Academic Press Inc., New York, 73-115.
16. Zuccarelli, A.J., Benbow, R.M. and Sinsheimer, R.L. (1972), 'Deletion mutants of bacteriophage φX174', *Proc. Nat. Acad. Sci., U.S.A.* **69**, 1905-1910.
17. Westmoreland, B.C., Szybalski, W. and Ris, H. (1969), 'Mapping

of deletions and substitutions in heteroduplex DNA molecules of bacteriophage λ by electron microscopy', *Science,* **163**, 1343-1348.

18. Judd, B.H., Shen, M.W. and Kaufman, T.C. (1972), 'The anatomy and function of a segment of the X chromosome of *Drosophila melanogaster', Genetics,* **71**, 139-156.
19. Crick, F. (1971), 'A general model for the chromosomes of higher organisms', *Nature,* **234**, 25-27.
20. Spudich, J.A., Horn, V. and Yanofsky, C. (1970), 'On the production of deletions in the chromosome of *Escherichia coli', J. Mol. Biol.,* **53**, 49-67.
21. de Serres, F.J. (1960), 'Genetic analysis of the structure of the *ad-3* region of *Neurospora crassa* by means of irreparable recessive lethal mutations', *Genetics,* **50**, 21-30.
22. Mohr, O.L. (1923), 'A genetic and cytological analysis of a section deficiency involving four units of the X-chromosome in *Drosophila melanogaster', Zeitsch. Ind. Abst. und Vereb.,* **32**, 108-232.
23. Wolff, S. (1959), 'Interpretation of induced chromosome breakage and rejoining' *Radiation Res. Suppl 1,* 453-462.
24. Snell, G.D. (1935), 'The induction by X-rays of hereditary changes in mice'. *Genetics* **20**, 545-567.

CHAPTER THREE

Classification of gene mutations

I. MUTATIONS AFFECTING SINGLE GENES

Most mutations are of this type. The main part of this chapter will be devoted to their classification both by effect and by molecular basis. At the end of the chapter we shall briefly consider the much smaller but very important class of mutations that affect simultaneously the action of several genes.

Classification by effect

In a lecture which Muller gave in 1932 at the 6th Congress of Genetics in Ithaca, he classified mutations by their dosage effect (1). He based himself on experiments carried out on sex-linked *Drosophila* genes, in which he manipulated dose with the aid of small deletions and duplications. He distinguished four main types of mutation, which fall into two pairs of mutually complementary members: *hypomorphs* and *hypermorphs, amorphs* and *neomorphs.* Members of the first pair carry out the same function as the normal allele, but hypomorphs do so less efficiently, while hypermorphs do so more efficiently. Thus, the normal allele is hypermorphic in relation to a hypomorphic mutation, but hypomorphic in relation to a hypermorphic one. *Amorphs* have completely lost the function of the normal gene, while *neomorphs* have acquired a new function, not carried out by the normal allele. Thus the normal allele acts as an amorph in relation to a neomorphic mutant gene. To these four main groups, he added *antimorphs* which act in ways that antagonize or impede the action of the normal allele. It is important to realize that this classification was based on rigorously defined experimental tests and not simply on phenotypic comparisons between flies with one, two, or three X-chromosomes. Moreover, Muller was careful to warn against a purely quantitative interpretation of the different types of gene action. A hypomorphic gene for example, does not necessarily produce less enzyme than its normal allele; it may equally well produce the same amount of a less efficiently working enzyme.

This classification of genes by dosage effect has never penetrated much beyond the field of *Drosophila* genetics. It seems useful to revive it now when the validity of the underlying concepts has been proved by biochemical studies, and when classification at the phenotypic level can be confirmed and extended by direct analysis of the gene products. We should, however, keep in mind that every classification by phenotype depends on the level of analysis and that Muller's classification was exhaustive only at the level available to him, that of correlation between gene dosage and phenotype. More recent enzyme studies have confirmed the reality of the classification, but they have also shown that there are mutants that remain outside it. Thus, *Neurospora* strains with very different amounts of the same enzyme may yet be indistinguishable from each other in growth rate (2,3). While these mutations might still be classified as hypo- or hypermorphic at the enzyme level, this is not so for mutations to different iso-enzymes in a polymorphic population, where the possession of one or the other enzyme need not affect the phenotype of the carrier. Such mutations can be classified only at the molecular level. *A fortiori* this is true for mutations that change an amino acid in a gene product without changing its electrophoretic mobility. Such mutations undoubtedly exist and, through heterotic effects or as stepping stones for further molecular changes, may play a role in evolution. Yet they will be missed by any method of ascertainment except sequencing of the amino acids in the gene product or of the nucleotides in messenger or template nucleic acid.

Hypomorphs

The first case of what Muller later called a hypomorph was discovered by C. Stern in 1929 (4). It concerns the sex-linked gene bobbed (*bb*) which, in contrast to other sex-linked genes of *Drosophila melanogaster*, has a homologue on the Y-chromosome. It is recessive, so that ♂♂ with *bb* on their X and $+^{bb}$ on their Y are phenotypically normal, while *bb*/*bb* ♀♀ have short bristles. Stern compared the effect of two doses of *bb* in homozygous ♀♀ with that of three doses in XXY ♀♀ that carried *bb* also on their Y. Contrary to what he had expected, three doses gave a less drastic effect than did two: in the XXY ♀♀ the bristles were more nearly normal than in the XX ones. Stern concluded that the mutant gene *bb* does not yield an abnormal product, but only less of the normal one; addition of an extra mutant gene therefore brings the phenotype nearer to normal.

One of the examples discussed by Muller (5) is white-apricot (w^a), a mutant allele at the white locus that produces apricot-coloured eyes. Like most sex-linked genes in *Drosophila melanogaster*, it is subject to dosage compensation, one recessive allele in the hemizygous male giving the same phenotypic effect as two alleles in the homozygous female. It was therefore necessary to restrict gene dosage comparisons to one or the other sex. When this was done, the following relation was obtained ($<$ stands for 'less normal than', Def for deficiency, Dp for duplication, + for the normal allele).

$$♀♀\ w^a/\mathrm{Def} < w^a/w^a < w^a/w^a/\mathrm{Dp}w^a < +/w^a = +/+$$

$$♂♂\qquad w^a < w^a/\mathrm{Dp}w^a < +$$

Males and females in the same column have the same phenotype in spite of different gene dosage; this is the phenomenon of dosage compensation, which cannot be discussed here. Females that carry one dose of w^a over a deficiency for the locus have lighter eye colour than homozygous w^a ♀♀; this is the phenomenon of exaggeration mentioned in the last chapter (p.27). In both sexes, adding w^a genes brings the phenotype nearer to normal; this stamps the mutant gene as hypomorph.

Another instance of hypomorphism, to which we shall come back later in different contexts, is provided by the cubitus interruptus gene (*ci*) on the small fourth chromosome (6). The mutant gene causes interruptions in one of the large wing veins, and the degree of effect can be measured by the quantity of vein substance present. Gene dosage can be manipulated by adding or subtracting whole chromosomes, since both haplo-IV and triplo-IV flies are viable. Using this technique, Curt Stern found the following relation

$$ci < ci/ci < ci/ci/ci < + < +/+ = +/+/+$$

Normal venation is approximated ever more closely as the dose of *ci* is increased, but even three doses of *ci* are still less effective than one dose of the normal allele. It is interesting to note that full saturation of gene effect is reached only with two normal alleles. One of them by itself still yields a certain proportion of flies with incomplete venation; *ci* is not 'haplo-sufficient' in regard to venation. We shall return to this concept later.

In recent years, the hypomorphic nature of many biochemical mutants has been shown to be due to the production of less enzyme or of an enzyme with reduced efficiency (7). Special interest attaches to 'conditional lethals', i.e. to mutants that behave normally under one set of circumstances, (the 'permissive conditions'), but act as lethals under another set (the 'non-permissive conditions'). The most important class among them are temperature-sensitive lethals, which grow normally at the permissive temperature, but fail to grow at the non-permissive (usually higher) one. They have been found in many organisms (8-13). In several cases, the *ts* mutant gene was shown to produce an enzyme with reduced heat tolerance. *ts* lethals may arise in any gene that codes for a heat-labile enzyme or other protein. Because they can be kept indefinitely at the permissive temperature, they afford excellent material for the genetical and biochemical analysis of mutations that otherwise would be lethal, such as defects in enzymes concerned with DNA-replication.

Hypermorphs

Any mutation that results in a more efficient gene is a hypermorphic mutation but, if the starting point is a hypomorph, the result is not necessarily hypermorphic in relation to wild-type. Muller pointed out that, in *Drosophila,* the saturation of the curve relating phenotype to gene dosage at the level of the normal allele makes it difficult or impossible to detect hypermorphs. In fact, the only examples of hypermorphic mutations cited by him are reverse mutations of hypomorphic genes. However, reverse mutations are exceedingly rare in *Drosophila,* and those that have been found may not have been true reversions. A better claim to be classed as hypermorphs can be made for some of the so-called wild-type iso-alleles that Curt Stern found at the *ci* locus (14). We have seen that the normal allele of *ci* is not haplo-sufficient, so that a certain proportion of individuals with incomplete venation appears among flies that have only one wild-type allele, either because they are haplo-IV or because they carry a deletion for the locus on one of their fourth chromosomes. Normal alleles differ in their degree of haplo-sufficiency and, since *ci* incompleteness of venation clearly is due to a hypomorphic effect, a wild-type allele that in single dose yields hardly any abnormal flies is a hypermorph of those that yield more of them. A somewhat similar situation has been analysed in species crosses of *Neurospora,* where the normal alleles of one particular mutant gene show different degrees of dominance to the same mutant gene when brought together with it in the same genetical background (15).

Enzyme studies have provided more direct proof for the reality of hypermorphic mutations. Reversions of biochemically deficient, hypomorphic mutations have repeatedly been shown to result in restoration of fully functional normal enzyme and, occasionally, in the production of an enzyme that functions better than the original one (2). In order to detect hypermorphs to wild-type genes, special selection techniques are required, and these have led to the detection of some genuine hypermorphs. In *Diplococcus pneumoniae,* selection for resistance to the antifolate amethopterin yielded hypermorphic mutations in the structural gene for the enzyme dihydrofolate reductase (16). Growth of *E.coli* on a normally not fermentable galactoside yielded mutants with increased β-galactosidase activity (17). While most of these had occurred in the regulator gene of the lactose operon, a few appear to have been true hypermorphic changes of the structural gene. In yeast, mutation of the gene for acid phosphatase yielded an enzyme with increased pH range (18).

Amorphs

In *Drosophila,* the lowest member of the allelic series at the white locus is white (*w*) itself, which yields no eye colour at all. The following relationship establishes it as amorph (1):

$$♀♀\ w/\mathrm{Def}w = ♀♀\ w/w = ♀♀\ \mathrm{Dp}w/w/w.$$

Thus, whether the mutant gene is present in single, double or triple dose, its effect is *nil.* While, in heterozygotes, the effect of a hypomorphic gene usually is exaggerated by the presence of a deficiency in the homologous chromosome, the effect of an amorphic gene has already reached the bottom of the scale and can no longer be exaggerated.

The difficulty of distinguishing between an intragenic mutation and deletion of a whole gene is greatest for amorphs and may be insuperable in many cases. To the previously discussed criteria for solving this problem (p.24), a further one can be added. An amorphic mutant, but never a deletion, often produces a protein that, while no longer enzymatically active, is immunologically related to the active enzyme (CRM = cross-reacting material) (19). Since, however, not every gene mutation produces CRM nor every small deletion can be recognized by the usual genetical, cytological or biochemical criteria, there remains a group of ambiguous amorphs for which a distinction between an intragenic change and deletion of the whole gene may not be possible.

Neomorphs

Since these are the converse of amorphs, they can be defined as mutant genes for which the normal allele acts as amorph. Thus, the effect of a given dose of a neomorph should be indifferent to the number of normal alleles with which it is associated in the same cell, while an increase in its own dose should result in increased abnormality. An example is the semi-dominant sex-linked gene hairy-wing (*Hw*) in *Drosophila* (1). A comparison between females carrying different doses of *Hw* and its normal allele (carried either on the X or on a small duplication), gave the following result (where < stands for less normal than).

$$Hw/Hw = Hw/Hw/\mathrm{Dp}+ < Hw/+ = Hw/+/\mathrm{Dp}+.$$

Thus, while increasing the dose of *Hw* made the flies more abnormal, the presence of 0, 1 or 2 normal alleles made no difference to the phenotype. Another gene that behaves in the same way is *B* (barshaped eyes). It is probably no coincidence that both *B* and *Hw* are known to be duplications. It has often been pointed out that a duplication creates a situation in which a gene is free to take over a new function because its previous function can still be carried out by the other member of the pair (see p. 24). In the two quoted cases, the whole duplication is required for the neomorphic action. One may imagine that this is an intermediate step in the development of a new function by one of the genes.

Good candidates for neomorphs that may have arisen in duplications are genes with immunological effects. Thus, the *A* and *B* blood group genes behave as neomorphs in relation to each other and in their biochemical action (20). There is, in fact, an indication that the locus may be a repeat, for recently a few crossovers who carry *A* and *B* in *cis* have been found (21). The genes coding for the α and β human haemoglobins seem to have arisen as neomorphs in duplicate genes which are now no longer linked. In *E.coli,* Campbell *et al.* have created a neomorphic mutation for the production of β-galactosidase by growing a strain lacking the gene for the enzyme on lactose medium (22). The neomorphic mutant gene is not linked to the lactose operon; its original function probably was concerned with lactose utilization, since it had a small amount of lactase activity *in vitro. In vivo,* however, it was unable to utilize lactose before this ability had been restored by the selected mutation (22a). It looks as though an evolutionary pathway that might eventually have led to a new function of a partially degenerated duplicate gene had been reversed by mutation. Whether the term 'neomorph' should be applied to this mutant is therefore doubtful.

Antimorphs

These form an additional class of mutant genes defined by Muller as genes 'having an opposite action to that of the normal allelomorph, competing with the latter when both are present'. The existence of antimorphic genes was inferred from cases where two doses of a recessive gene added to the normal allelomorph made the flies less rather than more normal, or where the mutant gene in combination with its normal allelomorph yielded a more abnormal phenotype than did heterozygosity for deficiency of the gene. Stern interpreted such cases in a way that does not require a special class of genes but can also be applied to hypomorphic genes. His analysis revealed several cases in which hypomorphic alleles of *ci* behaved like antimorphs. Particularly striking was the interaction between *ci* and one of the wild-type iso-alleles (see p. 34), $+^3$ (6). This is a fairly efficient allele, yielding very few flies with incomplete venation when present singly in haplo-IV flies or in combination with a deficiency for the locus. However, when $+^3$ was combined with *ci*, more than 40% of the flies had incomplete venation. Thus, $+^3/ci < +^3/\text{Def}\ ci$, the exact opposite of what is typically found for hypomorphs. Yet, previous studies left no doubt that *ci* is a hypomorph. Stern resolved this discrepancy by distinguishing between two aspects of a gene's efficiency: ability of its product to bind to substrate, and ability to convert substrate into a new product. In the above case, the *ci* allele is assumed to be more efficient than $+^3$ in binding ability but less efficient in substrate conversion. If substrate is limiting, then competition between *ci* and $+^3$ in the heterozygote could result in less of the final product being formed than when $+^3$ acts by itself. A different way of detecting antimorphic effects of mutant genes was derived from enzyme studies on fungi ('negative complementation', see p. 40).

Classification by molecular changes in nucleic acid

After it had been realized that genes consist of sequences of nucleotides carrying the four bases A, T, G, C (or HMC in certain phages), and that polypeptides are read off a cistron in the 5′ to 3′ direction according to a triplet code for amino acids, all possible gene mutations (intracistronic changes) could be fitted into a simple framework (Fig. 3.1). This was first provided by Freese (23) and subsequently enlarged and made comprehensive by Crick (24) and his collaborators. Since a mutation is by definition transmissible, and since the cellular machinery is adapted only to the synthesis and incorporation into DNA of the four naturally occurring nucleotides, mutant DNA cannot contain any new nucleotide base. This restricts the possible mutational changes severely. They may be listed like this. (*a*) Replacement of a base in a given position by a different base (base substitution). (*b*) addition or deletion of one or more bases within a cistron, (*c*) rearrangement – e.g. inversion – of bases in a cistron. No instance of (c) has come to my knowledge, and I therefore shall limit the discussion to the first two possibilities.

Base substitutions

These can be of two essentially different types, according to whether substitution has replaced a purine by a purine or a pyrimidine by a pyrimidine (transitions) or whether there has been substitution of a purine by a pyrimidine or vice versa (transversion). The four possible kinds of transition are $A \rightleftharpoons G$, $T \rightleftharpoons C$. The eight possible types of transversion are $A \rightleftharpoons T$, $A \rightleftharpoons C$, $G \rightleftharpoons C$, $G \rightleftharpoons T$. All have been found to occur. Sequencing of DNA is as yet restricted to very few cases, such as some genes coding for transfer-RNA (25,26); but where it is possible to obtain at least partial resolution of a

2nd position

1st position	U	C	A	G	3rd position
U	PHE	SER	TYR	CYS	U
	PHE	SER	TYR	CYS	C
	LEU	SER	CT	CT	A
	LEU	SER	CT	TRY	G
C	LEU	PRO	HIS	ARG	U
	LEU	PRO	HIS	ARG	C
	LEU	PRO	GLN	ARG	A
	LEU	PRO	GLN	ARG	G
A	ILU	THR	ASN	SER	U
	ILU	THR	ASN	SER	C
	ILU	THR	LYS	ARG	A
	MET	THR	LYS	ARG	G
G	VAL	ALA	ASP	GLY	U
	VAL	ALA	ASP	GLY	C
	VAL	ALA	GLU	GLY	A
	VAL	ALA	GLU	GLY	G

Fig. 3.1. The codon catalogue. CT = chain terminating codon. (Fig. 5.6, Drake, 1970 (bibl.). Courtesy Holden-Day)

polypeptide chain into peptides and amino acids (27-30), mutational changes in coding triplets can be inferred from changes in the corresponding amino acid.

Mis-sense mutations. In structural genes, most base-substitutions will give rise to mis-sense mutations, in which one amino acid has been replaced by a different one. Whether or not this will result in a phenotypic effect depends on the type and site of the change. Because of the degeneracy of the code, mutation in a codon does not always result in replacement of one amino acid by a different one. Where it does – and this will be the usual result – the importance of the change varies with its site in the cistron. It is a well-known observation that mutant sites tend to cluster in certain regions of a cistron, independent of whether the mutations are of spontaneous origin or have been induced by a mutagen. This qualification is important because it distinguishes these regions of generally high mutability from the mutagen-specific 'hot spots', which we shall discuss in Chapter 17. Langridge (31) has suggested that mutational clusters within a cistron code for regions in the corresponding polypeptide chain that are important for tertiary structure, e.g. in forming the binding site of an enzyme. He has started to apply this rationale to a study of tertiary conformation of β-galactosidase in *E. coli.*

The effects of mis-sense mutations on enzyme activity vary. Often, the mutant is 'leaky', i.e. it produces a less efficient enzyme or a smaller amount of the normal one. These mutants act as hypomorphs. If they should compete with the normal enzyme for substrate, they might act as antimorphs. Some mis-sense mutants produce enzymes that are fully functional under one set of conditions and non-functional or poorly functional under another. Temperature-sensitive mutants have already been mentioned (p. 34) (8–13). In fungi, 'osmotic-remedial' mutants, which survive only in hypertonic medium, are usually attributed to mis-sense changes (32,33). Mis-sense mutants that completely lack functional enzyme often yield an immunologically related protein, called cross-reacting material or CRM (p.36). Many mis-sense mutations are of the complementing type, i.e. they can form some functional enzyme when combined with certain other mis-sense mutations in the same cistron. Intracistronic complementation is restricted to enzymes that consist of two or more identical polypeptide chains. Polypeptide chains with localized errors at different sites may form aggregates in which the defect of one is complemented for or masked by the normal configuration of the other. Evidence for this model of intracistronic complementation has been obtained by *in vitro* complementation between mutant enzymes (34-37). Occasionally, the opposite has been found: the enzyme formed by a heterozygote for two mutant alleles is less active or more abnormal in properties than either of the enzymes formed by the alleles when acting alone (38-39). This has been called 'negative complementation'. It can be a source of antimorphic gene action (p. 38) (see also Fincham, Genetic Complementation; Bibliography).

In summary: mis-sense mutations may have any one or several of the following attributes: leakiness, conditional expression, formation of CRM, intracistronic complementation. Possession of one or more of these attributes makes it very likely that a particular mutation is of the mis-sense type, but exceptions have been found (see below).

Nonsense mutations. Instead of producing a mis-sense mutation, a base substitution may change a codon into one of the three 'nonsense' triplets which do not code for any amino acid (Fig. 3.1). Instead, the messenger triplets that they produce – UAG (amber), UAA (ochre), UGA (sometimes called opal) – are not translated and result in premature termination of the growing polypeptide chain. It is easy to see how they can arise from sense codons by base substitutions. For instance, a transition from G to A in the second position changes the triplet for tryptophan UGG into the amber triplet UAG.

The proportion of nonsense mutations among base substitutions is much higher then would be expected from the codon dictionary. In part this may be due to the fact that most mutagenic agents produce preferentially changes from GC into AT (see Chapters 16-18). In part it will result from the fact that nonsense mutations have much more drastic effects than mis-sense mutations and are therefore much more readily detected. While the amino acid changes that represent mutations may remain without noticeable effect, chain termination except very close to the end of the cistron destroys the function of the enzyme: nonsense mutations thus act usually as amorphs. This means that most of them are not leaky and cannot be made leaky by environmental conditions such as temperature or osmotic pressure. Neither do they usually yield CRM. Yet there are exceptions to all the rules that define nonsense mutations by *negative* criteria (33,40). A *positive* criterion is provided by their response to certain specific suppressors; this will be dealt with in the next chapter. In tests for intracistronic complementation, nonsense mutations either fall into the non-complementing class or, if the polypeptide fragment formed by them is sufficiently large, they complement only those alleles that produce a normal polypeptide chain from somewhere in front of the chain-terminating codon to the end of the gene. This results in a peculiar type of 'complementation pattern' which, in *Neurospora,* has been applied to the differentiation between mis-sense and nonsense base changes (41). We shall return to this in Chapter 11.

Both mis-sense and nonsense mutations have been used successfully for establishing co-linearity between the nucleotide sequence in DNA and the amino acid or peptide sequence in the polypeptide chain. In the tryptophan-A cistron of *E.coli*, Yanofsky has compared the genetically determined sites of mis-sense mutations in a

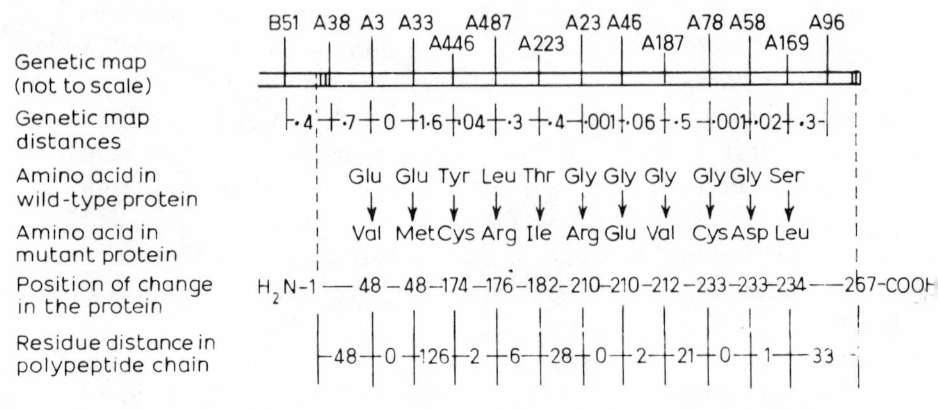

Fig. 3.2. Co-linearity of gene and protein as shown with the aid of mis-sense mutations. (Fig. 2, Yanofsky (42). Courtesy Nat. Acad. Sci. U.S.A.)

given region with the position of the amino acid changes in the corresponding region of the polypeptide (42) (Fig. 3.2). In bacteriophage T4, Brenner and his co-workers have compared the length and composition of the fragments of head protein produced by a series of amber mutations with the genetically determined sites of the amber codons in the DNA of a cistron (43) (Fig. 3.3). In both cases, co-linearity was very satisfactory.

Deletions and additions of one or several bases: frameshift (or sign) mutations

Deletion of one or more codon triplets will result in the loss of one or more amino acids from the corresponding polypeptide. The phenotypic effect depends on the importance of the missing amino acids for the functioning of the protein. An example of a DNA segment in which deletion of whole codons can be tolerated is the beginning of the B cistron of the rII gene in bateriophage T4 (44). The B cistron follows immediately on the A Cistron, but the two are functionally separate as shown by their ability to complement each other's mutations. (This is a case of intercistronic complementation, not to be confused with the intracistronic or interallelic complementation discussed on p. 40). A deletion is known that comprises the beginning of the B cistron together with the end of the A cistron. While the deletion mutant has lost the A function, it has retained the B function. This indicates that the remaining part of the B cistron is translated into the normal amino acid sequence.

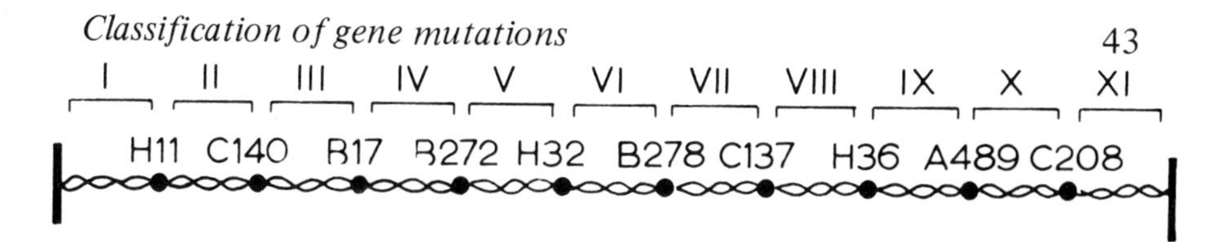

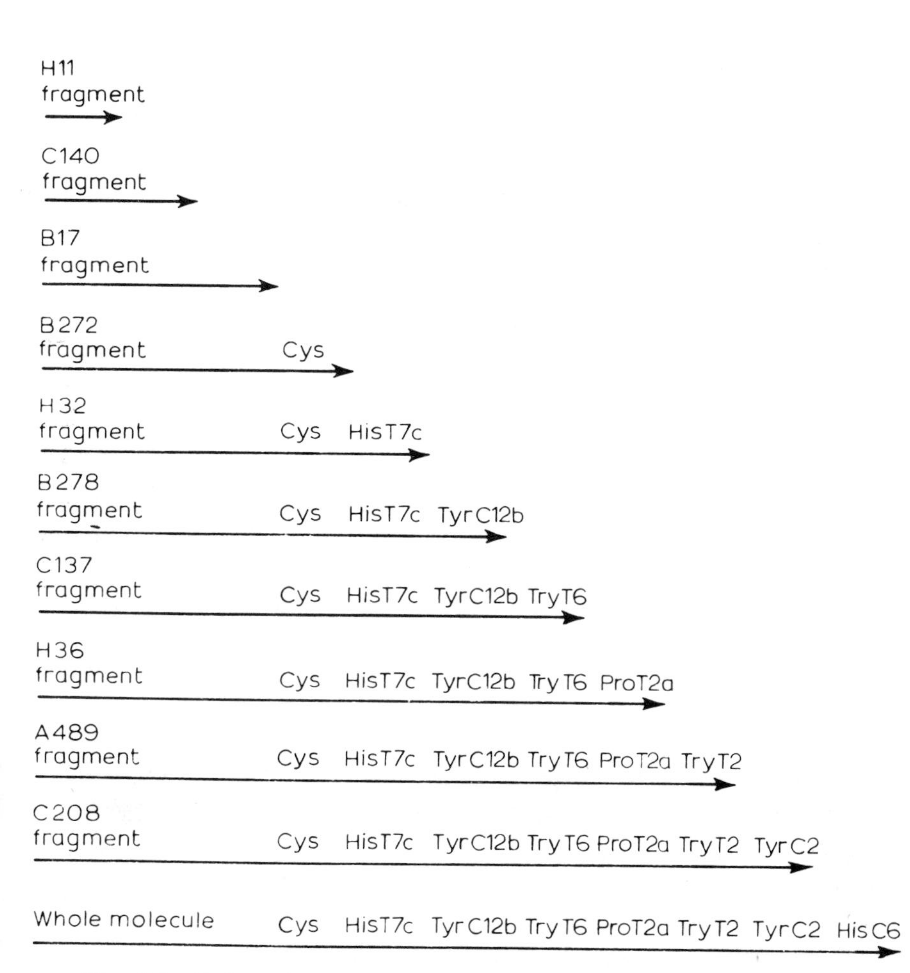

Fig. 3.3. Co-linearity of gene and protein as shown with the aid of nonsense mutations. Top: Sites of amber mutations in the head protein cistron of phage T4D. Bottom: Peptide fragments found in the mutants. (Fig. 3, Sarabhai (43). Courtesy Macmillan Journals.)

This would *not* be the case if the deletion, instead of removing one or more whole codon triplets, would have left in place one or two bases out of a complete triplet; for then translation, starting with the truncated triplet and continuing according to the triplet code, would have yielded a string of amino acids every one of which was 'wrong' in relation to the normal one.

Such changes in the 'reading frame' are produced by all deletions or insertions of bases except those that are multiples of three. Their mode of origin is still under debate and will be discussed in Chapter 21. There is, however, no doubt that frameshifts do occur both spontaneously and under the influence of mutagens. Their properties, as would be expected from their drastic effect, resemble those of nonsense mutations. Like these, frameshift mutants are usually not leaky (for an exception see (45)). Their effect is not dependent on temperature or other environmental conditions, and they do not produce CRM. Many of them are, in fact, nonsense as well as frameshift mutations because most frameshifts are likely to produce a nonsense codon somewhere further down the line of DNA. If this happens, the polypeptide chain terminates at this point but, since the preceding portion consists of a sequence of wrong amino acids, there is no suppression by those extracistronic supersuppressors that suppress simple nonsense mutations (Chapter 4). This is one of the main criteria by which a frameshift mutation can be distinguished from a nonsense mutation. Instead, frameshift mutations may revert by a second frameshift in the same cistron, which restores the register of reading (Fig. 3.4). The rules of codon translation determine the kind of secondary mutation that is complementary to, and suppressive of, a given primary one. Thus, deletion of a single base can be suppressed either by addition of a single base (or of 4 or 7 etc.), or by deletion of two (or 5 or 8 etc.). The stretch of DNA between the primary and secondary mutation will, of course, retain its garbled message. Whether or not the second mutation can effectively suppress the first depends on the length of this useless segment of polypeptide and on its importance in the tertiary structure.

The nature of frameshift mutations was at first inferred from mutational data on a small region of the rII B cistron of T4 (46). Starting with a mutation which, for theoretical reasons (Chapter 17), they assumed to be an insertion or deletion of a base, Crick and his group collected a large sample of spontaneous mutations which either suppressed the original one or each other. They showed that these could be divided into two different groups, called plus and minus, in such a way that suppression could occur only between a plus and a minus but not between two plus or two minus mutations.

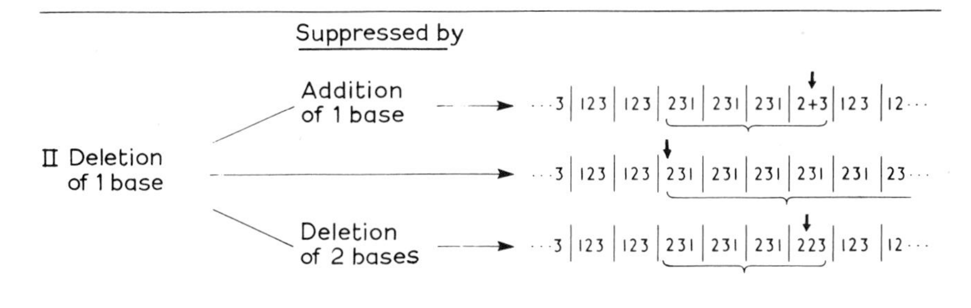

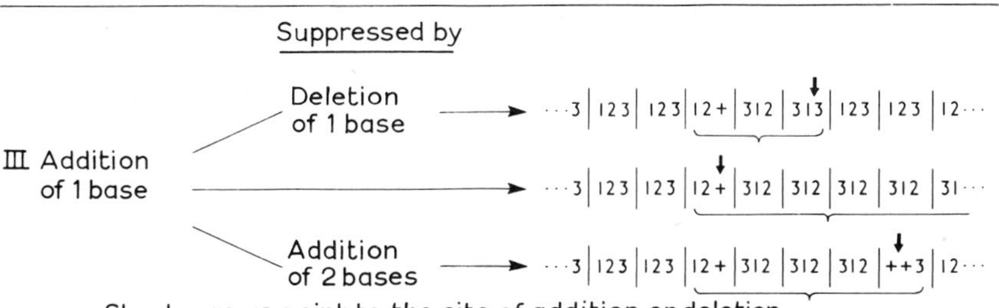

Short arrows point to the site of addition or deletion.
The brackets show the extent of the garbled codescript

Fig. 3.4. Frameshift mutations and their suppression.

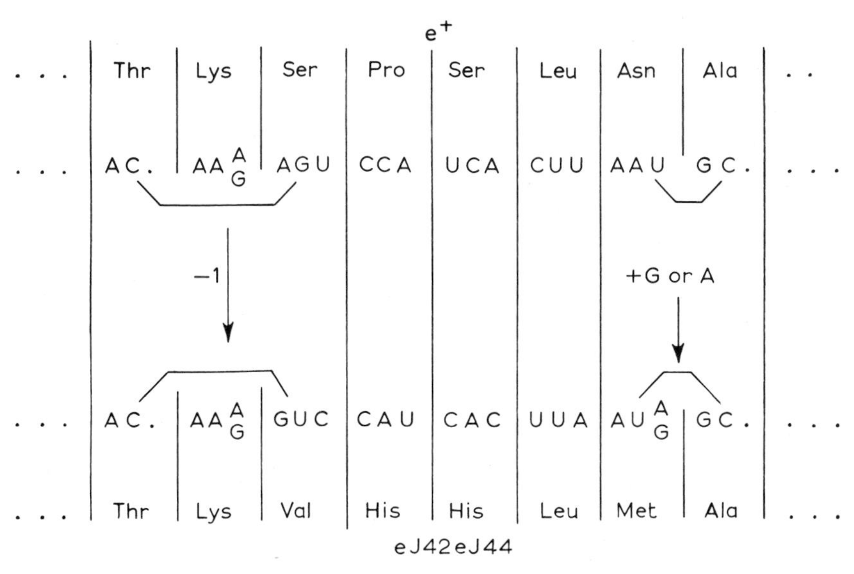

Fig. 3.5. Example of a suppressed frameshift in phage T4. Part of the amino acid sequence of the lysozyme gene in wild-type (e^+) and a suppressed frameshift (double mutant *e*J42*e*J44).(Fig. p.504, Terzaghi (68). Courtesy Nat. Acad. Sciences, U.S.A.)

A model of deletions and insertions of bases fitted these data, although there was no means to decide which of the two signs stood for insertion, which for deletion. Several years later, Streisinger and his group and, soon, other groups showed the correctness of the picture for genes whose products could be sequenced (47). They found that the polypeptides formed by a suppressed frameshift mutation carried the correct amino acid sequences at the beginning and the end of the gene, while a sequence of wrong amino acids was intercalated between two points which were presumed to be the sites of the primary and secondary mutation (Fig. 3.5). By reference to the genetic code, it was also possible to show that in each case the sequence of wrong amino acids could be attributed to one particular kind of primary frameshift. In these cases, it was therefore possible to allot plus signs to addition mutations and minus signs to deletion mutations. The whole story of the discovery of frameshifts is a beautiful example of brillant deduction being verified at first inferentially and finally by direct means.

II MUTATIONS THAT AFFECT SEVERAL OR MANY GENES

The majority of mutations have pleiotropic effects. This is the expected consequence of the fact that the biochemical pathways starting from different genes intersect in many places, reinforcing, inhibiting, deflecting and variously modifying each other. Sometimes the chain of events leading to pleiotropy is very short. Such, for example, is the case for a mutation that makes the activity of many different t-RNA's of *E.coli* temperature sensitive by interfering with the formation of a rare base common to all of them (48); for a mutant leucyl-transfer-RNA-synthetase in *E.coli* that affects five different leucyl-t-RNA's (49); for a permease mutation in *Streptococcus faecalis,* that simultaneously inhibits the uptake of four different amino acids (50). Even here, however, pleiotropy is due to effects on already completed gene products or established biochemical pathways. Such cases are topics for developmental or epigenetic studies and will not further concern us. We shall limit ourselves to the consideration of cases in which pleiotropy occurs because a mutant gene influences a number of other genes at the transcription or translation stages. Five main classes will be distinguished.

Supersuppressors

In strains of micro-organisms that carry several mutant genes, mutations may arise that appear to revert all these genes simultaneously. This is such an important class that we shall consider it separately and fully in the next chapter.

Polarity mutations (51)

These are restricted to genes that form part of an operon. They result in complete inactivation of one cistron and, simultaneously, in reduced activity of all cistrons that are operator-distal to the primary affected one. They are usually caused by chain-terminating (nonsense) mutations (52,53). When frameshifts act as polarity mutants, they may do so by creating a nonsense codon further along the cistron (54). There is, however, at least one case in which the frameshift occurred so close to the end of the cistron that it may have acted in its own right by a carry-over of the wrong reading-frame into the following cistron (55). In many cases, the degree of polarity depends on the position of the nonsense codon within its cistron: the nearer its position to the operator, the stronger the induced polarity (56,57). The detailed mechanism by which chain termination produces polarity has been much discussed and is still under debate. It is, however, certain that translation is involved. This has been shown convincingly in experiments in which the RNA of phage f2 was used as template for protein formation *in vitro* (58). It seems that the whole of the f2 RNA yields one polycistronic message and thus corresponds to an operon in the bacterial DNA. Amber mutations in the gene for coat protein showed polarity as well as a polarity gradient. An amber triplet replacing the sixth amino acid within the cistron prevented translation of the non-coat proteins and the formation of polyribosomes. An amber triplet replacing the seventieth amino acid had neither of these effects. Polarity may be suppressed either by supersuppressors that prevent premature chain termination (59) or by specific polarity suppressors acting in different ways (60,61).

Extreme polarity mutations in the galactose operon of *E.coli* were shown to be due to insertions of up to 1800 nucleotides into one of the cistrons; some with especially strong polar effects were located to the control region of the operon (62-64). Polarity was also caused by insertion of the mutagenic phage Mu-1 into the galactose-operon (65). Polarity mutations that are due to insertion of pieces of DNA differ in some observational respects from those due to chain-termination and may differ from them also in the mechanism by which they create polarity. We shall return to insertion mutations in Chapter 21.

Position effect mutations (66)

The changes due to the so-called 'variegated position effect' resemble polarity mutants in their action on several contiguous genes; they differ from them in a number of essential features. Their effects are not restricted to one operon but may spread over many genes and very long chromosomal distances: thus, inactivation of genes through a transposition in the mouse spreads over 15 centimorgans (67). While polarity mutants affect all operator-distal genes to the same extent, the position effect shows a gradient of decreasing gene inactivation with distance from the primary change. Finally, while polarity mutations are due to intragenic changes, position effects always seem to result from structural changes. For a long time, position effects were considered a unique type of interaction between neighbouring chromosome regions. To-day, there is a strong tendency for fitting them into the general picture of the eukaryotic genome as a dual system, consisting of structural genes and 'controlling elements'. We shall return to this in Chapter 22.

Mutator and antimutator genes

These are genes that influence spontaneous mutation frequencies through being concerned with steps in the replication of DNA or with processes that in other ways affect the fidelity of DNA replication or the degree to which it is buffered against environmental and cellular hazards. We shall deal with them more fully in Chapter 21. Some mutator genes affect sensitivity to the lethal and mutagenic effects of radiation or chemicals; these genes are usually involved in repair processes that play a role in induced as well as spontaneous mutability. Other genes of this category modify responses to certain mutagens while leaving spontaneous mutability unchanged. Mutations in any of these genes have pleiotropic effects on the mutability of all or certain genes.

References

1. Muller, H.J. (1932), 'Further studies on the nature and causes of

gene mutations', *Proc. 6th Int.Cong. Genetics, Brooklyn Botanic Gardens, U.S.A.* Vol. 1, 213-255.

2. Giles, N.H. (1958), 'Mutations at specific loci in *Neurospora', Proc. 10th Int. Congr. Genetics,* Vol. 1, 261-279.
3. Pateman, J.A. and Fincham, J.R.S. (1964), 'Complementation and enzyme studies of revertants induced in an *am* mutant of *N. crassa', Genet. Res. Camb.,* **6**, 419-432.
4. Stern, C. (1929), 'Über die Additive Wirkung Multipler Allele', *Sonderdruck aus dem 'Biologischen Zentralblatt',* **49**, 261-290.
5. Muller, H.J. (1950), 'Evidence of the precision of genetic adaptation', *Harvey Lecture Series XLIII,* 1947-1948, **1**, 165-229.
6. Stern, C. (1948), 'The effects of changes in quantity, combination and position of genes', *Science,* **108**, 615-621.
7. Bonner, D.M., Yanofsky, C. and Partridge, C.W.H. (1952), 'Incomplete genetic blocks in biochemical mutants of *Neurospora', Proc. Nat. Acad. Sci., U.S.A.,* **38**, 25-34.
8. Edgar, R.S. and Lielausis, I. (1964), 'Temperature-sensitive mutants of bacteriophage T4D: their isolation and genetic characterization', *Genetics,* **49**, 649-662.
9. Langridge, G. (1968), 'Thermal responses of mutant enzymes and temperature limits to growth', *Molec. Gen. Genetics,* **103**, 116-126.
10. Hartwell, L.H. (1967), 'Macromolecular synthesis in temperature-sensitive mutants of yeast' *J. Bacter.,* **93**, 1662-1670.
11. Horowitz N.H. and Fling, M. (1953), 'Genetic determination of tyrosinase thermostability in *Neurospora', Genetics,* **38**, 360-374.
12. Suzuki, D.T. (1970), 'Temperature-sensitive mutations in *Drosophila melanogaster', Science,* **170**, 695-706.
13. Roberds, D.R. and de Busk, A.G. (1973), 'Cold-sensitive mutants of *Neurospora crassa', J. Bacter.,* **115**, 1121-1129.
14. Stern, C. and Schaeffer, E.W. (1943), 'On wild-type iso-alleles in *Drosophila melanogaster', Proc. Nat. Acad. Sci. U.S.A.* **29**, 361-367.
15. Srb, A.M. and Basl, M. (1972), 'Evidence for the differentiation of wild-type alleles in different species of *Neurospora*', *Genetics,* **72**, 759-762.
16. Sirotnak, F.M. Hachtel, S.L. and Williams, W.A. (1969),

'Increased dihydrofolate reductase synthesis in *Diplococcus pneumoniae* following translatable alteration of the structural gene. 11. Individual and dual effects on the properties and rate of synthesis of the enzyme', *Genetics,* **61**, 313-326.

17. Langridge, J. (1969), 'Mutations conferring quantitative and qualitative increases in β-galactosidase activity in *Escherichia coli', Molec. Gen. Genetics,* **105**, 74-83.
18. Francis, J.C. and Hansche, P.E. (1972), 'Directed evolution of metabolic pathways in microbial populations. 1. Modification of the acid phosphatase pH optimum in *S. cerevisiae', Genetics,* **70**, 59-73.
19. Suskind, S.R., Yanofsky, C. and Bonner, D.M. (1955), 'Allelic strains of *Neurospora* lacking tryptophan synthetase: a preliminary immunochemical characterization; *Proc. Nat. Acad. Sci., U.S.A.* **41**, 577-582.
20. Ingram, V.M. (1961), 'Gene evolution and the haemoglobins', *Nature,* **189**, 704-708.
21. Salmon, C. (1971), 'Le complex cis AB et le locus ABO', *Proc. 4th Int. Congr. Human Genetics, Paris in Excerpta Medica. Int. Congress Series,* **233**, 10.
22. Campbell, J.H., Lengyel, J.A. and Langridge, J. (1973), 'Evolution of a second gene for β-galactosidase in *Escherichia coli', Proc. Nat. Acad. Sci., U.S.A.,* **70**, 1841-1845.

22a. Hartl, D.L. and Hall, B.G. (1974), 'Second naturally occurring β-galactosidase in *E.coli.' Nature,* **248**, 152-153.

23. Freese, Ernst (1963), 'Molecular mechanism of mutations', in *Molecular Genetics Part 1,* Ch. V., ed., J.H. Taylor, Academic Press, New York.
24. Crick, F.H., Barnett, L., Brenner, S. and Watts-Tobin, R.J. (1961), 'General nature of the genetic code for proteins', *Nature,* **192**, 1227-1232.
25. Holley, R.W., Apgar, J., Everett, G.A., Madison, J.T. *et al.* (1965), 'Structure of a ribonucleic acid', *Science,* **147**, 1462-1465.
26. Agarwal, K.L., Büchi, H., Caruthers, M.H., Gupta, N., Khorana, H.G. *et al.* (1970), 'Total synthesis of the gene for alanine transfer ribonucleic acid from yeast', *Nature,* **227**, 27-34.
27. Wittmann, H.G. and Wittmann-Liebold, B. (1966), 'Protein chemical studies of two RNA viruses and their mutants', *Cold Spring Harbor Symp. Quant. Biol.,* **31**, 163-172.

28. Yanofsky, C. (1965), 'Gene structure and protein structure', *The Harvey Lecture Series*, **61**, 145-168.
29. Prakash, L. and Sherman, F. (1973), 'Mutagenic specificity: reversion of Iso-1-cytochrome *c* mutants of yeast', *J. Mol. Biol.*, **79**, 65-82.
30. Gilmore, R.A., Stewart, J.W. and Sherman, F. (1971), 'Amino acid replacements resulting from super-suppression of nonsense mutants of Iso-1-cytochrome *c* from yeast', *J. Mol. Biol.*, **61**, 157-173.
31. Langridge, J. (1968), 'Genetic evidence for the disposition of the substrate binding site of β-galactosidase', *Proc. Nat. Acad. Sci. U.S.A.*, **60**, 1261-1267.
32. Hawthorne, D.C. and Friis J. (1964), 'Osmotic-remedial mutants. A new classification for nutritional mutants in yeast', *Genetics*, **50**, 829-839.
33. Jones, E.W. (1972), 'Fine structure analysis of the *ade* 3 locus in *Saccharomyces cerevisiae*', *Genetics*, **70**, 233-250.
34. Case, M.E., Burgoyne, L. and Giles, N.H. (1969), '*In vivo* and *in vitro* complementation between DHQ synthetase mutants in the *arom* gene cluster of *Neurospora crassa*', *Genetics*, **63**, 581-588.
35. Foley, J.M., Giles, N.H. and Roberts, C.F. (1965), 'Complementation at the adenylosuccinase locus in *Aspergillus nidulans*', *Genetics*, **52**, 1247-1263.
36. Glassman, E. (1962), 'Complementation between non-allelic *Drosophila* mutants deficient in xanthine dehydrogenase', *Proc. Nat. Acad. Sci., U.S.A.*, **48**, 1491-1497.
37. Schwartz, D. (1960), 'Genetic studies on mutant enzymes in maize: synthesis of hybrid enzymes by heterozygotes', *Proc. Nat. Acad. Sci., U.S.A.*, **46**, 1210-1215.
38. Nashed, N.G., Jabbur, G. and Zimmerman, K.F. (1967), 'Negative complementation among the *ad*2 mutants in yeast', *Molec. Gen. Genetics*, **99**, 65-75.
39. Garen, A. and Garen, S. (1963), 'Complementation *in vivo* between structural mutants of alkaline phosphatase from *E. coli*', *J. Mol. Biol.*, 7, 13-22.
40. Lin, S. and Zabin, I. (1972), 'β-Galactosidase: rates of synthesis and degradation of incomplete chains'. *J. Biol. Chem.* **247**, 2205-2211.
41. De Serres, F.J. (1964), 'Mutagenesis and chromosome structure', *J. Cell. Comp. Physiol.*, **64**, supp. 1, 34-42.

42. Yanofsky, C., Drapeau, G.R., Guest, J.R. and Carlton, B.C. (1967), 'The complete amino acid sequence of the tryptophan synthetase A protein (α subunit) and its colinear relationship with the genetic map of the A gene'. *Proc. Nat. Acad. Sci., U.S.A.,* **57**, 296-298.
43. Sarabhai, A.S., Stretton, A.O.W., Brenner, S. and Bolle, A. (1964), 'Co-linearity of the gene with the polypeptide chain', *Nature,* **201**, 13-17.
44. Champe, S.P. and Benzer, S. (1962), 'An active cistron fragment', *J. Mol. Biol.,* **4**, 288-292.
45. Atkins, J.F., Elseviers, D. and Gorini, L. (1972), 'Low activity of β-galactosidase in frameshift mutants of *Escherichia coli', Proc. Nat. Acad. Sci., U.S.A.* **69**, 1192-1195.
46. Crick, F.H.C., Barnett, L., Brenner, S. and Watts-Tobin, R.J. (1961), 'General Nature of the Genetic Code for proteins', *Nature,* **192**, 1227-1232.
47. Streisinger, G., Okada, Y., Emerich, J. *et al.* (1966), 'Frameshift mutations and the genetic code', *Cold Spring Harbor Symp. Quant. Biol.,* **31**, 77-84.
48. Yamamoto, M., Endo, H. and Kuwano, M. (1972), 'A temperature-sensitive mutation in *Escherichia coli* transfer RNA', *J. Mol. Biol.,* **69**, 387-396.
49. Kan, J. and Sueoka, N. (1971), 'Further evidence for a single leucyl transfer ribonucleic acid synthetase capable of changing five leucine transfer ribonucleic acids in *E.coli', J. Biol. Chemistry,* **246**, 2207-2210.
50. Ashgar, S., Levin, E. and Harold, F.M. (1973), 'Accumulation of neutral amino acids by *Streptococcus faecalis', J. Biol. Chemistry,* **248**, 5225-5233.
51. Section on 'Polarity', in 'Symposium on The Genetic Code', (1966), *Cold Spring Harbor Symp. Quant. Biol.,* **31**, 181-249.
52. Morse, D.E., Mosteller, R.D. and Yanofsky, C. (1966), 'Dynamics of synthesis, translation and degradation of trp operon messenger RNA in *E.coli', Cold Spring Harbor Symp. Quant. Biol.,* **34**, 725-740.
53. Imamoto, F., Kano, Y. and Tani, S. (1970), 'Transcription of the tryptophan operon in nonsense mutants of *Escherichia coli', Cold Spring Harbor Symp. Quant. Biol.,* **35**, 471-490.
54. Martin, R.G. (1967), 'Frameshift mutants in the histidine operon of *Salmonella typhimurium', J. Mol. Biol.,* **26**, 311-328.
55. Rechler, M.M. and Martin, R.G. (1970), 'The intercistronic

divide: translation of an intercistronic region in the histidine operon of *Salmonella typhimurium*', *Nature,* **226**, 908-911.

56. Newton, W.A., Beckwith, J., Zipser, D. and Brenner, S. (1966), 'Nonsense mutants and polarity in the *Lac* operon of *Escherichia coli*', *J. Mol. Biol.,* **14**, 290-296.
57. Fink, G.R. and Martin, R.G. (1967), 'Translation and polarity in the histidine operon, 2. Polarity in the histidine operon', *J. Mol. Biol.,* **30**, 97-107.
58. Engelhardt, D.L., Webster, R.E. and Zinder, N.D. (1967), 'Amber mutants and polarity *in vitro*', *J. Mol. Biol.,* **29**, 45-58.
59. Yanofsky, C. and Ito, J. (1966), 'Nonsense codons and polarity in the tryptophan operon', *J. Mol. Biol.,* **21**, 313-334.
60. Carter, T. and Newton, A. (1971), 'New polarity suppressors in *Escherichia coli:* Suppression and messenger RNA stability', *Proc. Nat. Acad. Sci., U.S.A.,* **68**, 2962-2966.
61. Morse, D.E. and Guertin, M. (1972), 'Amber *suA* mutations which relieve polarity', *J. Mol. Biol.,* **63**, 605-608.
62. Jordan, E., Saedler, H. and Starlinger, P. (1968), 'O^0 and strong-polar mutations in the *gal* operon are insertions', *Molec. Gen. Genetics,* **102**, 353-363.
63. Saedler, H., Besemer, J., Kemper, B., Rosenwirth, B. and Starlinger, P. (1972), 'Insertion mutations in the control region of the *gal* operon of *E.coli.* 1. Biological charaterization of the mutations', *Molec. Gen. Genetics,* **115**, 258-265.
64. Hirsch, H.J., Saedler, H. and Starlinger, P. (1972), 'Insertion mutations in the control region of the galactose operon of *E.coli,* 11. Physical characterization of the mutations', *Molec. Gen. Genetics,* **115**, 266-276.
65. Taylor, Austin L. (1953), 'Bacteriophage-induced mutation in *Escherichia coli*', *Proc. Nat. Acad. Sci. U. S. A.* **50**, 1043-1051.
66. Baker, Wm. K. (1968), 'Position-effect variegation' in '*Advances in Genetics*', Vol. 14, 133-169, ed. E.W. Caspari, Academic Press, New York-London.
67. Cattanach, B.M. (1968), 'Incomplete inactivation of the Tabby locus in the mouse X-chromosome', *Genetics,* **60**, 168.
68. Terzaghi, E., Okada, I., Streisinger, G., Emrich, J., Inouye, M. and Tsugita, A. (1966), 'Change of a sequence of amino acids in phage T4 lysozyme by acridine-induced mutations', *Proc. Nat. Acad. Sci., U.S.A.,* **56**, 500-507.

CHAPTER FOUR

Reverse mutations. Suppressors

Operationally, reverse mutations are defined as mutations that fully or partially restore the activity of a mutant gene. In the early days of mutation research, the occurrence of reverse mutations was used as an argument against the presence-absence theory (Chapter 1). Subsequently, the same argument was used to prove that X-rays can produce intragenic changes. However, in *Drosophila* the evidence for reverse mutation is meagre, and in maize – the only other object of mutation studies at that time – the then best authenticated case of reverse mutation appears to be due to loss of a controlling element rather than to true gene mutation (see Chapter 22). Moreover, it was soon realized that apparent reversions may be due to suppressor mutations in other genes. The distinction between true reversion and reversion by a suppressor mutation can be made easily when the mutant gene and the suppressor are not closely linked, for then the mutant gene will again segregate out unsuppressed in crosses to wild-type. If, however, linkage between mutant gene and suppressor is close, separation between them by crossing-over will be correspondingly rare. Large experiments are necessary to detect linked suppressors, and failure of the mutant gene to reappear in the progeny cannot usually exclude the possibility of very close linkage. A definite distinction between suppressors and true revertants has become possible only recently by analysis at the molecular level in some microbial systems.

While suppressors thus created difficulties in the interpretation of apparent reverse mutations, they were seen to be of interest in their own right as pointers to interactions between developmental pathways. With the advent of biochemical and molecular genetics, the role of suppressors for the study of gene action has become increasingly important. In addition, analysis of suppressors has been of great use in elucidating the nature of the genetic code and in allotting codons to chain-terminating triplets. Suppressors do not usually, as the term seems to imply, counteract the harmful effects of other genes. Much more frequently, they compensate in some way for deficiencies of function in the suppressed gene.

True reverse mutations

These have become a major tool for inferring the action of chemical mutagens on DNA; we shall deal with this in Chapter 17. Here, we shall consider reverse mutations only in contrast to suppressor mutations. By definition, true reversions restore the original base sequence in DNA. For base substitutions this requires replacement of the mutant base pair by the one present in wild-type; for frame-shift mutations it requires deletion of one particular base from one particular site, or insertion of one particular base into one particular site. It is obvious that true reverse mutations must be less frequent than forward mutations, which may arise at many sites within a gene and in a variety of ways. The only convincing way to prove that a true reversion has occurred is to show, by sequencing of the gene product, that the mutant amino acid has been replaced by the original one. The first analysis of this kind was carried out by Yanofsky and his co-workers on mutants of the tryptophan-synthetase A protein of *E.coli* (1). It was then found that even reverse mutations occurring inside the mutant cistron are not of necessity true reversions.

Intracistronic suppression

Suppression of a frameshift

The most frequent cause of a frameshift reversion is a second frame-shift in the same cistron (Fig. 3.4). In fact, this is one of the characteristics by which frameshift mutations can be identified. In phage T4, cases have been described in which a frameshift mutation was suppressed by a nearby base substitution, which created a signal for reinitiation of translation (2). Suppression of a frameshift by mutations outside the gene will be discussed in a later section of this chapter.

Suppression of a base substitution

The following case, taken from Yanofsky's work, shows that activity of a mutant enzyme can be restored when a second base substitution in the same codon results in production of an amino acid that is more acceptable than the original mutant one (3). Starting with a mis-sense mutation from glycine to arginine, four types of reversion were obtained, two of which produced fully active enzyme (full revertants), while the other two had reduced enzyme activity (partial revertants). Amino acid sequencing showed that one of the former was a true reversion; the other three had reverted by intracodon suppression. The amino acids in the wild-type, the mutant and the full and partial revertants were as follows.

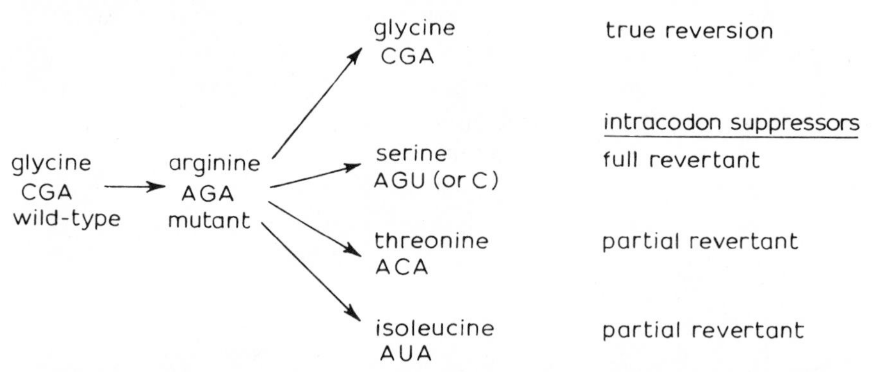

In these experiments, the amino acid replacements in the revertants were only a fraction of those that could have occurred theoretically as the result of single intracodon base changes. This was tentatively attributed to the fact that the missing amino acids were unable to restore the function of the gene. By the ingenious trick of selecting under the shelter of certain extragenic suppressors (see below), this was indeed found to be true (4). It illustrates the important truth that only a fraction of potentially mutagenic changes in DNA give rise to detected mutations.

The following example shows that also a nonsense mutation can be partially reverted by an intracodon suppressor. In bacteriophage T4, Koch (5) studied reversions of nonsense mutations at a site of the rII locus at which true reversions (to glutamine) could be phenotypically distinguished from reversion by mutation at a second site within the codon (to tryptophan).

Type of mutation	*Codon*	*Amino acid*	*Phenotype*
	CAG	glutamine	wild-type
primary: transition C → U	UAG	(amber)	chain termination
true reversion: transistion U → C	CAG	glutamine	wild-type
secondary: transition A → G	UGG	tryptophan	mutant

More difficult of interpretation are cases in which the effect of a base substitution in one codon can be suppressed by a second base substitution in a different codon of the same cistron. One such case, again taken from Yanofsky's work (6), is the following. In a particular region of the tryptophan synthetase gene, substitution of the normal amino acid glycine by glutamic acid (GGA → GAA) had resulted in loss of enzymatic activity; but this could be restored by substitution of cysteine (UGU) for tyrosine (UAU) in a codon that was separated from the mutant one by 36 amino acid residues. In a second case of this kind, the sites of both the primary and secondary mutations were shifted by two codons in relation to those involved in the first case, so that the two mutant amino acids were again separated by 36 residues. This points to a connection between these two regions in the tertiary structure of the polypeptide chain. Mutational analyses like this may help elucidate the tertiary structure of enzymes (see also p. 40).

Phenotypic suppression

As a preliminary to certain important types of extracistronic suppressors, it will be useful to review some instances of phenotypic suppression, i.e. of cases in which expression of a mutant phenotype can be prevented by special means. Many cases of phenotypic suppression are trivial results of palliative treatment, such as suppression of death from auxotrophy by provision of the required nutrilite, or suppression of hereditary diabetes through insulin. Others, such as suppression of auxotrophy at a certain temperature, are restricted to specific gene products that are temperature-sensitive or in other ways dependent on environmental conditions (p. 34). There are, however, some cases where phenotypic suppression acts on a variety of mutations at different loci which are not functionally related to each other. Thus, in *E.coli*, addition of sublethal doses of streptomycin to the medium allows a number of auxotrophs to grow even in the absence of their required supplement (7,8). That this is not due to the antibiotic standing in physiologically for the missing growth factor is shown by the fact that suppression is restricted to certain alleles at the responding loci, and that these loci may have quite diverse functions. In such a situation, it seems likely that suppression acts on a process which is required for the action of all genes whatever their function. Processes of this general nature are transcription and translation. For phenotypic suppression by streptomycin, translation has been implicated as the relevant step. Experiments *in vitro* (9) showed that ribosomes from streptomycin-resistant cells committed certain coding errors that were not committed by ribosomes from sensitive cells. It is easy to see how such coding errors may compensate for mutational faults in DNA by occasionally mis-reading the wrong code word for the correct one. Whether or not a mutant responds to this phenotypic 'curing' will depend on the type of error it carries in its DNA and on the kind of amino acid that may be inserted through mis-reading. The fact that the ionic environment plays an important role in determining the degree of mis-reading *in vitro* and of phenotypic suppression *in vivo* emphasizes the complexity of the processes that lead to the expression of mutant genes. Genetically, resistance to streptomycin and some other antibiotics and the concomitant ability to 'cure' certain auxotrophs is due to mutations in genes that code for the ribosomal proteins (10). Two mutations, *strA* and *ram,* control the degree of misreading in opposite ways: while *ram* (ribosomal ambiguity) increases the ambiguity of misreading, *strA* (restricted ambiguity) restricts it (11). When combined in the same cell, *ram* and *strA* counteract each other. Both mutations interact with the effects of streptomycin. In cells carrying *ram,* phenotypic suppression occurs even in the absence of streptomycin; in cells carrying different alleles of *strA,* suppression requires unusually high amounts of streptomycin or does not occur at all. We shall see later in this chapter that these two ribosomal mutations interact also with the *genotypic* suppression of auxotrophs. The possibility that the 'leakiness' of many auxotrophs may in part be due to occasional mis-reading even by normal ribosomes was suggested by Gorini and borne out by experiments in which the degree of leakiness in frameshift mutants of *E.coli* was shown to be controlled by streptomycin in the medium and by mutations in the *ram* and *strA* loci (12).

A different type of phenotypic suppression involves messenger-RNA (13,14). In bacteria and bacteriophage, 5-fluorouracil (5-FU) was found to suppress an array of nonsense mutations. The suppression was only phenotypic and was not carried over into the progeny. The following mechanism was postulated to account for these findings. FU, like U, pairs with A. At transcription, it can be incorporated into m-RNA instead of U. At a site where this has happened, translation becomes ambiguous, for FU may be read either 'correctly' as U or 'incorrectly' as C. In the latter case, the mutational error in DNA may be compensated for by the translational error; the amino acid formed is that of the wild-type, and the phenotype will be wholly or partially wild-type.

DNA (reading strand)	*m-RNA*	*t-RNA*	*protein*
G	C	G	wild-type
A	U	A	mutant
A	FU	A	mutant
		G	wild-type

The model makes two predictions, both of which were borne out by the results. (1) Only mutations which, on other grounds, are presumed to have an AT-pair at the mutant site, will be suppressed. (2) Of these, only one half will respond to suppression, namely those that have A on the reading strand of the double helix. In *E.coli*, amino acid sequencing of the alkaline phosphatase in a phenotypically suppressed amber mutant has proved the correctness of the inferred mis-translation process (15a).

UV, too, can produce phenotypic suppression by means of altered messenger-RNA (15b). One of its photoproducts (Chapter 13), hydrated uracil, results in the miscoding of UAG as CAG, and this in turn leads to the incorporation of glutamine at the site of the nonsense triplet. Amber mutants of genes that can function with glutamine at the nonsense site are suppressed by this mechanism. It is of interest in this context that UV is a universal mutagen and that fluorouracil has been found mutagenic in fungi (Chapter 18). The ability of mutagens to act as phenotypic suppressors via manipulation of m-RNA opens the intriguing possibility that a mutagenic treatment may at the same time *produce* mutations and selectively *prevent* some of these from becoming manifest. We shall return to this in Chapter 20.

By no means all amber and ochre mutations respond to suppression by 5-FU. Some are very strongly suppressed, others not at all, and in between there is a whole range of suppressibility. Part of this must be due to the fact that the amino acid inserted through misreading is not equally acceptable at all sites of the polypeptide chain; but even when this source of differences is excluded, variability remains enormous (16). Table 4.1 shows the efficiency with which 5-FU suppresses seven ochre mutations in phage T4, in all of which mis-reading inserts the wild-type amino acid.

Table 4.1

Differences in the efficiency of suppression by 5-FU acting on seven ochre mutants in the rIIB cistron of phage T4. (In all cases, mis-reading by 5-FU inserts glutamine at sites at which the normal amino acid is glutamine). *After* Table 1 in Salser (16).

Mutant	*AP53*	*N24*	*SD160*	*N17*	*N7*	*N12*	*N29*
Efficiency* of suppression	0.62	0.4	0.21	7.8×10^{-2}	5×10^{-2}	1.4×10^{-2}	9×10^{-5}

*measured as phage yield when the wild-type yield is taken as 1.

When the position of these and other nonsense mutations in the rIIB cistron was plotted against their response to suppression by 5-FU, no correlation whatsoever was found. Thus, if some kind of 'position effect' is responsible for their differences in suppressibility, it must be restricted to immediately adjacent nucleotide sequences. Salser has put forward a hypothesis according to which the efficiency of nonsense suppression depends on a nucleotide sequence that is longer than the nonsense codon itself. This hypothesis also covers site-specificities in extracistronic suppression, with which we shall deal in the next section.

Extracistronic suppression

We have seen that, in *E. coli,* streptomycin-induced ribosomal modifications produce phenotypic suppression of certain auxotrophs, and that genes concerned with ribosomal structure affect the efficiency of this process. In *Salmonella,* a suppressor of histidine-auxotrophs has been attributed to mutation in a gene of this type (17).

While these suppressors are genotypical analogues of phenotypic suppression by antibiotics, other suppressors act as genotypical equivalents of suppression by 5-FU. The same sets of rII mutants in phage T4 and of alkaline phosphatase mutants in *E.coli* that are suppressed phenotypically by 5-FU respond to genotypical suppression by external suppressors (18), and they show a similar site-specificity of response (16). This indicates that suppression in both cases acts by the same mechanism, i.e. through compensatory errors of translation. However, while in phenotypic suppression misreading is due to incorporation of a wrong base into m-RNA, in genotypic suppression a gene-controlled step must be involved. This suggested at once that these suppressors arise by mutation of genes that control the formation of either a t-RNA or an aminoacyl transferase. The latter possibility is excluded for nonsense suppression, since there are no t-RNA's that can read nonsense triplets. So far, there has also been no case of mis-sense suppression via a mutant aminoacyl transferase. If this is not due to an insufficient number of analysed cases, it may be explained by the fact that most or all transferases appear to be represented only once in the bacterial genome; this must make mutations to ambiguity in these genes very damaging or lethal. In contrast, the suppressor action of mutant t-RNA has been verified both for mis-sense and nonsense mutations. We shall consider these in turn.

Mis-sense suppression

Suppressed mis-sense mutants often contain not only the CRM (p. 36) that is formed by the mutant gene but also a small amount of functional protein. This results in dominance or semi-dominance of the suppressor. One of the best analysed cases refers to a mis-sense mutation in the tryptophan synthetase gene of *E.coli* (19). The mutant produces a CRM substance in which a glycine residue is replaced by arginine. The suppressed mutant produces the same CRM but also a small amount of the normal enzyme. *In vitro*, wild-type t-RNA translates poly AGAG... only into arginine (AGA) and glutamic acid (GAG), but addition of t-RNA from a suppressor strain also allows incorporation of a small amount of glycine (GGA). Obviously it does this by introducing a reading ambiguity. This might happen in various ways. A change in the anticodon of glycine-t-RNA might result in the occasional misreading of the arginine codon as glycine. Or a change in the specificity of the amino acid accepting end of arginine-t-RNA might result in its occasionally getting charged with glycine. Or finally, some change elsewhere in either the glycine- or arginine-t-RNA molecules might affect their specificity of reading or accepting. In this particular case, the evidence showed that the change had occurred in a minor species of several iso-acceptors for glycine, such that glycine (GGA) was inserted instead of arginine (AGA). The assumption that the change had occurred in the anticodon (from CCU to UCU) was supported by experiments (20) with glycine-t-RNA that had been treated with nitrous acid, a treatment that is known to change C into U (Chapter 17). *In vitro*, the treated t-RNA behaved like the suppressor, i.e. it inserted a small amount of glycine into the polypeptide chain coded for by poly (A-G). The presumed mechanism of suppression is shown in the following scheme.

Codon in m-RNA	*Anticodon*	*Amino acid inserted*	*Enzyme*
	normal gly-t-RNA		
GGA	CCU	glycine	normal
	normal arg-t-RNA		
AGA	UCU	arginine	mutant
	mutant gly-t-RNA		
AGA	UCU	glycine	normal

A similar mechanism was shown to operate in the suppression of a mis-sense mutation in which the mutant site of the protein carried cysteine instead of glycine (21). Although bacterial strains differ in the efficiency with which a given combination between a mutant mis-sense gene and its suppressor can produce normal enzyme (22), mis-sense suppression is generally not very efficient. Indeed, it could hardly be otherwise, for replacement of one amino acid by another at high frequency must lead to errors in practically all gene products. Only cells in which suppression is carried out by a minor t-RNA species, or in which mis-translation by a mutant major species has a low probability, will be able to survive. The possibility of producing translational suppressors without too much interference with the function of the normal genes may be one of the reasons for the over-representation of t-RNA loci in many species. In *Drosophila* (23), it has been estimated that each of 60 t-RNA's is, on the average, coded for by 13 structural genes. In yeast (24), one particular species of tyrosine-t-RNA is coded for by at least eight different genes. In a strain of *E.coli* that carried a suppressor for a mis-sense mutation, there was strong selection for cells with a duplication for the relevant t-RNA locus (25). This is a good example for the role of duplications in evolution.

Nonsense suppression (26)

Nonsense mutations, too, may be suppressed by complementary mutations in elements of the translation machinery. In the suppressed strains, a chain-terminating codon is read as an amino acid, which is inserted into the polypeptide chain; often, this results in restoration of gene function. One of the best analysed cases may serve as example (27). An amber mutation in the β-galactosidase gene of *E.coli* was suppressed by mutation in a gene for a minor species of tyrosine-t-RNA. Sequencing of t-RNA showed that the anticodon in the normal gene was AUG and attached to the UAC codon for tyrosine, while the anticodon in the suppressor gene was AUC and attached to the amber codon UAG. Since the amino acid accepting specificity had remained unchanged, the mutant t-RNA carried tyrosine to the point of chain termination and inserted it into the polypeptide chain. Another amber mutation of the β-galactosidase gene was refractory to the action of this suppressor, because functioning of the enzyme required glutamine rather than tyrosine at this particular site. When the acceptor specificity of the suppressor was changed from tyrosine to glutamine by a mutation at the amino acid acceptor end, the previously refractory mutant became responsive to the suppressor (28). Mutations elsewhere in the t-RNA molecule may affect its specificity of amino acid acceptance or of codon-reading. The UGA codon was shown to be suppressed by a mutation outside the normal anticodon for tryptophan (29).

By allowing the formation of at least a certain amount of functional enzyme, nonsense suppressors are dominant or semi-dominant in heterozygotes with the normal t-RNA gene. Recessive supersuppressors of bacteria (31,37) are assumed to act in a different way, probably through some protein that is involved in translation. Nonsense suppressors are more or less specific for one of the three nonsense codons and suppress mutations due to this particular codon wherever they occur, not only in the genome of the cell but also in that of the episomes or phages the cell may harbour. As a result, nonsense suppressors are allele-specific but not gene specific. Thus, an efficient ochre suppressor may restore the function of many auxotrophic genes, while being unable to restore the function of these very same genes when they have been made non-functional through a mutation other than ochre. This peculiar range of action distinguishes nonsense suppressors (and, to some extent, mis-sense suppressors) from other kinds of suppressor. Because of their pleiotropic action (p. 47), nonsense suppressors are often called supersuppressors. Their pleiotropic specificity can be used for recognizing nonsense mutations: any mutation that responds to a nonsense suppressor is classified as nonsense, and its type of nonsense codon may be determined by the use of specific suppressors. Conversely, reversion via supersuppression can be distinguished from true reversion by the ability of the revertant cell to suppress known nonsense mutations. In bacteria, phages carrying nonsense mutations are often used for classifying reversions in the host cell (32).

In contrast to most mis-sense suppressors, many nonsense suppressors allow good growth survival. Since nonsense codons are not found in non-mutant genes except as termination signs, the ill effects of suppression will be limited to the possible running together of the message for two adjoining polypeptide chains. It seems that genes may be protected against this contingency by being separated from their neighbours not only by termination and initiation codons (sometimes several of the former), but also by short strings of untranslatable nucleotide sequences (33-35). The fact that suppressed ambers and opals often grow very well, while suppressed ochres usually are weak strains, points to UAA as the most important codon for normal chain termination.

A cell carrying a nonsense suppressor may be disadvantaged not only by the new function that the mutant t-RNA carries out but also by the absence or inefficiency of its old function. A t-RNA that has been mutationally altered so as to carry its specific amino acid to the polypeptide chain in response to a nonsense codon, usually will no longer do this in response to the normal codon for this amino acid. Unless there are other t-RNA's that share in the transport of this particular amino acid, growth of the cell will be severely impaired and may cease altogether. The same applies to mis-sense mutations; on p.63 we have mentioned that, in a bacterial strain with a mis-sense mutation, a duplication for the relevant t-RNA gene was strongly selected for (25). Nonsense suppressors that act as lethals have been successfully selected in bacterial strains that carried the relevant chromosomal region in duplicate on an episome (36,37a). Under cover of the duplication, suppressors of auxotrophic nonsense mutations could survive; because of their dominance to the normal t-RNA allele (see above), they could be detected. In one case, the same change in the anticodon that allowed it to suppress amber mutations was found to be accompanied by a change in the amino acid-accepting specificity from tryptophan to glutamine. Since *E.coli,* in which this observation was made, has only one structural gene for t-RNA^{tryp}, the suppressor acts as recessive lethal (37b).

Evidence for supersuppression has been found also in fungi (38). In yeast, cells carrying a supersuppressible allele of the tryptophan synthetase gene together with the suppressor produce complete enzyme molecules as well as the fragments to be expected from precocious chain termination (39). Likewise in yeast, amino acid sequencing of the *iso*-1-cytochrome *c* in suppressed strains led to identification of the nonsense codons and of the bases inserted by suppressors (40-42). Possibly, the mutant t-RNA's carry an alteration outside the anticodon. At least, this has been suggested for an amber suppressor which failed to act on an *in vitro* system in which an *E.coli* suppressor with a proven change in the anticodon was effective (43). As in bacteria, the action of some completely recessive supersuppressors in yeast has been attributed to mutations in a protein that is concerned with translation (43a, 43b). In *Neurospora,* supersuppressors seem to be rare, but a few have been reported (44).

Frameshift suppression. Suppression of base changes by frameshifts

Unexpectedly, there have been several reports of extracistronic frameshift suppressors. These must act differently from the supersuppressors of base changes. It is true that frameshifts often create nonsense codons, but insertion of an amino acid in the place of any of these will not 'cure' the garbled polypeptide chain that extends from the site of the frameshift all the way to the end of the gene. One possibility for extracistronic frameshift suppression would be the creation, by mutation, of a quadruplet anticodon in t-RNA, able to read the original codon plus the inserted base as one amino acid. This has been assumed in several cases (45-47) and has been proved in one (48), in which the anticodon of a suppressor t-RNA was found to have an additional nucleotide (CCCC instead of CCC). If this were the general mode of frameshift suppression, all suppressible frameshifts should have an added rather than a deleted base, and this agrees with what is known about their origin and response to mutagens (see Chapter 17). It is probably no accident that so far all suppressible frameshifts that could be analysed at the molecular level had the inserted amino acid next to a triplet of three identical nucleotides (CCC for proline, GGG for glycine). It suggests that suppression might also occur when a non-mutated anticodon, through some outside change in the t-RNA molecule, acquires a tendency to slip along the message, especially when this consists of a run of identical bases. It will be interesting to see whether a search for frameshift suppressors of lysine (AAA) and phenylalanine (UUU) will be successful.

In bacteria and, particularly, in yeast, there has been indirect evidence for the occurrence of the converse situation: suppression of nonsense mutations by frameshifts in other genes, and this has been tentatively attributed to the creation, by frameshift, of a new anticodon in t-RNA (49,50). In T4, suppression of base substitutions in the rII cistrons by an extracistronic frameshift is limited to slightly leaky mutations, or to non-leaky ones in the presence of a weak supersuppressor; suppression here seems to act at some step following translation (51). We shall now consider some of the cases in which this was definitely so.

Suppressors acting in other ways

Analysis of the nature of supersuppression has been such an outstanding success that it has for some time channelled interest away from other types of suppression. There has been a tendency to use the model of supersuppressors for the interpretation of less well analysed cases of suppression. This is a risky procedure in biological systems with their enormous resources of mechanisms. An interesting case, which is still being investigated, is that of the vermilion suppressor in *Drosophila,* discovered as early as 1932. Vermilion flies lack tryptophan pyrrolase, one of the enzymes involved in the formation of brown eye pigment; as a result, their eyes are bright red instead of dark red. A suppressor gene, $su(s)^2$, allows partial restoration of enzyme activity and pigment formation when it is combined with certain *v*-alleles; other alleles are recalcitrant to its action (see Lindsley and Grell, Bibliography). On the other hand, $su(s)^2$ acts also on a number of other eye colour mutants. In its allele specificity, coupled with lack of locus specificity, $su(s)^2$ conforms to the operational definition of a supersuppressor. It differs from microbial supersuppressors in being fully recessive. Biochemical analysis (52a,b,c) revealed that in suppressor strains one species of tyrosine-t-RNA is modified in quantity, composition and reaction with tyrosyl-tRNA transferase. The recessivity of the suppressor action makes it doubtful whether suppression occurs via a change in codon recognition as in true supersuppression.

A biochemically better understood case of suppression in *Drosophila* is that of the wing-mutant rudimentary by mutation in an unlinked gene (53). Rudimentary is one of the few morphological mutants of *Drosophila* for which the basic biochemical defect – a requirement for pyrimidine – is known. The suppressor, by blocking a step in the catabolic pathway of uracil, provides the rudimentary mutants with precursors that they can utilize.

In *Neurospora,* mutations at the pyrimidine-3 locus, while creating a requirement for pyrimidine, lift the requirement for arginine; the effect is due to linking of the two pathways by a common precursor, carbamyl phosphate (54). Also in *Neurospora,* a more complex interaction of biochemical pathways is responsible for the suppression of mutations in the *arom*-1^+ gene by mutations in the unlinked *qa*-4^+ gene (55). In *Aspergillus,* suppressors of a proline deficiency act by the opening up of an alternative pathway (56). The same appears to be true for the repeatedly found locus-specific but not allele-specific suppressors of mutations to isoleucine-valine deficiency in bacteria and fungi (57). Even deletions can be suppressed metabolically; in *Salmonella,* proline deletion mutations were suppressed by mutations in an arginine locus (58). Conversely, certain auxotrophic mutations of *Salmonella* can be suppressed not only by mutations in an unlinked locus, but also by deletions of the locus (59).

References

1. Yanofsky, C., Ito, J. and Horn, V. (1966), 'Amino acid replacements and the Genetic Code', *Cold Spring Harbor Symp. Quant. Biol.,* **31**, 151-162.
2. Sarabhai, A. and Brenner, S. (1967), 'A mutant which reinitiates the polypetide chain after chain termination', *J. Mol. Biol.,* **27**, 145-162.
3. Guest, J.R. and Yanofsky, C. (1965), 'Amino acid replacements associated with reversion and recombination within a coding unit', *J. Mol. Biol.,* **12**, 793-804.
4. Berger, H. and Yanofsky, C. (1967), 'Suppressor selection for amino acid replacements expected on the basis of the genetic code', *Science,* **156**, 394-397.
5. Koch, R.E. (1971), 'The influence of neighboring base pairs upon base-pair substitution mutation rates', *Proc. Nat. Acad. Sci., U.S.A.* **68**, 773-776.
6. Yanofsky, C., Horn, V. and Thorpe, D. (1964), 'Protein structure relationships revealed by mutational analysis', *Science,* **146**, 1593-1594.
7. Gorini, L. (1966), 'Antibiotics and the Genetic Code', *Scient. Am.,* **214**, 102-109.
8. Gorini, L., Jacoby, G. and Breckenridge, L. (1966), 'Ribosomal ambiguity', *Cold Spring Harbor Symp. Quant. Biol.,* **31**, 657-664.
9. Davies, J., Gilbert, W. and Gorini, L. (1964), 'Streptomycin, suppression and the code', *Proc. Nat. Acad. Sci., U.S.A.,* **51**, 883-890.
10. Apirion, D. and Schlessinger, D. (1969), 'Functional interdependence of ribosomal components of *Escherichia coli,* *Proc. Nat. Acad. Sci., U.S.A.,* **63**, 794-799.
11. Gorini, L. (1969), 'The contrasting role of *strA* and *ram* gene products in ribosomal functioning', *Cold Spring Harbor Symp. Quant. Biol.,* **34**, 101-111.
12. Atkins, J.F., Elseviers, D. and Gorini, L. (1972), 'Low activity of β-galactosidase in frameshift mutants of *Escherichia coli',* *Proc. Nat. Acad. Sci., U.S.A.* **69**, 1192-1195.
13. Champe, S.P. and Benzer, S. (1962), 'Reversal of mutant phenotypes by 5-fluorouracil: an approach to nucleotide sequences in messenger-RNA', *Proc. Nat. Acad. Sci., U.S.A.,* **48**, 532-546.

14. Benzer, S. and Champe, S.P. (1962), 'A change from nonsense to sense in the genetic code', *Proc. Nat. Acad. Sci., U.S.A.*, **48**, 1114-1121.
15a. Rosen, B., Rothman, F. and Weigert, M.G. (1969), 'Miscoding caused by 5-fluorouracil', *J. Mol. Biol.*, **44**, 363-375.
15b. Ottensmeyer, F.P. and Whitmore, G.F. (1968), 'Coding properties of ultraviolet photoproducts of uracil. II Phenotypic reversion of the amber mutation: implication of the uracil hydrate', *J. Mol. Biol.* **38**, 17-24.
16. Salser, W., Fluck, M. and Epstein, R. (1969), 'The influence of the reading context upon the suppression of nonsense codons, *III', Cold Spring Harbor Symp. Quant. Biol.*, **34**, 513-520.
17. Topisirovič, L., Metlaš, R. and Kanazir, D.T. (1973), 'A slow-growing, ribosomal mutant of *Salmonella typhimurium', Molec. Gen. Genetics* **123**, 135-142.
18. Garen, A. and Siddiqi, O. (1962), 'Suppression of mutations in the alkaline phosphatase structural cistron of *E.coli', Proc. Nat. Acad. Sci., U.S.A.*, **48**, 1121-1127.
19. Carbon, J., Berg, P. and Yanofsky, C. (1966), 'Studies of missense suppression of the tryptophan synthetase A-protein mutant *A36', Proc. Nat. Acad. Sci., U.S.A.*, **56**, 764-771.
20. Carbon, J. and Curry, J.B. (1968), 'Genetically and chemically derived missense suppressor transfer RNA's with altered enzymic aminoacylation rates', *J. Mol. Biol.*, **38**, 201-216.
21. Gupta, N.K. and Khorana, H.G. (1966), 'Missense suppression of the tryptophan synthetase A-protein mutant *A78', Proc. Nat. Acad. Sci., U.S.A.*, **56**, 772-779.
22. Carbon, J., Berg, P. and Yanofsky, C. (1966), 'Missense suppression due to a genetically altered t-RNA', *Cold Spring Harbor Symp. Quant. Biol.*, **31**, 487-497.
23. Ritossa, F.M., Atwood, K.C. and Spiegelman, S. (1966), 'On the redundancy of DNA complementary to amino acid transfer RNA and its absence from the nucleolar organizer region of *Drosophila melanogaster', Genetics*, **54**, 663-676.
24. Gilmore, R., Stewart, J. and Sherman, F. (1971), 'Amino acid replacements resulting from supersuppression of nonsense mutants of *iso*-1-cytochrome *c* from yeast', *J. Mol. Biol.*, **61**, 157-173.

25. Hill, C., Foulds, J., Soll, L. and Berg, P. (1969), 'Instability of a missence suppressor resulting from a duplication of genetic material', *J. Mol. Biol.,* **39**, 563-581.
26. Garen, A. (1968), 'Sense and nonsense in the genetic code', *Science,* **160,** 149-159.
27. Goodman, H.M., Abelson, J., Landy, A., Brenner, S. and Smith, J.D. (1968), 'Amber suppression: a nucleotide change in the anticodon of a tyrosine transfer RNA', *Nature,* **217**, 1019-1024.
28. Smith, J.D. and Celis, J.E. (1973), 'Mutant tyrosine transfer RNA that can be charged with glutamine', *Nature New Biology* **243**, 66-71.
29. Hirsh, D. (1971), 'Tryptophan transfer RNA as the UGA suppressor', *J. Mol. Biol.,* **58**, 439-458.
31. Reeves, R.H. and Roth, J.R. (1971), 'A recessive UGA suppressor', *J. Mol. Biol.,* **56**, 523-533.
32. Bridges, B.A., Dennis, R.E. and Munson, R.J. (1967), 'Mutation in *Escherichia coli* B/r WP2 try$^-$ by reversion or suppression of a chain-terminating codon', *Mutation Res.,* **4**, 502-504.
33. Nichols, J.L. (1970), 'Nucleotide sequence from the polypeptide chain termination region of the coat protein cistron in bacteriophage R17 RNA', *Nature,* **225**, 147-151.
34. Steitz, J.A. (1969), 'Polypeptide chain initiation: nucleotide sequences of the three ribosomal binding sites in bacteriophage R17 RNA', *Nature,* **224**, 957-964.
35. Rechler, M.M. and Martin, R.G. (1970), 'The intercistronic divide: translation of an intercistronic region in the histidine operon in *Salmonella typhimurium', Nature,* **226**, 908-911.
36. Soll, L. and Berg, P. (1969), 'Recessive lethals: a new class of nonsense suppressors in *Escherichia coli', Proc. Nat. Acad. Sci., U.S.A.,* **63**, 392-399.
37a. Miller, C. and Roth, J. (1971), 'Recessive-lethal nonsense suppressors in *Salmonella typhimurium', J. Mol. Biol.,* **59**, 63-75.
37b. Yaniv, M., Folk, W.R., Berg, P. and Soll, L. (1974), 'A single mutational modification of a tryptophan-specific transfer RNA permits aminoacylation by glutamine and translation of the codon UAG'. *J. Mol. Biol.* **86**, 245-260.
38. Mortimer, R.K. and Gilmore, R.A. (1968), 'Suppressors and suppressible mutations in yeast', *Adv. Biol. Med. Physics,* **12**, 319-331.

39. Manney, T.R. (1968), 'Evidence for chain termination by supersuppressible mutants in yeast', *Genetics,* **60**, 719-733.
40. Gilmore, R.A., Stewart, J.W. and Sherman, F. (1971), 'Amino acid replacements resulting from supersuppression of nonsense mutants of *iso*-1-cytochrome *c* from yeast', *J. Mol. Biol.,* **61**, 157-173.
41. Stewart, J.W., Sherman, F., Jackson, M., Thomas, F.L.X. and Shipman, N. (1972), 'Demonstration of the UAA ochre codon in bakers' yeast by amino acid replacements in *iso*-1-cytochrome *c*', *J. Mol. Biol.,* **68**, 83-96.
42. Stewart, J.W. and Sherman, F. (1972), 'Demonstration of UAG as a nonsense codon in bakers' yeast by amino acid replacements in *iso*-1-cytochrome *c', J. Mol. Biol.,* **68**, 429-443.

43a. Kiger, J.A., Jr. and Brantner, C.J. (1973), 'The inability of transfer RNA to suppress an amber mutation in an *E.coli* system', *Genetics,* **73**, 23-28.

43b. Smirnov, V.N., Kreier, V.G., Lizlova, L.V., Andrianova, V.M. and Inge-Vechtomov, S.G. (1974), 'Recessive supersuppression in yeast', *Mol. Gen. Genetics* **129**, 105-121.

43c. Manthorne, D.C. and Leupold, U. (1974), 'Suppressor mutations in yeast', *Current Topics in Microbiology and Immunology,* **64**, 1-47.

44. Seale, T. (1968), 'Reversion of the *am* locus in *Neurospora:* evidence for nonsense suppression', *Genetics,* **58**, 85-99.
45. Yourno, J. and Kohno, T. (1972), 'Externally suppressible proline quadruplet CCC^{u}', *Science,* **175**, 650-652.
46. Riddle, D. and Roth, J. (1972), 'Frameshift suppressors II. 'Genetic mapping and dominance studies', *J. Mol. Biol.,* **66**, 483-493.
47. Riddle, D. and Roth, J. (1972), 'Frameshift suppressors III. 'Effects of suppressor mutations on transfer RNA', *J. Mol. Biol.,* **66**, 495-506.
48. Riddle, D.L. and Carbon, J. (1973), 'Frameshift suppression: a nucleotide addition in the anticodon of a glycine transfer RNA', *Nature New Biology*, **242**, 230-234.
49. Ames, B.N. and Whitfield, H.J. Jr. (1966), 'Frameshift mutagenesis in *Salmonella', Cold Spring Harbor Symp. Quant. Biol.,* **31**, 221-225.
50. Magni, G. (1969), 'Spontaneous mutations', *Proc. 12th Int. Congr. Genetics,* Vol. 3, 247-259.

51. Freedman, R. and Brenner, S. (1972), 'Anomalously revertible rII mutants of phage T4', *Genet. Res., Camb.*, **19**, 165-171.
52a. Twardzik, D.R., Grell, E.H. and Jacobson, K.B. (1971), 'Mechanism of suppression in *Drosophila:* a change in tyrosine transfer RNA', *J. Mol. Biol.*, **57**, 231-245.
52b. White, B.N., Tener, G.M., Holden, J. and Suzuki, D.T. (1973), 'Activity of a transfer RNA modifying enzyme during the development of *Drosophila* and its relationship to the su(*s*) locus'. *J. Mol. Biol.*, **74**, 636-651.
52c. Jacobson, K.B., Calvine, J.F. and Murphy, J.B. (1973), 'Tyrosyl-tRNA and its synthetase in a suppressor mutant of *Drosophila*' *Fed. Proc.* **32**, 654.
53. Bahn, E. (1972), 'A suppressor locus for the pyrimidine requiring mutant: rudimentary', *Drosophila Information Service*, **49**, 98.
54. Reissig, J.L. (1963), 'Spectrum of forward mutants in the *pyr-3* region of *Neurospora', J. Gen. Microbiol,* **30**, 327-337.
55. Case, M.E., Giles, N.H. and Doy, C.H. (1972), 'Genetical and biochemical evidence for further interrelationship between the polyaromatic synthetic and the quinate-shikimate catabolic pathways in *Neurospora crassa', Genetics,* **71**, 337-348.
56. Weglenski, P. (1967), 'The mechanism of action of proline suppressors in *Aspergillus nidulans', J. Gen. Microbiol,* **47**, 77-85.
57. Gundelach, E. (1973), 'Suppressor studies on *ilv-1* mutants of *S. cerevisiae', Mutation Res.,* **20**, 25-33.
58. Kuo, T. and Stocker, B. (1969), 'Suppression of proline requirement of *proA* and *proAB* deletion mutants in *Salmonella typhimurium* by mutation to arginine requirement', *J. Bacter.*, **98**, 593-598.
59. Dubnau, E. and Margolin, P. (1972), 'Suppression of promotor mutations by the pleiotropic *sup*^{X} mutations', *Molec. Gen. Genetics,* **117**, 91-112.

CHAPTER FIVE

The classical X-ray work. I: Mutations. The target theory

Soon after the discovery of X-rays by Röntgen at the turn of the century, the effects of this new powerful type of radiation on biological systems were studied. Embryological and cytological disturbances were observed in a variety of organisms, and this led to attempts at influencing heredity by radiation. Indications that X-rays may produce hereditary changes in germ cells were obtained quite early, but decisive proof had to wait for the development of systems in which nuclear mutations could be clearly distinguished from other types of damage to e.g., the spindle apparatus or the cytoplasm, and in which mutation frequencies could be objectively measured. In the 1920s such systems became available in *Drosophila* and maize. The first definite proof for the mutagenic action of X-rays was obtained in 1927 by Muller on *Drosophila* (1) to be followed a year later by Stadler's evidence for the production of mutations in maize (2). Even before this, *Drosophila* techniques had made it possible to discover X-ray effects on non-disjunction (3) and crossing-over (4,5). Since most of the classical X-ray work was done on *Drosophila,* this is the place to describe the two main tests used at that time and still used widely; the test for sex-linked lethals and the test for sex-linked visibles. I shall discuss the rationale of these tests rather than details of technique, which can be found in my book on the methods of mutation research (Bibliography).

Drosophila techniques

The test for sex-linked lethals

Recessive lethals are genes that kill the homozygous individual before a specified developmental stage, which in *Drosophila* is chosen as eclosion of the fly from the pupa. Recessive sex-linked lethals are due to mutations in the X-chromosome and affect genes that have no allele on the Y. They therefore kill males and homozygous females but can be carried hidden in heterozygous females. Among the progeny of such females only males with the lethal-free X will survive. If the two X's of the mother are distinguished by marker genes, a culture will be produced in which all males of one genotype are missing. The criterion for a sex-linked lethal is therefore absence of a whole class of flies, and this removes most of the subjective sources of error from the test. In addition, lethals have the great advantage that they are much more frequent than viable mutations with visible effects. This is only to be expected from the consideration that presently existing species have already incorporated most beneficial mutations into their genotype, so that the majority of newly arising ones are likely to be harmful by disturbing the integrated functions of the genome. In work with organisms that can be tested only on a moderate scale, tests for lethals can yield statistically meaningful results without undue or impossible demands on material and labour. The question whether recessive lethals can be considered a representative sample of genes was already raised by Muller (6) and is still under debate. On the whole, it seems reasonable to assume that lethals, which occur at a great number of loci, which can be produced by many different types of lesion in the genetic material, and which act in widely different ways will, in general, give better estimates of mutagenicity than visible mutations, the scoring of which is subject to unavoidable personal error, which are each restricted to one or a few loci, which are produced only by the milder types of genetic lesion, and which may show specificities in their response to mutagens (Chapter 20).

The essentials of the test are shown in Fig. 5.1. A population of spermatozoa is treated inside a ♂ (artificial insemination has never been successful in *Drosophila*). Each treated X-chromosome is introduced through fertilization into an F_1 ♀ where it multiplies into a population of identical chromosomes. Under the shelter of heterozygosity, this is possible also for X-chromosomes in which treatment has produced a recessive lethal. Each F_1 ♀ is then mated to a brother or any other suitable type of ♂; absence of ♂♂ with the treated X among her progeny (F_2) shows that the treated X carried a lethal. The same test can be applied to untreated control populations of

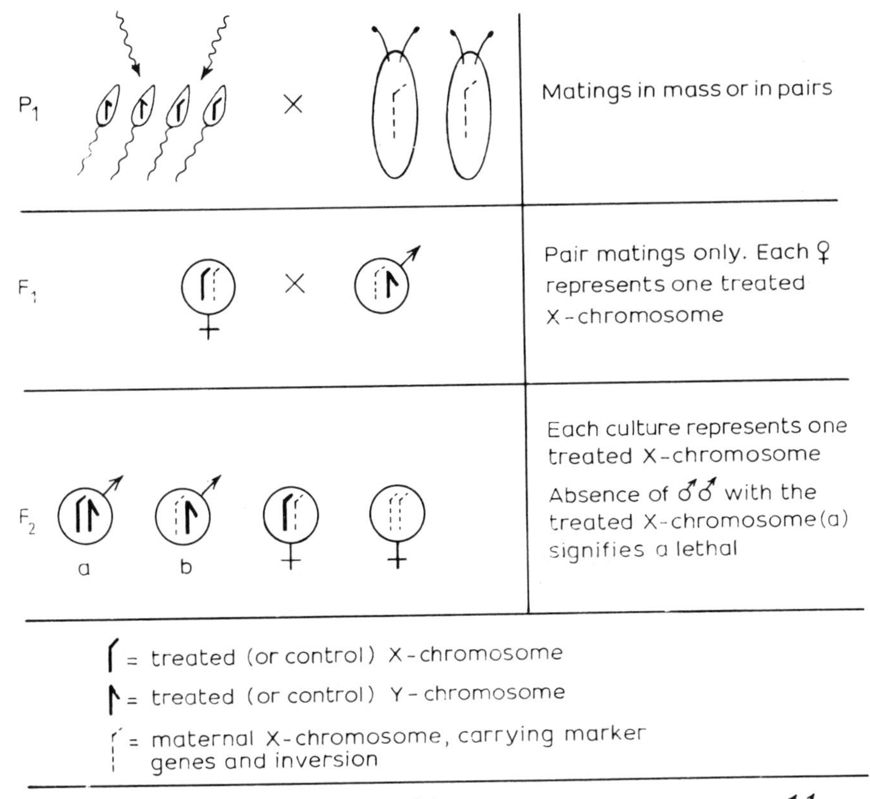

Fig. 5.1. Test for sex-linked recessive lethals in *Drosophila*. (Fig. 1, Auerbach, 1962 (bibl.). Courtesy author)

male germ cells. The X-chromosome of the tester ♀♀ has to carry a marker gene which, in F_2, distinguishes ♂♂ with this X from ♂♂ with the X under test. It also has to carry an inversion which prevents crossing-over of the marker gene into the homologous X. Various combinations of marker genes and inversions are in use. In the presently much used Muller-5 or Basc test, the X-chromosome of the tester ♀♀ carries a system of two inversions and the two marker genes apricot (eye colour; w^a) and Bar (shape of eye; B). Thus, an F_1♀ that carries no lethal on the treated wild-type X-chromosome has both wild-type and w^aB sons, while an F_1♀ with a lethal has only w^aB sons.

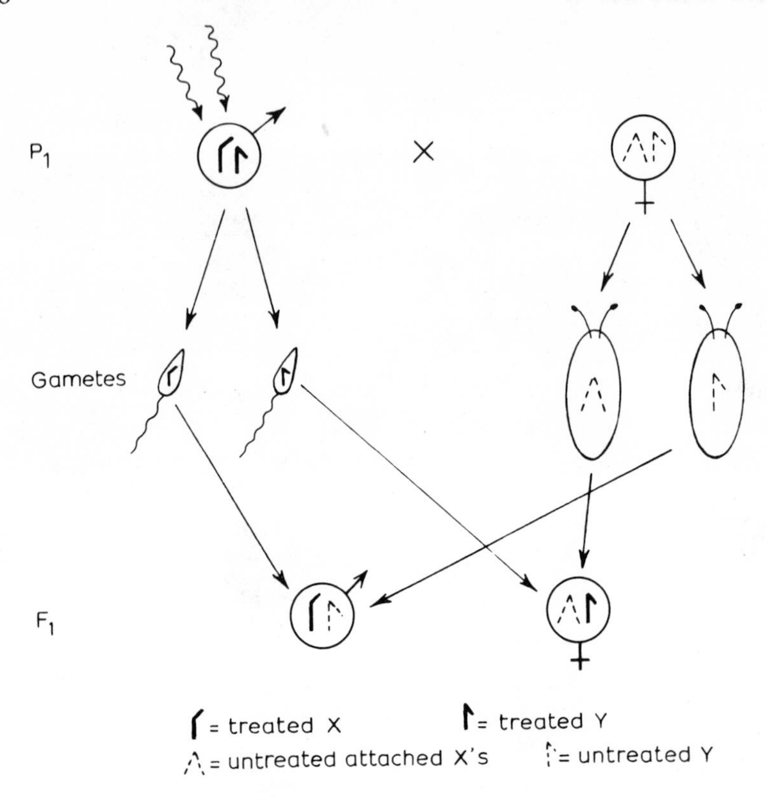

Fig. 5.2. Test for sex-linked visible mutations in *Drosophila* by the attached-X method. (Fig. 4, Auerbach, 1962 (bibl.). Courtesy author)

The testis of an adult *Drosphila* male contains all germ cell stages from spermatogonia to spermatozoa. By mating treated males to a succession of tester females one obtains a succession of broods representing germ cells that were successively younger at the time of treatment. This 'brood pattern analysis' provides information on the response of different germ cell stages to treatment.

The test can be applied to females by mating them to untreated tester males. In this case, it is necessary to take account of the fact that some of the treated females may be heterozygous for a pre-existing lethal. In tests for autosomal recessive lethals this precaution is required also for males. Since recessive autosomal lethals become manifest only when homozygous, an additional generation of inbreeding is required, and autosomal lethals can be scored only in F_3.

When Muller did his first tests for sex-linked lethals, no suitable inversion was available, and he had to use a tester stock in which the X-chromosome was marked by several well-spaced genes. Subsequently a suitable inversion arose in a stock that already carried the gene Bar (*B*) on its X-chromosome. The inversion was first detected by its inhibiting effect on crossing-over (crossing-over-inhibitor, *C*). An additional feature of this chromosome was that it also carried a recessive lethal (*l*) so that females heterozygous for the chromosome *always* lacked one class of males. If a treated spermatozoon introduced a lethal on the other X-chromosome, the F_2 culture was entirely maleless. This made the famous *ClB* test easy and attractive, and for decades it was used almost exclusively in tests for sex-linked lethals. It is described in most older textbooks of genetics, but is now only rarely used because it has some inherent disadvantages.

The attached-X test for sex-linked visibles and large deletions

Sex-linked visible mutations can be detected in the F_1 of treated males that have been mated to females with attached X-chromosomes (Fig. 5.2). Such females transmit their two X-chromosomes to their daughters and their Y-chromosome to their sons. The sons thus inherit their X-chromosome from their father, and any recessive visible carried in a spermatozoon can manifest itself in a son. Unavoidably this test is more open to personal error than the recessive-lethal test. In addition, since each mutated chromosome is represented by a single individual rather than by a whole culture, errors due to chance and selection play an important role. Both these disadvantages are avoided if visible mutations are scored together with lethals and by the same technique; for then each sex-linked visible will be represented by an F_2 culture in which *all* males with the treated X show the same visible abnormality. It is, however, very difficult to obtain sufficiently large numbers of tested chromosomes in this way. While it is not too difficult to test tens of thousands of X-chromosomes by the attached-X method, a sample of 2000 in a sex-linked lethal test is considered quite a big experiment.

Attached X-females can also be used for the detection of large deletions as shown in Fig. 5.3. The females are homozygous for several recessive markers. When a spermatozoon with a deleted

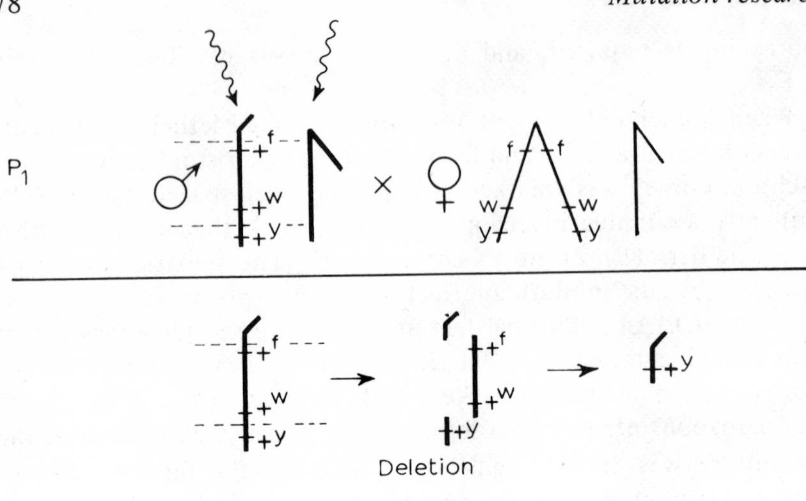

F_1

Hyperploid non-yellow, white forked ♀

= treated X and Y   = untreated attached X's and Y

w = white eyes   f = forked bristles   y = yellow body

Regular F_1 ♀♀ are y w f. F_1 ♀♀ with a deleted paternal X-chromosome have one or more of the marker genes covered

Fig. 5.3. The detection of large deletions by the attached-X method. (Fig. 6. Auerbach, 1962 (bibl.))

X-chromosome fertilizes an ovum with attached X's, the zygote develops into a hyperploid female in which one or several of the marker genes are covered by the supernumerary chromosome fragment.

First results. The target concept: preliminary tests for a direct and immediate action of X-rays on the genes

Muller's first experiments with X-rays yielded over 10% sex-linked lethals in irradiated males as compared with about 0.1% in non-irradiated ones. A few visible mutations were also found in the *ClB* tests and more in tests with attached-X females. There could be no doubt that X-rays are a powerful mutagenic agent for *Drosophila.* Soon this was shown to be true also for maize and barley. The next decade saw an enormous expansion of research on radiation mutagenesis. The experiments were extended to a variety of cell types and organisms, to different types of genetic effect and to different classes of ionizing radiation. This period culminated in the formulation of the 'target theory', which was a joint product of geneticists and biophysicists.

It should be realized that the powerful biological effects of X-rays presented physicists with a puzzling phenomenon, because the energies involved are so small. Thus, a man could be killed by the caloric content of a cup of tea if this energy was given as X-rays. Vice versa, a cell can be killed by an amount of radiation energy which, when expended as heat, would raise the temperature of the cell by less than 10^{-8} °C. This pointed to the existence in cells of small discrete regions or particles of special importance, in which absorption of quanta from the discontinuous distribution of radiation energy leads to disproportionally drastic overall effects. This idea already implied the existence of 'targets' for radiation. In the special case of mutations, the targets obviously were the genes. Before, however, a more rigorous biophysical analysis was attempted, it was necessary to settle certain fundamental questions. In particular, it had to be shown that it is really the impact of radiation on the genes themselves which leads to mutation. If, instead, radiation acted via the production of a diffusible substance somewhere else in the cell or organism, analysis at the chemical rather than at the physical level would have been required. It was equally important to establish whether all X-ray induced mutations become 'fixed' at once or whether a proportion of them give rise to premutational lesions which at some later stage, perhaps after many mitoses, may result in mutation proper. If this were the case, the mutations scored in the sex-linked lethal test could not be used as a reliable measure of the mutagenic strength of a given treatment for, as a glance at Fig. 5.1 will show, these mutations must already be present in the treated spermatozoa in order to be detected in F_2.

Thus, as a preliminary to a biophysical analysis, tests were carried out to decide whether the action of X-rays on the genes is *direct* and *immediate.*

The experiments that proved the direct effect of X-rays on the genes were again done on *Drosophila,* using the *ClB* test. Various methods were applied to test for the production of mutagenic substances in the cytoplasm. One consisted of exposing males to a very high dose of very soft X-rays, too soft to reach the gonads but sufficiently penetrating to enter the more superficially located parts of the animal. It was argued that, if mutagenically acting substances are produced outside the gonads, these might diffuse to the gonads and either produce mutations directly or predispose the chromosomes to give a stronger reaction with a subsequent dose of hard X-rays. A more direct answer to the question was attempted by irradiating females, inseminating them with sperm from untreated males, and looking for sex-linked lethals on paternal X-chromosomes which had spent the whole developmental cycle from inseminated zygote to the sexually mature adult in treated cytoplasm. As a control for the possibility that irradiation of females might be less effective than irradiation of males, spermatozoa were irradiated inside the seminal receptacles of females. The results of all these experiments are summarized in Table 5.1.

The following conclusions were drawn. (*a*) Radiation that does not penetrate to the gonads does not produce germinal mutations (III) nor does it make the chromosomes of the germ cells more likely to respond to the mutagenic action of penetrating radiation (IV compared with II). (*b*) Although spermatozoa that are irradiated inside the seminal receptacles of the ♀♀ (VI) yield at least as many mutations as do spermatozoa irradiated in the ♂♂ (II), no mutations are produced in unirradiated X-chromosomes that have undergone many mitoses in irradiated cytoplasm (V). Thus, the question of a direct action of X-rays was answered in the affirmative. We shall see in Chapter 7 that a more sophisticated way of testing it gave the same answer (p. 117).

The same was true for the question whether X-rays produce mutations immediately in the treated genes themselves; it, too, received a positive answer. Since tests for delayed mutations play an important role in research on chemical mutagens, I shall describe their essential features (Fig. 5.4).

Suppose that treatment produces a potentially mutagenic lesion in the X-chromosome of a spermatozoon and that this lesion gives rise to a mutation proper at some time during the development of the gonads in the F_1 female. The ovaries of this female will then

Table 5.1

Tests for a mutagenic effect of irradiated cytoplasm on unirradiated chromosomes in *Drosophila*. (After Tables 31 and 32, Timoféeff-Ressovsky and Zimmer, 1947 (Bibliography))

Series	*Chromosomes*		
	No. tested	*No. lethals*	*% lethals*
I controls	3708	7	0.2
II ♂♂ treated with hard X-rays (3000R)*	2239	198	8.8
III ♂♂ treated with soft X-rays (50,000R)	1883	5	0.3
IV ♂♂ treated first with soft, then with hard X-rays	1107	89	8.0
V Unirradiated chromosomes in irradiated cytoplasm	2163	3	0.1
VI Spermatozoa irradiated with hard X-rays in the seminal receptacles of the ♀♀	2631	244	9.3

* R = unit of dose; see below.

have two types of cell with different versions of the treated X: those with a lethal and those without a lethal. Such a female is called a gonadic mosaic, and she will produce two types of ova with the treated X: those that carry a lethal and those that do not. This will result in a shortage of sons with the treated X but, unless the shortage is very striking, it will not be detected in the quick routine-scoring of lethals in F_2. The trick for the detection of delayed lethals is to make the mosaic F_1 female produce daughters that carry the lethal *throughout* their ovaries and thus do not produce *any* sons with the treated X. In practice, females with the treated X are sampled from F_2 cultures that do not already have an immediately produced lethal, and the progeny of these females is then tested for lethals in the ordinary way. Delayed mutations are detected in F_3 in the same way as immediate mutations are detected in F_2. None were found in the F_3 of irradiated ♂♂.

It should be pointed out – and was realized by the early radiation geneticists – that none of these experiments excluded indirect effects that are very short-lived or extend over very short distances; nor did they exclude delayed effects that are confined to the period between irradiation and the first cleavage division in the F_1 zygote. We shall see in Chapter 7 that such effects do occur after irradiation. However, such as it stood in the middle thirties, the

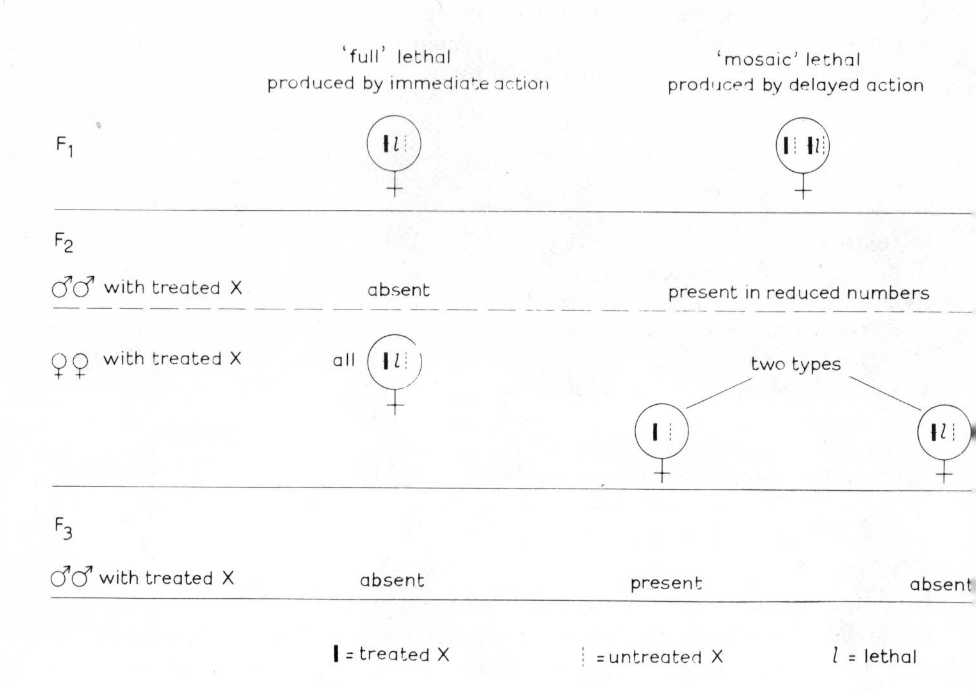

Fig. 5.4. Test for delayed production of a sex-linked lethal.

evidence for a direct and immediate action of X-rays on the genetic material was good enough to open the way to a biophysical interpretation. This was based on the kinetics of mutation induction and on comparisons between ionizing radiations of different wavelengths.

Biophysical analysis. Hit-theory and target-theory (7,8)

When X-rays pass through matter, they deliver their energy to the atoms with which they collide. The primary collision usually results in the ejection of an electron from the atom, leaving the atom behind as a positively charged ion. The ejected secondary electron produces further ion pairs along its track, until it is sufficiently slowed down to be captured by an atom or molecule. Most of the energy

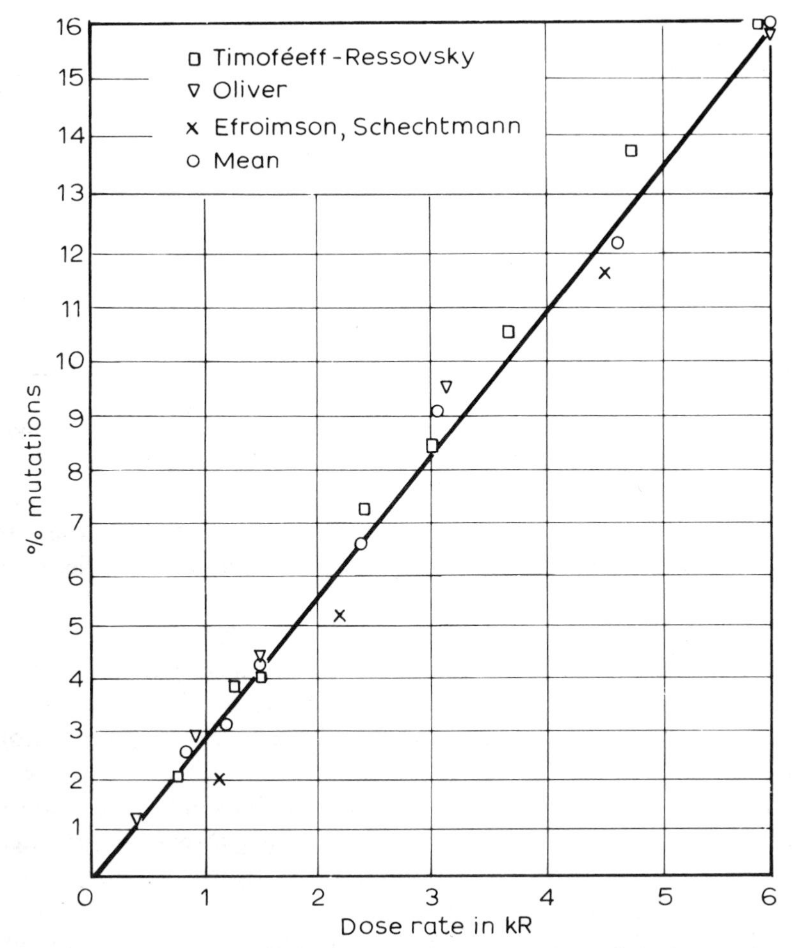

Fig. 5.5. Direct proportionality of the rates of sex-linked mutations (mainly recessive lethals) in *Drosophila melanogaster* to the X-ray dose (in kR) in *ClB* tests of three different authors. (Fig. 96, Timoféeff-Ressovsky and Zimmer 1947, (bibl.). Courtesy S. Hirzel-Verlag)

of the radiation is therefore dissipated in the formation of ion pairs, and this forms the basis for dosimetry. The unit of radiation is the R(Röntgen), which produces about 2×10^9 ion pairs in 0.001293g of air (= $1 cm^3$ at 0^0 and 760 mm). For practical purposes, one uses the rad as a measure of the absorbed energy; in air 1R is equivalent to 0.876 rad.

In *Drosophila* spermatozoa, the frequencies of sex-linked lethals, autosomal lethals and sex-linked visible mutations all increased linearly with dose (Fig. 5.5). Moreover, a given dose produced the

Table 5.2.
Frequency of sex-linked lethals in the progeny of males that had been exposed to 3600R delivered in three different ways. (After Table 28, Timoféeff-Ressovsky (16))

Time of treatment	*R min^{-1}*	*No. F_2 Cultures*	*No. Lethals*	*% lethals*
15 min	240	493	54	11.0
6 h	10	521	60	11.5
5 min daily, 6 days	120	423	47	11.1
Controls	–	1827	2	0.1

same frequency of mutations whether it was given over a short time at high intensity or over a long time at low intensity i.e. it was independent of dose rate (= number of R per s). Finally, when the dose was fractionated into several periods separated by radiation-free intervals, mutation frequency was proportional to the total dose (Table 5.2).

This led to the conclusion that mutation is a 'one-hit' event, a hit being a localized energy-yielding process in or near a gene. At first, it was not clear what to consider as a hit; the absorption of an energy quantum by the first atom to collide with a ray, or any ionization produced in the track of secondary electrons. This question was decided by comparing the efficiency of radiations of different wavelengths. The energy of a quantum of radiation is inversely correlated with wavelength and increases continuously from soft (long-wave) to very hard X-rays or γ rays. Hard X-rays, therefore, need to deliver fewer quanta than soft ones in order to produce a given total amount of energy as measured by the dose in R. If the frequency of mutations were determined by the number of quanta, then – dose for dose – hard X-rays should produce fewer mutations than soft ones. This was found not to be so. In fact, within a very wide range of wavelengths, mutational yield depended only on dose as measured in R. This, then, established an ionization as the 'hit' that produced a mutation.

The linear dose-effect curve for mutations has played a crucial role in estimates of genetical hazards to the human population from X-rays, fallout, radioactive waste, and other sources of ionizing radiation in our environment. If the frequency of mutations is, at all doses, proportional to dose, then there will be no threshold to the mutagenic action of radiation, and even the lowest dose will produce mutations in exact proportion to its magnitude. Barring nuclear warfare, nuclear accidents and certain medical practices (now abandoned), the environmental doses to which mankind is exposed are far below those used in the classical X-ray work. Therefore it became important to extend the experiments to lower and lower doses. This has been done repeatedly over the years and on a variety of objects. In *E.coli* and *Drosophila,* the linear dose-effect relationship has been shown to hold down to 8R (9,10). In *Tradescantia*, white cells were scored in stamen hairs that were heterozygous for red and white (11). This technique confounds somatic crossing-over with mutations, but since both are localized nuclear effects, this is no serious objection to its use. The dose-effect curve was linear down to 0.25R. On the other hand, there has been some evidence that in mouse oocytes and *Drosophila* oogonia very low doses of X-rays produce fewer mutations than would be predicted by linear extrapolation from the results of higher doses. (see Chapter 9).

The hit theory had been developed in the early twenties in order to account for the killing curves of small organisms or cells by X-rays. It was at first not much more than a formal description of a certain kind of kinetics but gained concrete content with its modification into the target theory by Crowther (12). In this theory, a reaction is produced whenever an ionization takes place within a well-defined small volume, the target. On the assumption that X-ray induced ionizations are distributed at random, and that every ionization within the target but none outside it produces a reaction, the volume of the target can easily be calculated. When Crowther did this for mitosis inhibition in tissue cells, he found that the calculated target volume agreed fairly well with the volume of the centriole as estimated from stained slides. Subsequent experiments on viruses and enzymes likewise yielded good agreement between size as determined directly, e.g. by microfiltration, and size as calculated from target considerations. We shall see in Chapter 7 that the conditions in these experiments (irradiation in the dry state or in broth) were in large part responsible for this good agreement. When the target theory was applied to mutations, the question arose whether the size of the gene could be determined from the mutational yield of given radiation doses. This was indeed attempted by several workers but others, especially Muller (13), raised both theoretical and observational objections to the method. On the theoretical side, Muller pointed out that neither of the two underlying assumptions – every ionization within a gene produces a mutation, none outside the gene does so – is likely to be true. Against the first speaks the consideration that among the many chemical changes that may occur in a gene, only a proportion will lead to an observable mutation; we now have good evidence for this at the molecular level. Against the second hypothesis stands not only the known fact of energy transfer over submicroscopic distances, but also the observation – discussed in the next chapter – that the frequency of very small X-ray induced deficiencies follows linear kinetics and therefore must be due to energy from a single ionization that is transferred either along a short stretch of the chromosome or, from a point outside the chromosome, to two points on it. Moreover, it had already been found in early experiments on *Drosophila* (14) that mutation frequencies are higher in spermatozoa than in spermatogonia; since it is most unlikely that the gene should expand during the progress of the spermatogonium into the spermatozoon, one must assume that for some reason the probability that an ionization will produce a mutation is higher for genes in spermatozoa than for genes in spermatogonia, making the former more 'sensitive' to the mutagenic effects of radiation. Similarly, Stadler (15) had found that a given dose of X-rays produced eight times as many mutations in the genes of germinating as of dormant seeds. In order to avoid identifying the target with the gene, the term 'sensitive volume' was introduced for a volume surrounding and including the gene; within this volume an ionization has a high probability of producing a mutation. The early experiments had already shown that the sensitive volume cannot be very much larger than the gene and that both its exact size and the probability of mutation induction within it are subject to a certain degree of variation depending on cell type and experimental conditions. In this less rigid form the target theory has retained its validity and usefulness and is still widely applied to genetical data from experiments with radiation and even with chemicals. In Chapters 7-9, we shall deal with the processes that blur the rigid outline of the target theory; but first we shall see how far the target theory could account for the production of chromosome rearrangements by ionizing radiation.

References

1. Muller, H.J. (1927), 'The problem of genic modification',

Fifth Int. Genetics Congress, Berlin; *Z. ind. Abst. Vererb. Lehre,* I (suppl.), 234-260.
2. Stadler, L.J. (1928), 'Genetic effects of X-rays in maize', *Proc. Nat. Acad. Sci. U.S.A.* **14**, 69-75.
3. Mavor, J.W. (1924), 'The production of non-disjunction by X-rays', *J. Exp. Zo-ol,* **39**, 381-432.
4. Mavor, J.W. and Svenson, H.K. (1924), 'An effect of X-rays on the linkage of Mendelian characters in the second chromosome of *Drosophila melanogaster', Genetics,* **9**, 70-89.
5. Friesen, H. (1937), 'Artificial release of crossing-over in meiosis and mitosis', *Nature*, **140**, 362.
6. Muller, H.J. (1928), 'Mutation rate in *Drosophila' Genetics,* **13**, 279-357.
7. Zimmer, K.G. (1961), *Studies on Quantitative Radiation Biology* Oliver and Boyd, Edinburgh and London.
8. Gray, L.H. (1946-47), 'Comparative studies of the biological effects of X-rays, neutrons and other ionizing radiations', *British Medical Bulletin,* **4**, 11-18.
9. Demerec, M. and Sams, J. (1958), 'Dose-effect relationships for X-ray induction of mutations in three genes of *Escherichia coli', Science,* **127**, 1059.
10. Shiomi, T., Inagaki, E., Inagaki, H. and Nakao, Y. (1963), 'Mutation rates at low dose level in *Drosophila melanogaster', J. Radiation Research,* **4**, 105-110.
11. Sparrow, A.H., Underbrink, A. and Rossi, H. (1972), 'Mutations induced in *Tradescantia* by small doses of X-rays and neutrons; analysis of dose-response curves', *Science,* **176**, 916-918.
12. Crowther, J.A. (1924), 'Some considerations relative to the action of X-rays on tissue cells', *Proc. Roy. Soc. Lond. B,* **96**, 207-211.
13. Muller, H.J. (1950), 'Some present problems in the genetic effects of radiation', *J. Cell. Comp. Physiol.,* **35**, suppl. I., 9-70.
14. Serebrovskaya, R.J. and Shapiro, N.J. (1935), 'The frequency of mutations induced by X-rays in the autosomes of mature and immature germ cells of *Drosophila melanogaster* males', *Dokl. Acad. Sci. U.S.S.R.,* **2**, 421-428.
15. Stadler, L.J. (1928), 'The rate of induced mutation in relation to dormancy, temperature and dosage', *Anatomical Record,* **41**, 97.
16. Timoféeff-Ressovsky, N.W. (1939), *Genetica Moderna.* ed. Ulrico Hoepli, Milano.

CHAPTER SIX

The classical X-ray work II: Chromosome breaks and rearrangements. Do X-rays produce true gene mutations?

The detection of breaks and rearrangements in mitotic cells

Already in Muller's first experiments various types of chromosome rearrangement were discovered by genetical tests (1). Soon this was followed by cytological evidence for the occurrence of chromosome breaks and rearrangements in irradiated plant cells (2). More recently, these effects have been scored in cell or tissue cultures of higher animals by essentially the same methods as in mitotic plant cells, and some of these results will be included in this chapter.

It should be realized that different criteria are used for the assessment of chromosome damage in mitotic cells of plants or animals and in germ cells of *Drosophila* (methods for detecting aberrations in germ cells of maize will be discussed in Chapter 10). In mitotically dividing cells, the mitosis following irradiation can be scanned, so that potentially lethal changes are not lost before scoring. In fact, assessment of radiation damage is mainly based on aberrations that will lead to cell death at a later mitosis: dicentrics formed from asymmetrical rejoining between two broken chromosomes, and rings formed from deletions (Fig. 6.1). Single unrestituted breaks can be seen as such in the irradiated cells themselves, or as bridges at the next anaphase (p. 17). In *Drosophila,* on the contrary, inviable rearrangements are lost as dominant lethals and only viable ones are detected by the genetical or cytological methods described in Chapter 2.

Whether or not a rearrangement acts as a lethal depends, not on the type of damage to the chromosomes, but on accidents of rejoining and their consequences for mitosis and cell viability. At the level of the chromosome, viable and inviable aberrations

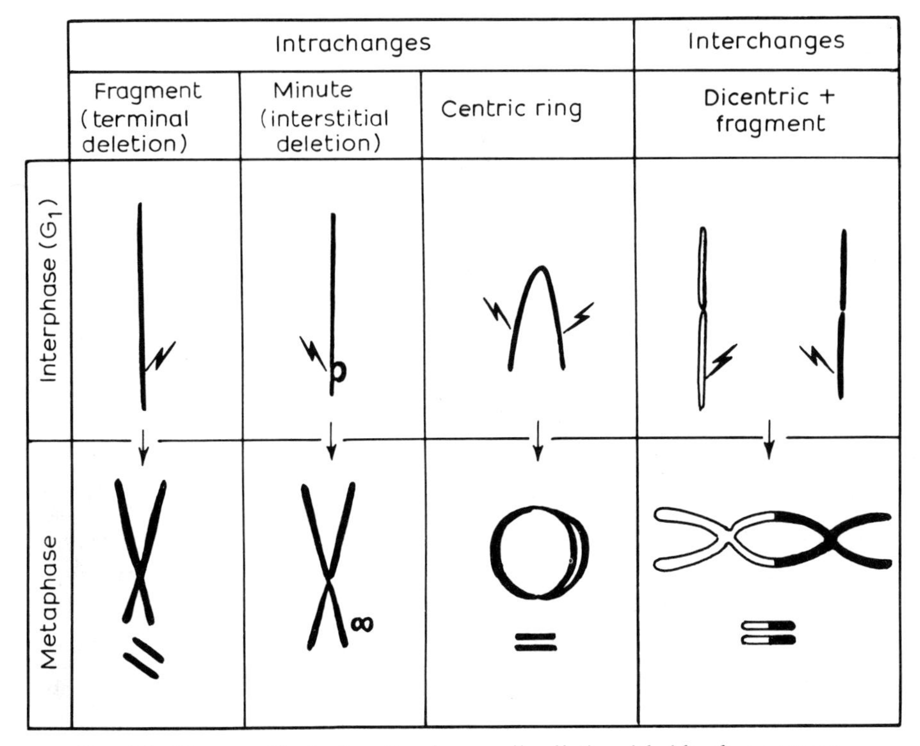

Fig. 6.1. Diagram illustrating the four easily distinguishable chromosome-type aberrations. (Fig. 2, Evans 1967 (37). Courtesy Athlone Press)

measure the same type of damage, and both have to be considered for a complete picture of the way a mutagenic agent affects the chromosomes. Indeed, a recent study of X-ray effects on mammalian cell cultures has led to the conclusion that 'there is little difference between the behavior of the chromosomes of normal human cells and those of *Tradescantia* microspores' (3). It is extremely unlikely that the chromosomes of *Drosophila* should react differently. At the moment there is, unfortunately, a great deal of compartmentalization between mutation research on plants and animals. In particular, cytologists often seem suspicious of rearrangements that have been detected only by genetical means. Yet, linkage between *Drosophila* genes on different chromosomes (Fig. 2.4) is fully as valid evidence for a translocation as is a cytologically detected quadrivalent in meiosis or a dicentric in mitosis.

Kinetics of break production

Let us now look at some of the experiments by which the production of chromosome breaks and rearrangements was fitted into the target theory. The underlying assumption that radiation energy has to be delivered in the chromosome or very close to it has more recently been confirmed by microbeam experiments, which showed that only beams hitting the chromosome itself produce breakage (4). Sax was the first to study the kinetics of chromosome breakage in plant cells (5). He found that, in *Tradescantia* microspores, the frequency of chromosome breaks (bridges and fragments at anaphase) increased linearly with dose and was independent of dose rate and fractionation. This was confirmed in many experiments on plant and animal cells and, in spite of some deviant results (6), it seems safe to conclude that chromosome breaks usually arise from single hits in or close to the chromosome. During the first part of interphase, whole chromosomes are broken. Later, when chromatids have been formed, each of them is broken independently of the other. There is an intermediate stage when the two newly formed chromatids are still closely apposed, so that both can be broken by the same hit.

Kinetics of rearrangement formation. 'Breakage-first' and 'contact' hypotheses

In order to account for the origin of chromosome rearrangements, two hypotheses were put forward in the thirties: the contact hypothesis of Serebrovsky (7), and the breakage-first hypothesis of Stadler (8). Serebrovsky assumed that fortuitous close contact between non-homologous chromosome points is a prerequisite for rearrangement formation. In a point of contact, a hit was presumed to cause a rearrangement by a kind of unorthodox crossing-over, resulting in an inversion or a deletion when the contact was formed between points in the same chromosome, a translocation when it was formed between points in different chromosomes. In contrast to this, the breakage-first hypothesis assumed that a rearrangement arises from two broken ends which are brought together by random chromosome movement and rejoin in the wrong order. The two hypotheses lead to different expectations for the kinetics of rearrangement production. On the contact hypothesis a rearrangement requires a single hit, and the frequency of rearrangements should increase linearly with dose. On the breakage-first hypothesis, each rearrangement requires two independently produced hits. If the probability for one hit to be produced within a given nucleus by a given dose is p, then the probability of having two hits produced by the same dose in the same nucleus is p^2. Since p is linearly related to dose, the breakage-first hypothesis predicts that the frequency of rearrangements should increase as the square of the dose, four times as many rearrangements being formed when the dose is doubled, nine times as many when it is trebled etc.

These predictions were so clear that it is surprising to find how much work and dispute had to go into establishing which of them was correct. This was in part due to the interruption of research and communication through the second world war; but there were also scientific obstacles. These were created by contradictory results obtained in different laboratories. In particular, there were essential differences between data on *Drosophila* and on plants. Fortunately, research workers at this period had not yet succumbed to compartmentalization, and they felt that these contradictions had to be resolved before a generally valid interpretation could be formulated. This has led to a better insight into the processes of breakage and reunion.

Sax' data on *Tradescantia* microspores were in good agreement with the breakage-first hypothesis: while the frequency of chromosome breaks increased almost linearly with dose, that of rearrangements increased nearly as the square of the dose (Table 6.1) (9).

More recently, a dose-square law of response to X-rays has been found also for aberrations in human leukocytes that had been irradiated *in vitro* (10), and for multilocus deletions in *Neurospora* (11).

In Sax' experiments, the frequency of aberrations decreased when a given dose was spread over a longer time (applied at 'lower intensity'; applied at a lower 'dose rate') (Fig. 6.2). The same was true for multilocus deletions in *Neurospora* (11). This is what would be expected if, of two independent breaks in the same nucleus, one had time to rejoin before the second was produced. In fact, it was possible to use fractionated exposure for calculating the time over which a break remains available for rejoining. We shall see in Chapter 8 that this method has played an important role in analysing the mechanism of rearrangement formation.

Table 6.1

Dose-response of the frequencies of breaks and rearrangements in irradiated microspores of *Tradescantia*. (The dose exponents in columns 4 and 6 are those that give the best fit to the observed data.) (After Table 3, Sax (9))

Dose in R	*n*	*% breaks*	*Expected for dose exponent 1.1*	*% rearrangements*	*Expected for dose exponent 1.9*
10	6330	0.2	0.2	0.03	0.02
20	6930	0.4	0.5	0.06	0.08
40	8610	0.9	1.1	0.4	0.3
80	7368	2.0	1.9	1.5	1.4
120	4962	2.9	2.9	2.5	3.0
160	8202	4.4	4.0	5.1	5.3
200	8508	5.1	5.1	8.1	8.1

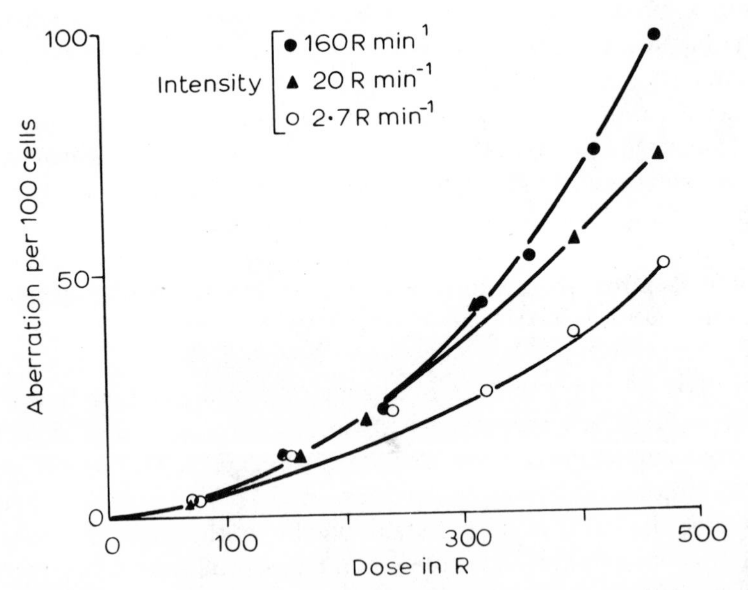

Fig. 6.2. The frequency of X-ray induced translocations in *Tradescantia* microspores in relation to dose rate (intensity). abscissa: dose in R, ordinate: aberrations per 100 cells. (Fig. 58, Rieger and Michaelis (38). Courtesy Verl. Gustav Fischer.

In *Drosophila,* the frequency of single chromosome breaks cannot be measured directly. Unrestituted breaks give rise to dominant lethals, mainly through loss of chromosomes by bridge formation (p. 17); but there are additional causes of dominant lethality, and dose effect curves for dominant lethals include these as well. A better measure of unrestituted single breaks is furnished by the so-called 'viable chromosome losses', e.g. loss of an X-chromosome from an XX zygote, which transforms a potential female into a male. The frequency of X-chromosome losses is greatly amplified in strains with a ring-shaped X-chromosome. Here, not only open breaks but also a high proportion of restituted ones lead to loss, provided they are still present when the chromosome has split into two chromatids. For if these twist round each other before rejoining, separation at anaphase is prevented by the formation of either a double-sized ring or two interlocked rings. This can be easily verified by cutting longitudinally through a twisted paper ring (Moebius strip). When it was found that, at low and moderate doses, the frequencies of dominant lethals and viable chromosome losses followed approximately linear kinetics (12), this was taken as indirect confirmation of the single-hit interpretation of chromosome breakage.

The kinetics of rearrangement frequency could be tested directly in *Drosophila,* and it is here that the difficulties arose. Two essential differences were found from the results obtained on plants. First, the dose exponent for the frequency of rearrangements was variable and, though it was usually greater than 1, it hardly ever reached the 2 expected from the dose-square law of the breakage-first model. Secondly, there was no effect of either dose rate or fractionation on translocation frequency. Thus, when Muller irradiated spermatozoa inside irradiated females, he found that a given dose of radiation produced the same frequency of translocations whether it was administered in a single concentrated treatment at 100R min^{-1}, in four fractions separated by intervals of one week, or spread over a whole month (12).

The second difference was more easily interpreted than the first. It was taken to mean – and this agrees with subsequent experiments – that breaks in the condensed and metabolically inactive chromosomes of the spermatozoa can neither restitute in the old order nor rejoin in a new one, and thus remain available for rearrangement formation until after insemination.

Table 6.2

The dose dependence of translocation frequency in *Drosophila*. (Genetically scored translocations between the two large autosomes.) (After Table 5, Muller (12))

Dose in R	No. of cultures	Frequency of translocations (%)	Ratio between frequencies: Observed	Expected for dose exponent 2	Expected for dose exponent 1.5
4000	1 373	8.3	$\frac{8.3}{1.2} = 6.9$	16	$\sqrt{(4^3)} = 8$
1000	10 196	1.2			

The problem of the dose exponent was more difficult to solve. In genetical tests carried out by Muller and his associates shortly before the outbreak of the Second World War (12), the dose exponent for translocations was about 1.5, halfway between the 1 expected for linearity and the 2 expected from the square law (Table 6.2). (A dose exponent of 1.5 means that the frequency of translocations increases by a factor of $\sqrt{(k^3)}$ if the dose is increased by a factor of k.)

Simultaneously, Bauer in Germany had carried out a large-scale experiment on the dose-response of cytologically scored rearrangements in *Drosophila* (13). The results are shown in Table 6.3.

When these data together with others obtained by the same author were fitted to a dose-response curve, it was again found that an exponent of 1.5 gave the best fit. It is, however, interesting to see from Table 6.3 that the dose exponent for the comparison between the two lowest doses was 2, as expected from the square law for two-hit rearrangements. We shall return to this presently.

While the '3/2 power' law gives a good approximation to most results obtained with moderate X-ray doses, it is an empirical rule rather than a theoretical expectation based on a biophysical model. All it states is that, over a fairly wide range of doses, the frequency of rearrangements falls short of expectation. The explanation must be looked for at the biological level. Either, a certain proportion of theoretically possible rearrangements are not formed, or they are not detected. Muller assumed the second explanation to be correct. He thought that, with increasing dose, more and more viable translocations would be drawn into complex inviable rearrangements and would be lost as dominant lethals. It agrees with this assumption

Table 6.3

The dose dependence of cytologically scored rearrangements in the salivary glands of F_1 progeny from irradiated *Drosophila* males. (After Table 2, Bauer (13))

Dose in R	*No. of F_1 larvae*	*Frequency of rearrangements (%)*
1000	329	2.74
2000	377	10.61
3000	463	18.14
4000	309	33.66
5000	523	45.70

Ratio of frequencies	*Observed*	*Expected for dose exponent*	
		2	*1.5*
4000R/1000R	33.7/2.7 = 12.4	16	$\sqrt{(4^3)} = 8$
4000R/2000R	33.7/10.6 = 3.2	4	$\sqrt{(2^3)} = 2.8$
2000R/1000R	10.6/ 2.7 = 3.9	4	$\sqrt{(2^3)} = 2.8$

that data obtained with low doses could be fitted to the square law. Subsequently, results with mustard gas provided indirect evidence for the hypothesis. Mustard gas produces much fewer rearrangements in relation to sex-linked lethals than do X-rays (Chapter 15). It is therefore possible to give doses that, in terms of the overall frequency of translocations, are comparable to very low X-ray doses; for these, the dose response follows the square law closely (p. 94).

Dose response to neutrons. Role of LET. Rejoining distance and rejoining sites

While in *Drosophila* the inviability of complex rearrangements appeared a satisfactory explanation for the low dose exponent of rearrangement formation, this was not so for plants; for here the dicentrics and rings that are scored are in any case inviable. Indeed, as Table 6.1 shows, good approximations to the square law have been obtained for plant material. Later work, however, has shown that these are the exception rather than the rule. Also in plants the dose exponent is usually less than 2 and often lies around 1.5 as in *Drosophila* (14). A satisfying explanation, which is of general importance for the theory of rearrangement formation, was provided by S. Wolff. It rests on recalculation of the distance within which two separately induced chromosome breaks can rejoin with each other ('rejoining distance'). In order to understand it, we have to give a more thorough look at the consequences of the breakage-first mechanism for rearrangement formation.

It is, of course, to be expected that spatial arrangements within the nucleus will to some extent determine the way in which chromosome fragments join with one another. Experimental evidence shows that this is true. *Tradescantia,* for example, has six chromosomes, each with two equal arms. If rejoining were wholly at random, one would expect dicentrics (interchanges) to be about ten times as frequent as rings (intrachanges), since a break in one chromosome arm should be ten times as likely to join up with one in the other five chromosomes than with one in the same chromosome. Yet, the actually observed ratio is about 3:1 (14). Similarly, in *Drosophila* the ratio of X-ray induced translocations to inversions and deletions is lower than would be expected from random rejoining. Obviously, then, not any two breaks in the same nucleus can form a rearrangement; they have to be within 'rejoining distance' from each other. A means for estimating this distance was provided by a comparison between the effects of X-rays and fast neutrons. Since the importance of such a comparison goes far beyond its bearing on the special question of rejoining distance, we shall digress for a moment to deal with it (15).

X-rays and neutrons differ from each other in LET (linear energy transfer) i.e. the ion density along the tracks they produce. X-rays, (except very soft ones), γ rays and β rays have low LET's, i.e. the ionizations along their tracks are far apart and each acts independently of the others. This is the basis of the square law for rearrangement frequencies. Only towards the end of its track, when an electron has lost most of its energy and is slowed down, does it produce a 'tail' of densely clustered ionizations. Within these tails, the ionizations no longer act independently, and two breaks may be produced by ions of the same tail. This adds a linear component to the equation relating aberration frequency to dose; the correct equation reads $A = a + bD + cD^2$, where A stands for aberration frequency, D for dose, and a, b and c are constants. At all but very low doses, so many electron tracks traverse the cell nucleus that the linear term becomes insignificant. The situation is different for radiation of high LET – neutrons and α rays – which yield densely ionized tracks throughout. For these radiations, the chance of producing two breaks in the same track is very high. It has indeed been found that neutrons produce aberrations by single-hit kinetics even at doses when several or many different proton tracks traverse the cell nucleus. Now this implies that a broken chromosome rejoins more readily with another broken chromosome in its neighbourhood than with one farther away. If this were not so, breaks that have been produced close together might still rejoin with breaks produced in other tracks, and the dose-effect curve would again contain a dose-squared component.

Lea (see Bibliography) used the different kinetics of rearrangement production by X-rays and neutrons for an estimate of the rejoining distance; he arrived at a value of roughly 1μ. Wolff, from a study of the interaction between neutrons and X-rays in the same cell, arrived at the much smaller value of between 0.1 and 0.3μ (16). He calculated that the number of 'sites' in which chromosome breaks are within this distance from each other is about two in *Vicia* and about four in *Tradescantia*. These values are small enough to result in early saturation, so that the originally quadratic curve soon tends to become convex. When all points are fitted by the same expression, this will often be best approximated by a 3/2 power exponent. The site concept, although still only a working hypothesis, fits not only the data from which it was derived, but also others that were obtained later by other workers. Naturally, neither rejoining distance nor site are fixed parameters, even for a given kind of cell. They vary with the type of rearrangement scored, with the geometrical arrangement of the chromosomes, and with physiological conditions such as degree of hydration; they have to be calculated separately for any given set of conditions.

Summary of the target theory

The classical target theory and its extension to densely ionizing radiation may now be summarized in the following three statements:

(1) Gene mutations, very small deletions, and chromosome breaks arise from single hits. Their frequency increases linearly with dose and is independent of dose rate and fractionation.

(2) Large rearrangements arise from rejoining between two chromosome breaks. After X- or γ radiation, their frequency increases essentially as the square of the dose. Saturation of the available 'sites' for rejoining and, in *Drosophila,* the lethality of complex rearrangements, lead to a bending down of the curve at higher doses, so that the dose exponent over the whole curve usually is about 1.5. Neutrons, because of their high LET, produce rearrangements by single-hit kinetics.

(3) Lowering the dose rate or splitting the dose into fractions reduces the frequency of X-ray induced rearrangements except in the special case of *Drosophila* spermatozoa in which breaks are incapable of restitution or rejoining before fertilization. The frequency of neutron-induced rearrangements is independent of dose rate and fractionation.

The biological objects that furnished the data for these conclusions were particularly well suited to a purely biophysical interpretation of the kinetic data. *Drosophila* spermatozoa and *Tradescantia* microspores are synchronized populations. In addition, *Drosophila* spermatozoa are metabolically inactive. So are plant seeds, in which the cells are arrested in G1, the period in the cell cycle that precedes DNA-synthesis (S). In dividing tissues, all stages of the cell cycle are present, but synchrony can be imposed on the sample for scoring by pulse-labelling with thymidine immediately prior to irradiation. Cells that were in G1 or S at the time of irradiation can be recognized by the possession of label, and scoring is restricted to unlabelled cells, all of which must have been irradiated in G2, the period between S and the next mitosis. Recently, it has been shown that even in this case the dose-effect curves cannot be interpreted wholly on target principles (17). Radiosensitivity changes during G2, and the duration of the various sensitivity phases varies between cells. Radiation, by perturbing the cell cycle, introduces an additional source of asynchrony, and one that differs between radiation doses. It is thus not surprising to find that frankly asynchronous populations such as mouse spermatogonia often fail to show the responses to dose, dose rate and fractionation that would be expected from target theory. We shall come back to this in Chapter 9, where we shall also deal with the role of metabolism in causing deviations from target expectations.

Do X-rays produce true gene mutations?

In the first chapter I mentioned the lively argument between those scientists that assumed X-rays could produce true gene mutations, and others that believed all X-ray-induced mutations to be due to deletions or position effects. *Drosophila* geneticists, with Muller as the leading protagonist, defended the first opinion; maize geneticists, in particular Stadler, the second. The difference between the two-hit kinetics for the formation of rearrangements and the one-hit kinetics for the production of point mutations seemed to provide a means for distinguishing between these two effects. This belief became untenable when it was found that, in maize as in *Drosophila*, the frequency of very small deletions increased linearly with dose like that of gene mutations (18,12). This, incidentally, was the first indication that energy delivered to a chromosome may spread over a short distance. The nature of this 'indirect' action of X-rays will be dealt with in the next chapter.

In recent years, the problem has been attacked in different organisms, by a variety of techniques, and with varying results. In flowering plants, even small deletions are easily detected by their effect on the haplophase (p. 27). Stadler and Roman (19) analysed 415 mutations from *A* (purple aleurone colour) to *a* (colourless aleurone) and found that all but two segregated defective pollen and, by this token, were deletions*. The remaining two, with partially viable pollen, fulfilled other criteria for deletions (shortening of map distance between outside markers; cell lethality in homozygous condition). Mottinger has now repeated this work for the bronze locus with the same result (20).

In *Drosophila* and mice, the proportion of cytologically or genetically detectable deficiencies among X-ray induced point mutations is high in spermatozoa but very low in spermatogonia (21,22). For spermatozoa alone, the data from *Drosophila* are conflicting. Lifschytz and Falk (23) studied X-ray induced lethals in a small region of the X-chromosome for which a duplication was available. Under cover of the duplication, the lethals could be carried in males, and pairwise crosses between all of them could be carried out. Lethals that yielded a viable heterozygote were presumed to have occurred in different genes; those that yielded a lethal heterozygote were supposed to be allelic. Deletions covering two or more

* The aleurone surrounds the seed and determines the colour of the maize kernel.

genes were recognized by the fact that they yielded lethal heterozygotes with every one of these. In this way, a 'complementation map' for a whole chromosome region was constructed on the same principle on which intracistronic complementation maps are constructed. It showed that most lethals covered two or more segments, and the authors came to the conclusion that most, if not all, of the lethals studied by them were deficiencies of various sizes and that X-rays hardly, if ever, produce true gene mutations. It has been objected (24) that the mutational type (lethal) and the region (partially heterochromatic) favoured the production of deletions over that of point mutations; but among lethals produced in the same region by a chemical mutagen (ethylmethane-sulphonate) about 50% covered only one 'unit' of the complementation map and were classified as point mutations (25). Possibly there are regional differences in response to X-rays. Judd and his collaborators (26), who analysed lethals in a short distal section of the X-chromosome both by complementation and by cytological study of the salivary chromosomes, found that most X-ray induced lethals registered as single-unit mutations without cytologically visible deficiencies. This was also found to apply to lethals in the fourth chromosome (27). At two different loci (rosy and rudimentary), the majority of X-ray induced mutations were capable of interallelic complementation (p. 40) and thus were due to intracistronic changes (28,29). While these observations show that X-rays can and often do induce point mutations, they do not exclude the possibility that all of these are intracistronic deletions. An indication that this is not so comes from the similarity of the complementation patterns for X-ray induced and chemically induced mutations, and from Suzuki's finding (30) that some γ ray induced sex-linked lethals were temperature-sensitive.

Short of a molecular analysis of the gene product, revertibility of a mutation is the best evidence that it is not due to a deletion. Malling and De Serres found that all of 33 X-ray induced adenine-auxotrophs in *Neurospora* reverted either spontaneously or under the influence of chemical mutagens, and thus were due to intragenic changes (31). At higher doses and with a different scoring technique, deletions were also obtained (32). Strong evidence for the ability of ionizing radiation to produce intragenic changes comes from experiments in which extracellularly irradiated bacteriophage or its isolated DNA were shown to mutate after irradiation (33-35). It is highly improbable that these mutations should have been deletions, especially as those induced in phage T4 included forward and reverse mutations at the rII locus. The argument was finally clinched by analysis at the molecular level, which showed that ionizing radiation can produce base changes in micro-organisms; we shall deal with this in Chapter 9.

Thus, Muller's contention that X-rays can produce true gene mutations has been fully vindicated in a variety of organisms. On the other hand, Stadler's finding that all X-ray induced mutations in maize are in reality due to deletions has been confirmed. This may be due to the extreme sensitivity of plant chromosomes to X-ray breakage; probably, breakage events predominate to such an extent that they mask the rare gene mutations. A few gene mutations seem to have occurred in experiments in which meiotic cells of maize were exposed to cobalt-γ radiation (36). Apparently, the proportion of true gene mutations and rearrangements in X-rayed cells depends on many factors, among which are organism, cell type, chromosome region, state of condensation of the chromosome, type of mutation scored, and X-ray dose. It is well to keep in mind that the enormous variability which is characteristic of living processes applies also to the mutation process. Even if the effects of a given mutagen on DNA should be the same in all cells and all organisms – and this will often be true – differences in accessibility of DNA and in the cellular processes that are required for fixation, expression and survival of a potential mutation will create a diversity of observed responses. X-ray mutagenesis is no exception to this rule.

References

1. Muller, H.J. (1930), 'Types of visible variations induced by X-rays in *Drosophila*', *J. Genetics,* **22**, 299-335.
2. Sax, K. (1938), 'Chromosomal aberrations induced by X-rays', *Genetics,* **23**, 494-516.
3. Bender, M.A. and Gooch, P.C. (1962), 'Types and rates of X-ray-induced chromosome aberrations in human blood irradiated *in vitro*', *Proc. Nat. Acad. Sci. (U.S.A.),* **48**, 522-532.
4. Zirkle, R.E. and Bloom, W. (1953), 'Irradiation of parts of individual cells', *Science,* **117**, 487-493.
5. Sax, K. (1941), 'Types and frequencies of chromosomal aberrations induced by X-rays', *Cold Spring Harbor Sym. Quant. Biol.,* **9**, 93-103.

6. Brewen, J.G. and Brock, R.D. (1968), 'The exchange hypothesis and chromosome-type aberrations', *Mutation Res.,* **6**, 245-255.
7. Serebrovsky, A.S. (1929), 'A general scheme for the origin of mutations', *Amer. Nature,* **53**, 374-378.
8. Stadler, L.J. (1932), 'On the genetic nature of induced mutations in plants', *Proc. 6th Int. Cong. Genetics,* **1**, 274-294.
9. Sax, K. (1940), 'An analysis of X-ray induced chromosomal aberrations in *Tradescantia*', *Genetics,* **25**, 41-68.
10. Bender, M.A. and Barcinski, M.A. (1969), 'Kinetics of two-break aberration production by X-rays in human leukocytes', *Cytogenetics,* **8**, 241-246.
11. De Serres, F.J., Malling, H.V. and Webber, B.B. (1967), 'Dose-rate effects on inactivation and mutation induction in *Neurospora crassa*', *Brookhaven Symposia in Biology,* **20**, 56-76.
12. Muller, H.J. (1940), 'An analysis of the process of structural change in chromosomes of *Drosophila*', *J. Genetics,* **40**, 1-66.
13. Bauer, H. (1939), 'Röntgenauslösung von Chromosomenmutationen bei *Drosophila melanogaster* I', *Chromosoma,* **1**, 343-390.
14. Giles, N.H., Jr. and Riley, H.P. (1949), 'The effect of oxygen on the frequency of X-ray induced chromosomal rearrangements in *Tradescantia* microspores', *Proc. Nat. Acad. Sci., U.S.A.,* **35**, 640-646.
15. Gray, L.H. (1946-47), 'Comparative studies of the biological effects of X-rays, neutrons and other ionizing radiations', *British Medical Bulletin,* **4**, 11-18.
16. Wolff, S. (1963), 'The kinetics for two-break chromosome exchanges and the 3/2 power rule', in *Repair from Genetic Radiation,* (ed. F. Sobels) pp.1-10 Pergamon Press, Oxford, London, New York and Paris.
17. Savage, J.R.K. and Papworth, D.G. (1973), 'The effect of variable G_2 duration upon the interpretation of yield-time curves of radiation-induced chromatid aberrations', *J. Theor. Biol.,* **38**, 17-38.
18. Rick, C.M. (1940), 'On the nature of X-ray induced deletions in *Tradescantia* chromosomes', *Genetics,* **25**, 466-482.
19. Stadler, L.J. and Roman, H. (1948), 'The effect of X-rays upon mutation of the gene *A* in maize', *Genetics,* **33**, 273-303.
20. Mottinger, J.P. (1970), 'The effect of X-rays on the bronze and shrunken loci in maize', *Genetics,* **64**, 259-271.

21. Ward, C.L. and Alexander, M.L. (1957), 'Cytological analysis of X-ray induced mutations at eight specific loci in the third chromosome of *Drosophila melanogaster*', *Genetics*, **42**, 42-54.
22. Russell, W.L. and Russell, L.B. (1959), 'The genetic and phenotypic characteristics of radiation-induced mutations in mice', *Rad. Research Suppl.*, 1, 296-305.
23. Lifschytz, E. and Falk, R. (1968), 'Fine structure analysis of a chromosome segment in *Drosophila melanogaster;* analysis of X-ray-induced lethals', *Mutation Res.*, **6**, 235-244.
24. Schalet, A. and Lefevre, G. (1974), 'The proximal region of the X chromosome', in *The Genetics and Biology of Drosophila*, Vol. I, (ed. M. Ashburner and E. Novitski) in press.
25. Lifschytz E. and Falk, R. (1969), 'Fine structure analysis of a chromosome segment in *Drosophila melanogaster;* analysis of ethylmethanesulphonate-induced lethals', *Mutation Res.*, **8**, 147-155.
26. Judd, B.H., Shen, M. and Kaufman, T.C. (1972), 'The anatomy and function of a segment of the X-chromosome of *Drosophila melanogaster*', *Genetics*, **71**, 139-156.
27. Hochman, B., Gloor, H. and Green, M.M. (1964), 'Analysis of chromosome IV in *Drosophila melanogaster.* I Spontaneous and X-ray-induced lethals', *Genetica*, **35**, 109-126.
28. Chovnick, A., Schalet, A., Kernaghan, R.P. and Talsma, J. (1962), 'The resolving power of genetic fine structure analysis in higher organisms as exemplified by *Drosophila*', *Am. Nat.*, **94**, 281-296.
29. Carlson, P.S. (1971), 'A genetic analysis of the rudimentary locus of *Drosophila melanogaster*', *Genet. Res., Camb.*, **17**, 53-81.
30. Suzuki, D.T., Piternick, L.K., Hayashi, S., Tarasoff, M., Baillie, D. and Erasmus, U. (1967), 'Temperature-sensitive mutations in *Drosophila melanogaster.* I. Relative frequencies among X-ray and chemically-induced sex-linked recessive lethals and semilethals', *Proc. Nat. Acad. Sci., U.S.A.*, **57**, 907-912.
31. Malling, H.V. and De Serres, F.J. (1967), 'Identification of the spectrum of X-ray-induced intragenic alterations at the molecular level in *Neurospora crassa*', *Radiation Res.*, **31**, 637.

32. Webber, B.B. and De Serres, F.J. (1965), 'Induction kinetics and genetic analysis of X-ray induced mutations in the *ad*-3 region of *Neurospora crassa', Proc. Nat. Acad. Sci., U. S. A.*, **53**, 430-437.
33. Van der Ent, G.M., Blok, J. and Linckens, E.M. (1965), 'The induction of mutations by ionizing radiation in bacteriophage ϕX174 and its purified DNA', *'Mutation Res.*, **2**, 197-204.
34. Bleichrodt, Y.F. and Verley, W.S.D. (1973), 'Reversion of cistron *A* amber mutants of bacteriophage ϕX174 by ionizing radiation', *Mutation Res.*, **18**, 363-365.
35. Brown, D.F. (1966), 'X-ray-induced mutation in extra-cellular bacteriophage T4', *Mutation Res.*, **3**, 365-373.
36. Caspar, A. and Singleton, W.R. (1957), 'Induced "gene" mutation in maize', *Genetics,* **42**, 364-365.
37. Evans, H.J. (1967), 'Actions of radiations on chromosomes', *The Scientific Basis of Med. Ann. Rev.*, 321-339.
38. Rieger, R. and Michaelis, A. (1967), 'Chromosomenmutationen', *Verl Gustav Fisher Verl, Jena.*

CHAPTER SEVEN

Direct and indirect action of X-rays. Radio-isotopes. Radiation-sterilized food

Indirect action of X-rays

The strict target theory left no room for dependence of X-ray-induced genetic changes on environmental conditions. Yet already in the earliest experiments on *Drosophila,* it had been found that more lethal and visible mutations are produced in spermatozoa than in spermatogonia (1), while Stadler had observed that X-rays are genetically more effective in soaked sprouting seeds of barley than in dry dormant ones (2). The latter difference might conceivably have been due to a shift in cell stage at the time of irradiation. These experiments had shown that at least the physiological environment of the chromosomes or, perhaps, their state of contraction, movement or neighbourhood relationship, could modify the results of X-radiation. In the late thirties and during the forties, increasing evidence was provided for the modification of X-ray effects by a variety of external factors, such as temperature, oxygen tension, hydration, and the presence of various chemical substances in irradiated cells (3-5). The impact of these observations on the theory of radiation mutagenesis was great. They led research out of the narrow confines of a purely biophysical model into the complexity of the living cell and organism. We shall see that similar observations have led the way out of the more recent, but equally narrow picture of mutagenesis as a branch of nucleic acid chemistry.

The first enlargement of the target theory stemmed from radiochemical observations and led only to minor modifications of the original concepts (6). It was due to the realization that, in wet systems, X-rays act on sensitive macromolecules not only by direct 'hits', but also indirectly through energy transfer by radiation products of water. The action of ionizing radiation on water is complex. It leads to the formation of a variety of chemical species which, in turn, by interacting with each other or with still unreacted molecules, produce new reactive species. Those most important for energy transfer to biological macromolecules are the short-lived, electrically neutral, highly reactive radicals $H^{\cdot}$ and $OH^{\cdot}$. In the presence of oxygen, these give rise to the radical $O_2H^{\cdot}$ and the molecule H_2O_2 (hydrogen peroxide). Another highly reactive radiation product of water is the hydrated electron, which in the absence of oxygen acts as a reducing agent.

When target molecules, e.g. viruses or enzymes, are irradiated dry, the whole action is direct. When the target size is calculated from the radiation data, it agrees remarkably well with that obtained by other means (Fig. 7.1), although even in this case external conditions such

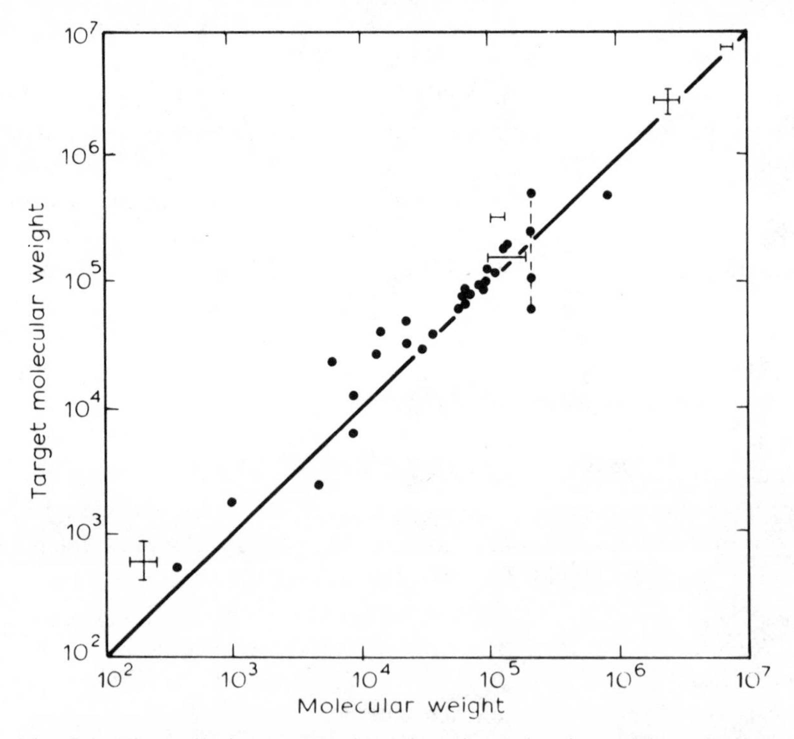

Fig. 7.1. The radiation target size, given in molecular-weight units for a variety of biological molecules irradiated in the dry state, plotted against the known physicochemical molecular weight. The straight line is the expected relation if the two molecular weights are equal. (Fig. 3, Hutchinson (7). Courtesy American Assoc. Advancement of Science)

as temperature may somewhat shift the result. The four points connected by a dashed vertical line represent the variation in sensitivity of the dry enzyme catalase at temperatures from $-180°C$ (bottom) to $+112°C$ (top). When these same macromolecules are irradiated in solution, their sensitivity is increased, i.e. the same degree of inactivation is achieved with a lower dose. Expressed in terms of the target theory, macromolecules irradiated wet offer a larger target than macromolecules irradiated dry. This is so because the target of wet molecules includes not only the molecule itself but also a sphere of water surrounding it within which water radicals can diffuse to the actual target. The difference between inactivation doses in the dry and wet states furnishes an estimate of the size of this sphere. For a number of enzymes, and for transforming DNA, its diameter was found to be no more than two to three times that of the macromolecule itself (Fig. 7.2) (7). Thus, inclusion of indirect

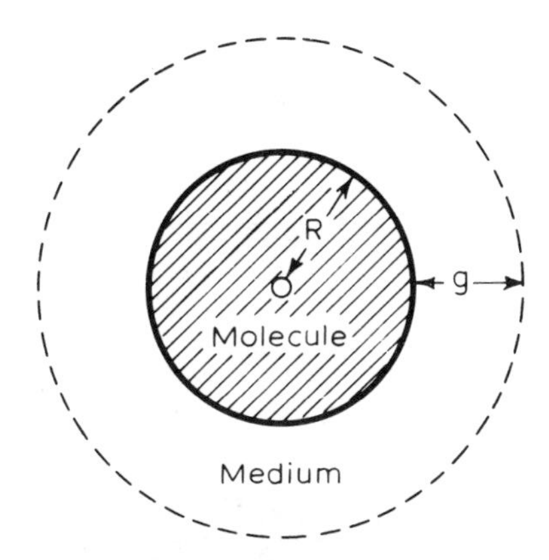

Fig. 7.2. A simple picture of radiation action on a molecule in a cell. The molecule, shown schematically as a sphere of radius R, is surrounded by a water-layer of average thickness *g*. Water radicals from this water layer contribute to the inactivation of the molecule. (Fig. 4, Hutchinson (7). Courtesy American Assoc. Advancement of Science)

radiation effects left the target concept essentially unchanged, but made it more flexible. In particular, it gave a more precise meaning to the conclusion that genetical effects can be produced by ionizations occurring 'in or very close to' the chromosome. They also made it understandable that physiological and environmental conditions could affect the target via effects on the formation and diffusion of radicals. Before going on to discuss this point, let us first look at the evidence for indirect radiation effects on biological molecules.

The dilution effect

When target molecules, e.g. enzymes, viruses, or transforming DNA are irradiated in solution, hits on target molecules are much rarer than hits on water molecules, and most or all of the damage is due to reaction of the target molecules with water radicals. A given dose of radiation produces a given number of radicals and these, in turn, inactivate a given number of target molecules. Radiation yield is measured as the percentage of inactivated molecules, and since a given *number* of molecules forms a larger *percentage* of all exposed ones at low than at high concentration, the radiation yield due to indirect action decreases with increasing concentration of the target molecules. In Table 7.1 this can be seen from the fact that, while the number of inactivated enzyme molecules per 1000 R is approximately the same at all dilutions, the dose required for a 25% inactivation decreases from 60 000 to 45 R when the original solution is diluted 884 times.

Table 7.1
25% inactivation of an enzyme at different dilutions

Dilution	*Dose in R*	*Enzyme units inactivated*	
		overall	*per 1000 R*
1x	60 000	2 600	43
62x	800	40	50
258x	145	8	55
884x	45	2.3	50

(Dale (6). Courtesy Springer-Verlag)

The protection effect

When target molecules are irradiated in the presence of other molecules that are capable of reacting with water radicals, these 'radical scavengers' protect the target molecules from the indirect action of radiation. They do this the more efficiently the greater their concentration relative to that of the target molecules and the higher their reactivity with radicals. At high concentrations of protective substances, e.g. in nutrient broth, practically the whole effect on the target molecules may be due to direct hits. This explains why there is good agreement between the directly determined sizes of bacteriophages and their target volume as calculated from irradiation in broth (8). Calculated from irradiation in pure water, the target volume is greater than the actual one. In addition to a general protective effect of organic molecules, which is approximately proportional to their molecular weight, there are atom groups with specific protective ability (9). Sulphur-compounds, for example, are outstandingly good protectors. Thus, thiourea ($NH_2CS.NH_2$) has 10 000 times the protective power of urea ($NH_2CO.NH_2$). Not all protective substances act as radical scavengers. They may also react with organic radicals formed in the macromolecule itself and 'repair' them by donation of a hydrogen atom before damage has become irremediable. Other means of protecting cells from radiation damage or, conversely, sensitizing them (10,11) to it cannot be discussed here; some will be mentioned in other contexts (p. 112/128).

Dilution and protection effects are features of indirect radiation action and can be used to distinguish it from direct action. Dose effect curves cannot. Inactivation of targets by single direct hits yields exponential survival curves; but so may inactivation via water radicals. In both cases, target molecules that were knocked out by a first hit or radical continue to serve as targets for further hits or radicals without, however, yielding an additional effect on survival. As a result, every increment of dose inactivates *the same proportion* of all targets that are still active (e.g. 1% or 10%). This yields an exponential curve in which the log of the surviving fraction is proportional to dose and which yields a straight line in a semi-log plot. For very rare events such as mutations, the number of twice affected targets is negligibly small, and the dose effect curve becomes linear in an arithmetic plot. Thus, exponential or linear dose-effect curves are no proof of direct action.

Neither is modifiability of the result by experimental conditions. It is true that this can often be attributed to differences in the numbers and types of radical formed and in their diffusion ranges, but even direct effects can be modified, as shown, for example, in Fig. 7.1. Moreover, both direct and indirect radiation effects are subject to modification at later levels of the mutation process (see Chapter 9). The complexities of biological processes are such that the same observable effect may have many different causes. We shall realize this best by looking in some detail at the most important single means of modifying biological radiation damage: variation in oxygen tension.

The oxygen effect

Ubiquity of oxygen effect

In most biological systems, X-rays are two to three times as effective when applied in air or oxygen than when given in nitrogen or an inert gas. The first evidence for this was provided by Thoday and Read, who studied chromosome breakage in root tip cells of *Vicia faba.* A given dose of X-rays produced three times as many bridges and fragments when applied in oxygen as when applied in nitrogen (12). On the contrary, the effectiveness of α-rays was not enhanced by oxygen (13). Giles and Riley studied the frequencies of dicentrics and rings in *Tradescantia* microspores (14,15). Starting from irradiation in anoxia, they found a steep increase with oxygen tension up to about 10% oxygen; after this, the increase was slower and levelled off at 21%, the normal oxygen content of air. The oxygen enhancement ratio (OER = effect in oxygen divided by effect in nitrogen) was between 2.5 and 3.5 for X-rays and γ-rays, but only 1.5 for neutrons. Thus, there is a negative correlation between increasing LET (X- and γ-rays, neutrons, α-rays) and decreasing oxygen effect (2.5-3.5; 1.5; 1). We shall come back to this presently. Oxygen

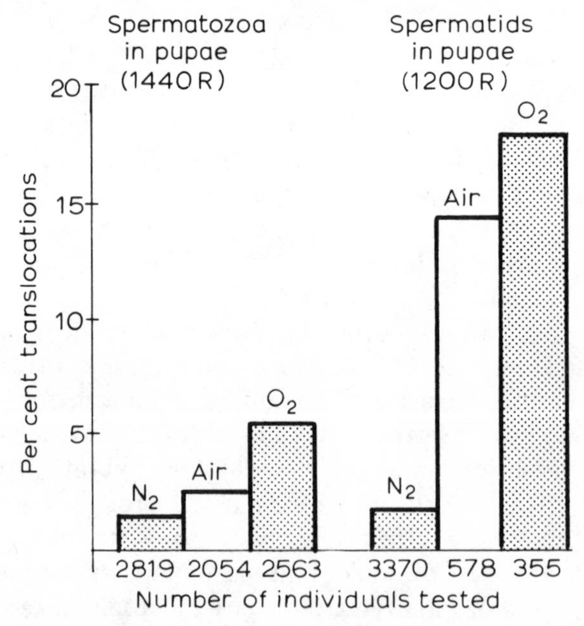

Fig. 7.3. Frequencies of translocations obtained by X-irradiating spermatozoa in pupae and spermatids in pupae in either nitrogen, air or oxygen. (Fig. 2, Oster (62). Courtesy Genetics Institut, Utrecht)

enhancement of radiation damage has been found also for translocations and recessive lethals in *Drosophila* (Figs. 7.3 and 7.4), and for mutations in micro-organisms (16-18).

In spite of the wide-spread occurrence of the oxygen effect, exceptions do exist. Thus, inactivation of transforming DNA (19) or of enzymes in solution is independent of the presence of oxygen. Isolated DNA of bacteriophage ϕX174 was even protected by oxygen against the killing and mutagenic action of X-rays (20), while the presence of oxygen during irradiation of the whole phage increased

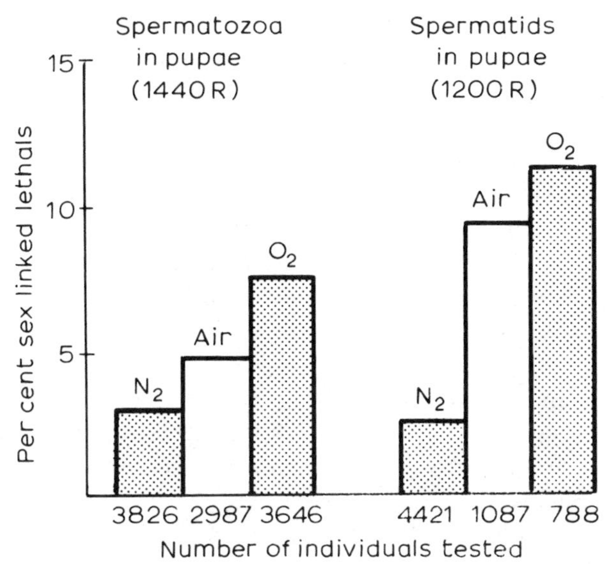

Fig. 7.4. Frequencies of recessive sex-linked lethal mutations obtained by X-irradiating spermatozoa in pupae and spermatids in pupae in either nitrogen, air or oxygen. (Fig. 3, Oster (62). Courtesy Genetics Institut, Utrecht)

the frequency of amber reversions (21). The frequency of dominant lethals in irradiated spermatozoa of the mouse is the same in oxygen as in nitrogen (22); probably, these cells are already so anoxic in their natural environment that nitrogen in the atmosphere has no further effect. In an auxotrophic strain of *Escherichia coli,* the frequency of reverse mutations was increased by the presence of oxygen during irradiation (23). Curiously enough, this was not the case for the frequency of mutations to streptomycin resistance in a different strain, although both strains had the same OER for survival. This observation was not followed up at the time. It gains renewed interest from the much more recent finding that these two categories of bacterial mutation differ also in their response to the conditions under which UV is used as mutagen (Chapter 14).

Depending on which gaseous environment is taken as baseline, anoxia may be considered as *radioprotective,* or oxygen as *radiosensitizing.* Certain chemical protectors act by producing anoxia. On the contrary, respiratory inhibitors, by promoting the accumulation of oxygen in the cells, may sensitize them to radiation damage. Kihlman (24) irradiated *Vivia faba* roots after treatment of the plants with a variety of respiratory inhibitors; he found a close correlation between the frequencies of chromosome aberrations and the degree of respiratory inhibition. Similarly, Sobels (25) could increase the frequency of X-ray induced translocations and recessive lethals in *Drosophila* spermatozoa and spermatids by pretreating the animals with sodium azide or potassium cyanide, both potent respiratory inhibitors. The finding that, in *Drosophila,* X-rays produce more recessive lethals at low than at high temperature (26,27) can also be attributed to the higher oxygen tension in cells kept cold.

Mechanism

There is no simple answer to the question how oxygen modifies radiation damage. It is, however, clear that it acts at many levels of the mutation process, from the primary delivery of radiation energy to repair mechanisms of various kinds. At the biophysical level, the influence of oxygen on *radical formation* is important. The oxidizing radical $HO_2^{\cdot}$ is formed only in the presence of oxygen; in systems that show an oxygen effect, this radical may have special importance as an energy carrier. On the contrary, reducing radicals, especially the $H^{\cdot}$ atom, may be more important in systems in which oxygen acts as radioprotector.

In addition to water radicals, ionizing radiation produces *organic ions, radicals and peroxides in the target molecules.* According to Alper and Howard-Flanders (28a), it is these that are responsible for the oxygen effect because reactions of these unstable groups with oxygen prevent their return to the normal state.

It has also been suggested that the oxygen effect is due to the formation of *hydrogen peroxide* from water radicals when oxygen is present. The fact that chromosome breakage by high LET radiation is not subject to an oxygen effect or only to a diminished one (p. 110) appeared to speak in favour of this assumption; for these radiations produce $OH^{\cdot}$ radicals so close together that they can form H_2O_2 even in the absence of oxygen by combining with each other. Alternatively, the generation of oxygen in the track of densely ionizing particles may reduce the effect of externally supplied oxygen. Recent results favour the second explanation (28b). Kimball (29) tested directly whether the addition of H_2O_2 increased the lethal and mutagenic action of X-rays on *Paramecium*. The results were negative, showing that at least in this system H_2O_2 does not act as mediator of radiation damage.

In certain viruses and bacteria, an oxygen effect on killing is, indeed, mediated by H_2O_2, but in a curiously indirect manner (30-33). The irradiated organisms become sensitized to the very small amount of H_2O_2 that is formed in irradiated oxygenated water. This leads to delayed killing after the cessation of radiation and, since survival values taken in the usual way do not distinguish between immediate and delayed killing, to increased killing by irradiation in oxygen. This effect disappeared when the organisms were transferred to pure water immediately after irradiation. Unirradiated controls were not killed by the small amounts of H_2O_2 in irradiated water. Obviously, no such situation exists in *Paramecium*; if it did, it would have shown up as an enhancement of killing and mutation by hydrogen peroxide. Whether sensitization to H_2O_2 plays a role in any other mutational system has not been investigated.

As a final complication, and one that certainly does play a role in the oxygen effect, the influence of oxygen on *repair systems* has to be considered. We shall do this in the next chapter.

In the present chapter we have seen that some of those observations that did not fit into the strict target theory could be accounted for by indirect radiation effects via water radicals or organic radicals in the genetic material itself. Others could not be so interpreted, and their number went on increasing. Especially intriguing were observations which showed that the frequencies of X-ray induced mutations and chromosome rearrangements can be influenced by conditions *after* irradiation, when action on the target has long been consummated. It was these observations that led mutation research beyond the target itself and into the cell, where target lesions become transformed into observable and transmitted genetic changes. This will form the topic of the next two chapters. Before going on to it, however, we shall consider two other possible ways in which ionizing radiation might produce genetic damage: one, from inside the gene by disintegration of incorporated radioisotopes; the other, from outside the gene, by the production of mutagenic substances in irradiated food.

Radioisotopes

Ionizing radiation from radioactive decay outside or inside cells may cause genetic damage in the same way as does ionizing radiation from other sources. If, however, an isotope that is subject to β-decay is incorporated into DNA, it may affect the chromosome in an essentially different way (34). β-decay *in situ* results in a change of atomic number, transforming e.g. ^{32}P into ^{32}S, or ^{14}C into ^{14}N. This is accompanied by atomic excitation and ionization and by nuclear recoil, which may be energetic enough to rupture chemical bonds. The whole complex of localized effects is called transmutation. It seems plausible to expect that transmutation of atoms in DNA will lead to genetic damage, but it has been difficult to obtain valid evidence for it. There is no doubt that incorporation of radioisotopes into cells can produce both lethality and mutation (35); the problem is to distinguish between true transmutation and the effects of β radiation from other parts of the cell or organism. Controls are usually run to compare outside radiation with radioactive decay inside the labelled cells. If the latter is considerably more effective than the former, this is presumptive evidence for transmutation; but the possibility remains that radiation produced inside the cell has a better chance than outside radiation of reaching the DNA. This possibility is greater for isotopes that produce β rays of low energy and short path length than for those that produce β rays of high energy and long path length. The problem of distinguishing between true transmutation and effects of distant radiation is therefore more severe for ^{3}H with a mean β path length of 0.5μ than for ^{32}P with a mean β path length of 2600μ. Since ^{32}P has also a convenient half-life (14.3 days) and is readily incorporated into DNA, it is not surprising that this isotope has been used in most attempts to prove genetic damage through transmutation.

For the killing of bacteriophage the evidence is convincing. Hershey and his collaborators (36) stored heavily labelled phage T2 for several weeks at 4°C and measured survival at daily intervals. They found that survival decreased exponentially with increasing decay of the isotope. Since this relationship was preserved even when the phage was stored at high dilutions, so that β radiation from outside was minimized, the conclusion was drawn that lethality was due to disintegration of ^{32}P atoms inside DNA. Comparison between radiochemical and biological data showed that approximately 1 in 10 disintegrations was lethal, and this was confirmed for other phages with double-stranded DNA. In the single-stranded phage ϕX174, on the contrary, practically every disintegration caused death (37). This was taken to mean that, in two-stranded DNA, only double-strand breaks produce death, and that the ratio of single-strand to double strand breakage is 10:1. Subsequently, Japanese workers (38,39) have shown for phage λ that this ratio depends on storage temperature and on a very rapidly acting repair mechanism that, *in vivo,* repairs the majority of breaks. In their system, too, all double strand breaks were lethal; but so were a proportion of the non-repaired single-strand breaks.

The fact that all double-strand breaks are lethal leads to the interesting conclusion that they cannot serve as sources of gene mutations, although they may well lead to chromosome breakage in eukaryotes. Intragenic mutations must trace back to some other type of primary radiation damage, such as single-strand breaks. We shall see in Chapter 9 that the same conclusion has been drawn for the effects of X-rays.

The production of mutations through radioactive decay of ^{32}P has been reported for a variety of organisms, prokaryotes as well as eukaryotes (40-42); but mutagenesis by non-localized effects of β radiation could not be rigorously excluded. The difficulty is greatest for multicellular organisms, where β radiation may reach the germ cells from radioactive decay in the surrounding tissues. This may conceivably have been responsible for the increased dominant lethal frequency in spermatogonia of mice that had been injected with strontium-90 (43). For *Drosophila,* Lee (44,45) overcame the difficulty by studying the decay of ^{32}P in spermatozoa that had been labelled in the male, but were stored in the seminal receptacles of untreated females. Because of the long mean pathway of β rays from ^{32}P, there is very little chance that the DNA in one spermatozoon will be exposed to a mutagenically effective dose of radiation produced in the other spermatozoa. Unexpectedly, 25 days of storage at 18°C did not increase the frequency of sex-linked lethals over that present before storage. When, however, scoring was extended to the F_3 (Fig. 5.4), a significant increase in mutation frequency over the controls was observed. Thus, all mutations had first occurred as gonadic mosaics and therefore had escaped detection in F_2. This indicates that ^{32}P decay produces mutations in *Drosophila* sperm only when the effect is restricted to one strand of DNA; double-strand scission presumably results in death and elimination of the affected spermatozoon.

For the reasons given above, the possibility of mutation through transmutation of isotopes other than ^{32}P has attracted less attention. Tritium decay in DNA certainly may lead to death of the cell, but it is difficult to decide how much of its killing action is due to transmutation. More decisive proof was obtained for its mutagenic action. This came from work in which different ^{3}H-labelled compounds were compared. Both in bacteria and *Drosophila,* more mutations were produced by uracil-5-^{3}H than by either uracil-6-^{3}H or thymidine-^{3}H (46-48). In bacteria, the analysis could be carried to the molecular level by classification of nonsense-revertants into amber-or-ochre-suppressors or true revertants. It was then found that uracil-5-^{3}H produced specifically transitions from cytosine to guanine, while the other two compounds gave the same spectrum of mutations as did ionizing radiation. This has been attributed to the incorporation into DNA of uracil-5-^{3}H as cytosine-5-^{3}H and the subsequent transmutation of the labelled cytosine *in situ.* Possibly, the position of the 5-C atom between two carbon atoms makes transmutation more capable of breaking the pyrimidine ring than does the position of the 6-C atom between a carbon and a nitrogen.

The production of mutations in *Neurospora* by ^{35}sulphur (63) cannot be attributed to transmutation within DNA. We shall return to it in Chapter 18 in connection with the mutagenic effects of amino acid analogues.

Is X-ray sterilized food mutagenic?

Ionizing radiation is used for breaking the dormancy of seeds, for inhibiting the sprouting of stored potatoes, for killing parasites and for the sterilization of food for human consumption. The latter has the great advantage over heat sterilization that it avoids heat-induced changes and can be applied to packed and sealed containers. Very high doses are used for it, of the order of megarads (1 Mrad = 10^6 rads). This raises the question whether radiation-sterilized food is mutagenic (49). The possibility that this might be so was suggested through the observation, to be discussed in Chapter 12, that ultraviolet radiation of organic media makes these media mutagenic for micro-organisms. Recent experiments (50) have indeed shown that bacteria are killed and induced to mutate when exposed to heavily irradiated medium, the active principle apparently being a radiolysis product of sugar, probably a peroxide. That this might be true also for mammals is suggested by experiments (51,52) in which chromosome breakage was shown to be increased in cells that themselves had not been irradiated but had been exposed to blood or blood plasma from irradiated individuals or had been cultured in irradiated medium or in medium containing irradiated sucrose.

It should be clearly understood that this kind of indirect effect of radiation is essentially different from the indirect effect due to water radicals, discussed above. The life time of water radicals is measured in fractions of milliseconds. It is inconveivable that radiation-produced radicals should still be present in food that reaches the consumer. If sterilization by radiation should, in fact, make the food mutagenic, it would have to do this by producing a stable mutagenic chemical. Early experiments, discussed in Chapter 5, had shown that doses of a few thousand R do not produce mutagenic substances in *Drosophila* ooplasm, for no mutations were induced in untreated spermatozoa that had been introduced into irradiated ova by fertilization. Subsequently (53), A.R. Whiting confirmed and extended these observations by experiments on the parasitic wasp *Habrobracon,* in which the peculiar type of sex determination permits tests of higher doses than in *Drosophila.* As in other Hymenoptera, haploid zygotes develop into patroclinoùs males, diploid ones into biparental females. When Whiting irradiated females before fertilization with high doses, a third type of progeny developed from zygotes in which all maternal chromosomes had been destroyed: patroclinous males, which carried only paternal chromosomes in irradiated cytoplasm (Fig. 7.5). The frequency of these males increased with dose from 2000 to 29 000 R; after this, it declined again, presumably because these high doses caused cytoplasmic injury to the developing embryo. A treatment of females that produced 132 visible mutations in 2414 directly irradiated maternal chromosomes (transmitted through biparental daughters and scored in their progeny), yielded none in 199 paternal chromosomes that had spent a generation in irradiated cytoplasm. The difference is highly significant and shows that, in *Habrobracon,* even cytoplasm exposed to doses that injure

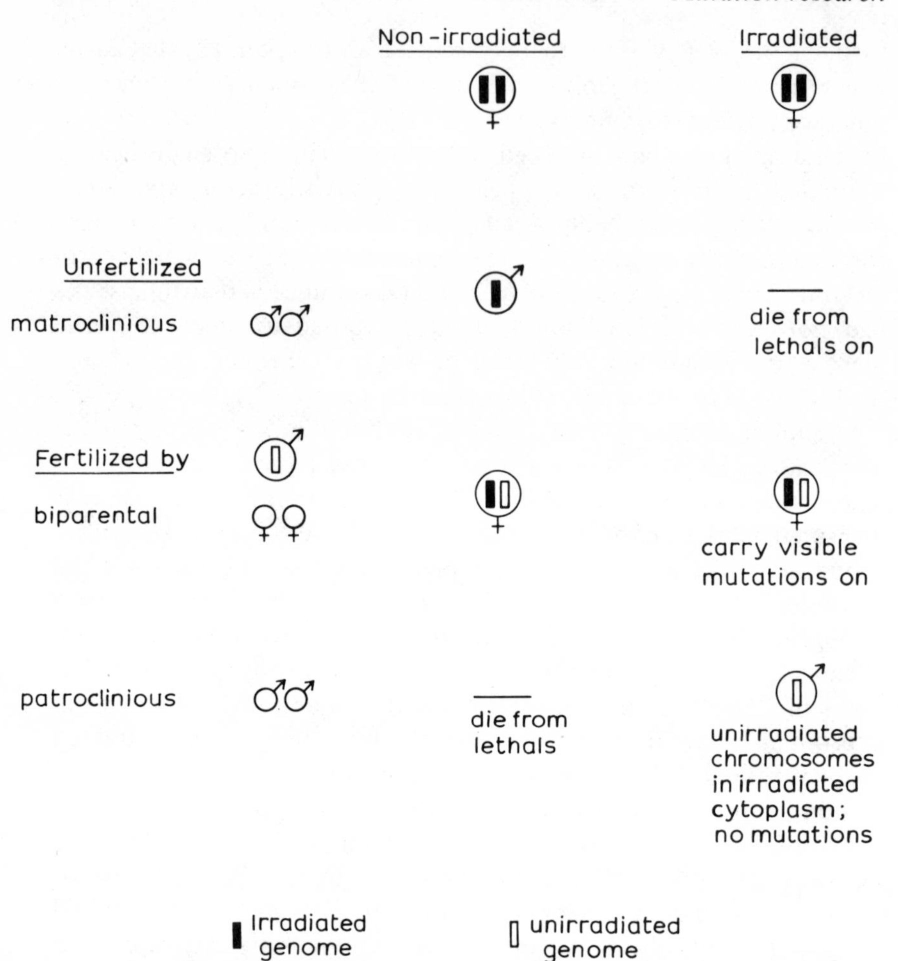

Fig. 7.5. Types of progeny produced by normal and by heavily irradiated *Habrobracon* females.

development fails to produce mutations in untreated chromosomes.

Because of the relatively low doses used, these experiments were not sufficient to prove the genetic harmlessness of radiation-sterilized food, and the problem was tackled again in different laboratories and on different organisms. The first to report genetical effects of irradiated medium were Indian workers (54,55) who, in the late fifties and early sixties, showed that chromosome aberrations are produced in root tip cells of plants grown in irradiated sugar solution or fruit juices, and in barley embryos cultured on irradiated potato mash. The active principle in these experiments, as in those on bacteria, seems to be a radiolysis product of sugar, for irradiated glucose produced chromosome breaks in plant (56) cells and in cultured human lymphocytes (57).

While these results leave little doubt that irradiated sugar – and perhaps other food constituents – can produce chromosome breaks in cells that are exposed to them, it would be rash to conclude that this may happen also after ingestion of irradiated food by animals. Indeed, it was found (58) that irradiated sucrose solutions, while inhibiting a variety of cellular processes in cultured rat cells, had no deleterious effects when given to live rats for 8 weeks, presumably because of rapid metabolism and excretion of the radiolysis products. Tests for genetic effects of ingested irradiated food were carried out on *Drosophila* in different laboratories and with different results. Rinehart (59) found a slight, but statistically significant, increase in mutation frequency, while other experiments gave negative results (60). It seems that the effect is at best marginal. Rats, however, fed through their life a food mixture that was made up to one half of heavily irradiated flour (5 Mrad), showed a significant increase in the frequency of chromosome aberrations in male meiosis (61). Obviously, radiation-sterilized food when taken in quantities and over long periods, can induce chromosomal aberrations. Whether it does so when taken in smaller quantities, and whether it can induce gene mutations, remains doubtful.

References

1. Serebrovskaya, R.I. and Shapiro, N.I. (1935), 'The frequency of mutations induced by X-rays in the autosomes of mature and immature germ-cells of *Drosophila melanogaster* males', *Compt. Rend. Acad. Sci. U.R.S.S.*, **2**, 421-428.
2. Stadler, L.J. (1928), 'Mutations in barley induced by X-rays and radium', *Science*, **68**, 186-187.
3. Gustafsson, Ake (1941), 'Mutation experiments in barley', *Hereditas*, **27**, 225-242.
4. D'Amato, F. and Gustafsson, A. (1948), 'Studies on the experimental control of the mutation process', *Hereditas*, **34**, 181-192.
5. Stapleton, G.E. (1955), 'Factors modifying sensitivity of bacteria to ionizing radiations', *Bacteriological Reviews*, **19**, 26-32.
6. Dale, W.M. (1966), 'Direct and indirect effects of ionizing

radiations', in *Encyclopedia of Medical Radiology,* Vol. II, part 1, pp. 1-38, Berlin-Heidelberg-New York: Springer.

7. Hutchinson, F. (1961), 'Molecular basis for action of ionizing radiations', *Science,* **134**, 533-538.
8. Lea, D.E. and Salaman, M.H. (1946), 'Experiments on the inactivation of bacteriophage by radiations, and their bearing on the nature of bacteriophage', *Proc. Roy. Soc. Lond. B,* **133**, 434-444.
9. Hollaender, A. (Ed.) (1960), 'Radiation Protection and Recovery', Pergamon Press: New York.
10. Bridges, B.A. and Munson, R.J. (1967), 'Radio-sensitization by N-ethylmaleimide: observations with sensitive and resistant strains of *Escherichia coli', Int. J. Rad. Biol.,* **13**, 179-181.
11. Bridges, B.A. (1969), 'Sensitization of organisms to radiation by sulfhydryl-binding agents', in *Advances in Radiation Biology,* **3**, 123-176. Academic Press; New York and London.
12. Thoday, J.M. and Read, J. (1947), 'Effect of oxygen on the frequency of chromosome aberrations produced by X-rays', *Nature,* **160**, 608.
13. Thoday, J.M. and Read, J. (1949), 'Effect of oxygen on the frequency of chromosome aberrations produced by α-rays', *Nature,* **163**, 133-134.
14. Giles, N.H. and Riley, H.P. (1949), 'The effect of oxygen on the frequency of X-ray induced chromosomal rearrangements in *Tradescantia* microspores', *Proc. Nat. Acad. Sci. U.S.A.,* **35**, 640-646.
15. Giles, N.H. and Riley, H.P. (1950), 'Studies on the mechanism of the oxygen effect on the radiosensitivity of *Tradescantia* chromosomes', *Proc. Nat. Acad. Sci. U.S.A.,* **36**, 337-344.
16. Hollaender, A., Baker, W.K. and Anderson, E.H. (1951), 'Effect of oxygen tension and certain chemicals on the X-ray sensitivity of mutation production and survival', *Cold Spring Harbor Symp. Quant. Biol.,* **16**, 315-326.
17. Deering, R.A. (1963), 'Mutation and killing of *E.coli* WP-2 by accelerated heavy ions and other radiations', *Radiation Res.,* **19**, 169-178.
18. Bridges, B.A. (1963), 'Effect of chemical modifiers on inactivation and mutation-induction by γ radiation in *Escherichia coli', J. Gen. Microbiol.,* **31**, 405-412.
19. Ephrussi-Taylor, H. and Latarjet, R. (1955), 'Inactivation, par

les rayons X, d'un facteur transformant du pneumocoque', *Biochim. Biophys. Acta,* **16**, 183-197.

20. van der Ent, G.M., Blok, J. and Linckens, E.M. (1965), 'The induction of mutations by ionizing radiation in bacteriophage ϕX174 and in its purified DNA', *Mutation Res.,* **2**, 197-204.
21. Bleichrodt, J.F. and Verheij, W.S.D. (1973), 'Reversion of cistron A *amber* mutants of bacteriophage ϕX174 by ionizing radiation', *Mutation Res.,* **18**, 363-365.
22. Russell, W.L., Kile, J.C. Jr. and Russell, L.B. (1951), 'Failure of hypoxia to protect against the radiation induction of dominant lethals in mice', *Genetics,* **36**, 574.
23. Anderson, E.H. (1951), 'The effect of oxygen on mutation induction by X-rays', *Proc. Nat. Acad. Sci. U. S. A.,* **37**, 340-349.
24. Kihlman, B.A. (1959), 'The effect of respiratory inhibitors and chelating agents on the frequencies of chromosomal aberrations produced by X-rays in *Vicia', J. Biophysic. Biochem. Cytol.,* **5**, 479-490.
25. Sobels, F.H. (1955), 'The effect of pretreatment with cyanides and azide on the rate of X-ray induced mutations in *Drosophila', Z. Vererbungslehre Bd.,* **86**, 399-404.
26. Mickey, G.H. (1939), 'The influence of low temperature on the frequency of translocations produced by X-rays in *Drosophila melanogaster', Genetica.* **21**, 386-407.
27. Novitski, E. (1949), 'Action of X-rays at low temperatures on the gametes of *Drosophila', Am. Nat.,* **83**, 185-193.

28a. Howard-Flanders, P. and Alper, T. (1957), 'The sensitivity of microorganisms to irradiation under controlled gas conditions', *Radiation Res.,* 7, 518-540.

28b. Alper, T. and Bryant, P.E. (1974), 'Reduction in oxygen enhancement ratio with increase in LET: a test of two hypotheses', *Int. J. Rad. Biol.,* **261**, 203-218.

29. Kimball, R.F., Hearon, J.Z. and Gaither, N. (1955), 'Tests for a role of H_2O_2 in X-ray mutagenesis II. Attempts to induce mutations by peroxide', *Radiation Res.,* **3**, 435-443.
30. Alper, T. (1952), 'Indirect inactivation of bacteriophage S13 during and after exposure to ionizing radiation', *Nature,* **169**, 183-184.
31. Alper, T. (1952), 'A new after-effect of X-rays on dilute aqueous suspensions of bacteriophage', *Nature,* **169**, 964-965.

32. Adler, H.I. and Stapleton, G.E. (1958), 'The role of radiation-produced hydrogen peroxide in inactivation of a catalase-negative bacterium', *Radiation Res.*, **9**, 84-206.
33. Johnson, C.D. (1965), 'The influence of previously X-irradiated aqueous solutions on the infectivity of the viruses of Foot-and-mouth disease and Vesicular Stomatitis', *J. Gen. Microbiol.*, **38**, 9-19.
34. Krisch, R.E. and Zelle, M.R. (1969), 'Biological effects of radio-active decay: the role of the transmutation effect', in *Advances in Radiation Biology* **3**, 177-213: Academic Press: New York and London.
35. Person, S. (1968), 'Lethal and mutagenic effects of tritium decay produced by tritiated compounds incorporated into bacteria and bacteriophage', in *Biological effects of transmutation and decay of incorporated radio-isotopes,* p. 29-64: Int. Atomic Energy Agency: Vienna.
36. Hershey, A.D., Kamer, M.D., Kennedy, J.W. and Gest, H. (1951), 'The mortality of bacteriophage containing assimilated radio-active phosphorous', *J. Gen. Physiol.*, **34**, 305-319.
37. Tessman, Irwin (1959), 'Some unusual properties of the nucleic acid in bacteriophages S13 and ϕX174', *Virology,* **7**, 263-275.
38. Ogawa, H. and Tomizawa, J. (1967), 'Breakage of polynucleotide strands by disintegration of radiophosphorus atoms in DNA molecules and their repair I. Single-strand breakage by transmutation', *J. Mol. Biol.*, **30**, 1-6.
39. Tomizawa, J. and Ogawa, H. (1967), 'Breakage of poly-nucleotide strands by disintegration of radiophosphorus atoms in DNA molecules and their repair II. Simultaneous breakage of both strands', *J. Mol. Biol.*, **30**, 7-15.
40. Rubin, B.A. (1950), 'The pattern and significance of delayed phenotypic expression of mutations induced in *E.coli* by absorbed ^{32}P, *Genetics,* **35**, 133.
41. Strigini, P., Rossi, C. and Sermonti, G. (1963), 'Effects of disintegration of incorporated ^{32}P in *Aspergillus nidulans'*, *J. Mol. Biol.*, **7**, 683-699.
42. Thompson, K.F., MacKey, J., Gustafsson, A. and Ehrenberg, L. (1950), 'The mutagenic effect of radiophosphorus in barley', *Hereditas,* **36**, 220-224.
43. Lüning, K.G., Frölen, H., Nelson, A. and Rönnback, C. (1963), 'Genetic effects of strontium-90 on immature germ-cells in mice', *Nature,* **199**, 303-304.

44. Lee, W.R., Oden, C.K., Bart, C.A., Debney, C.W. and Martin, R.F. (1966), 'Stability of *Drosophila* chromosomes to radioactive decay of incorporated phosphorus-32', *Genetics,* **53**, 807-822.
45. Lee, W.R., Sega, G.A. and Alford, C.F. (1967), 'Mutations produced by transmutation of phosphorus-32 to sulphur-32 within *Drosophila* DNA', *Proc. Nat. Acad. Sci. U.S.A.,* **58**, 1472-1479.
46. Olivieri, G. and Olivieri, A. (1965), 'The mutagenic effect of tritiated uridine in *Drosophila* spermatocytes', *Mutation Res.,* **2**, 381-384.
47. Kieft, K. (1968), 'Induction of recessive lethals by ^{3}H-uridine and ^{3}H-thymidine in *Drosophila',* in *Biological effects of Transmutation and Decay of Incorporated Radioisotopes,* p. 65-78: International Atomic Energy Agency: Vienna.
48. Person, S. and Osborn, M. (1968), 'The conversion of amber suppressors to ochre suppressors', *Proc. Nat. Acad. Sci. (U.S.),* **60**, 1030-1037.
49. 'Wholesomeness of irradiated food with special reference to Wheat, Potatoes and Onions'; *World Health Organization Technical Report Series* No. 451 Geneva, 1970.
50. Chopra, V.L. (1969), 'Lethal and mutagenic effects of irradiated medium on *Escherichia coli', Mutation Res.,* **8**, 25-33.
51. Kerkis, Ju.Ja and Jasnova, L.N. (1967), 'Distant mutagenic effects of ionizing radiation in mammals', *Nature,* **215**, 1403-1405.
52. Goh, K.O. and Sumner, H. (1968), 'Breaks in normal human chromosomes: are they induced by a transferable substance in the plasma of persons exposed to total body irradiation?', *Radiation Res.,* **35**, 171-181.
53. Whiting, A.R. (1950), 'Absence of mutagenic action of X-rayed cytoplasm in *Habrobracon', Proc. Nat. Acad. Sci., U.S.A.,* **36**, 368-372.
54. Swaminathan, M.S., Chopra, V.L. and Bhaskaran, S. (1962), 'Cytological aberrations observed in barley embryos cultured in irradiated potato mash', *Radiation Res.,* **16**, 182-188.
55. Chopra, V.L., Natarajan, A.T. and Swaminathan, M.S. (1963), 'Cytological effects observed in plant material grown on irradiated fruit juices', *Radiation Bot.,* **3**, 1-6.
56. Moutschen, J. and Matagne, R. (1965), 'Cytological effects of irradiated glucose', *Radiation Bot.,* **5**, 23-28.

57. Shaw, M.H. and Hayes, E. (1966), 'Effects of irradiated sucrose on the chromosomes of human lymphocytes *in vitro*', *Nature,* **211**, 1254-1256.
58. De, A.K., Aiyar, A.S. and Sreenivasan, A. (1969), 'Biochemical effects of irradiated sucrose solutions in the rat', *Radiation Res.,* **37**, 202-215.
59. Rinehart, R.R. and Ratty, F.J. (1965), 'Mutation in *Drosophila melanogaster* cultured on irradiated food', *Genetics,* **52**, 1119-1126.
60. Chopra, V.L. (1965), 'Tests on *Drosophila* for the production of mutations by irradiated medium or irradiated DNA', *Nature,* **208**, 699-700.
61. Bugyaki, L., Deschreider, A.R., Moutchen, J., Moutchen-Dahmen, M., Thys, A. and Lafontaine, A. (1968), 'Les aliments irradiés exercent-ils un effet radiomimétique? II. Essais d'alimentation de la souris avec une farine de froment irradiée à 5 megarad', *Atompraxis* **14**, 1-7.
62. Oster, I.I. (1958), 'Radiosensitivity', *Genen en Phaenen,* **3**, 53-66.
63. Hungate, F.P. and Mannell, T. (1952), 'Sulphur-35 as a mutagenic agent in *Neurospora*', *Genetics,* **37**, 709-719.

CHAPTER EIGHT

Repair of X-ray damage I: Repair of chromosome breaks

We have seen in Chapter 2 that chromosome breakage may have one of three consequences. (1) The breaks rejoin in the old order – restitution. (2) The breaks rejoin in a new, wrong order – formation of rearrangements. (3) The breaks remain open. This leads to either terminal deletion or chromosome loss via a breakage-fusion-bridge cycle. Since there is no essential difference between the types of rejoining that result in restitution or in the formation of rearrangements, we shall consider both as forms of repair. Sobels has introduced the useful term 'misrepair' for the production of rearrangements.

Fractionation experiments. Rejoining time

Chromosome rejoining was at first considered a natural consequence of the 'stickiness' of broken chromosome ends. Sheldon Wolff and his collaborators replaced this somewhat unbiological concept by that of rejoining as a process that requires energy and protein synthesis. They measured the time during which X-ray-produced breaks remain open by challenging the cells with two doses in succession. If the interval between irradiations is short, the breaks produced by the first fraction will still be available when those of the second fraction make their appearance. Rearrangements will be formed from this joint pool of breaks, and their frequency will be the same as if both fractions had been given as a single dose. If, on the other hand, the interval between the two fractions is sufficiently long for all breaks produced by the first to have restituted or formed rearrangements before the second is given, then each fraction will produce rearrangements independent from the other, and the total rearrangement frequency will be the sum of the two individual ones. The first situation – full interaction between all breaks – results in a much higher frequency of rearrangements than does the second – additivity. If the square law of dose-response for rearrangement formation (Chapter 6) holds strictly, interaction between two equal fractions will give twice as many rearrangements as additivity. Suppose for example, that a given dose has produced 4% translocations. These are two-hit rearrangements and a dose yielding 4% of them must have produced 20% breaks within 'sites' (see p. 97) ($0.2^2 = 0.04$). A dose of twice the strength will yield 40% breaks within sites and 16% translocations ($0.4^2 = 0.16$). Additivity, on the other hand, will yield only 4% + 4% = 8% translocations. In general, if a dose yielding $2p$ breaks within sites is split into two equal fractions, additivity of effect will yield $p^2 + p^2 = 2p^2$ two-hit rearrangements, while full interaction between breaks will yield $(2p)^2 = 4p^2$. Saturation of sites may reduce the interaction yield somewhat, but it will always be considerably above the additive value. This is also true when the two dose fractions are of unequal magnitude. If one starts with an interval between fractions that is short enough to allow full interaction, gradual lengthening of the interval will reduce the rearrangement yield until it reaches the low level of additivity. Using this rationale, Sax (1), in 1939, had estimated that breaks in *Tradescantia* microspores remain open for about one hour. In *Vicia faba* chromosomes, Wolff found rejoining times that varied between 30 min and 2 h depending on the conditions of the experiment (see below). For plants as well as *Drosophila,* there is evidence that X-rays produce also a second type of chromosome break, which rejoins much more rapidly (2-5).

In order to obtain insight into the nature of the rejoining process, Wolff tested a number of conditions that modified the time taken for the rejoining of chromosome breaks in *Vicia faba* seeds and *Tradescantia* microspores (6). Table 8.1 shows some of his data for *Vicia.*

The table shows that rejoining of breaks is slowed down by all conditions that inhibit oxidative metabolism: low temperature, potassium cyanide, CO in the dark (light breaks the complex between CO and cytochromoxidase). The authors speculated that oxidative metabolism might influence rejoining via the formation of energy-rich ATP molecules. Dose-rate experiments were in agreement with this idea. Lowering the dose-rate decreases aberration frequency

Table 8.1

Post-treatment conditions that determine whether chromosome breaks produced in *Vicia faba* seeds by a dose of 600R will be still open when a second dose of 400R is given 75 min later. (After Table 1, Wolff (7))

Post-treatment conditions	*No. of rings and dicentrics*			*Rejoining has taken place*
	Observed	*Expected for*		
		Additivity	*Interaction*	
Water, room temperature	0.220	0.193	0.380	yes
Water, 0^{0}C	0.367	0.193	0.380	no
2×10^{-3} M KCN*	0.367	0.236	0.440	no
95% CO + 5% O_2, dark	0.347	0.193	0.380	no
95% CO + 5% O_2, light	0.210	0.193	0.380	yes

* KCN given as pretreatment before the second fraction slightly increased aberration frequency; hence the higher expected values.

through providing a chance for the early produced breaks to rejoin before the late-produced ones become available (p.92). Wolff found that the decrease in aberration frequency was more rapid when the seeds had been pre-soaked in ATP (7). More recently, it has been shown (8a,b) that the rejoining of chemically produced chromosome breaks also requires ATP. Evidently, the inability of broken chromosomes in *Drosophila* spermatozoa to rejoin before entry into the ovum (p.93) is due to insufficient oxidative metabolism in mature sperm. Wolff tested whether the role of ATP in the repair of chromosome breaks is to allow protein synthesis. If this is correct, then protein synthesis inhibitors should delay rejoining. Table 8.2 shows that this is indeed the case. Chloramphenicol which inhibits protein synthesis, delayed rejoining time beyond 75 min. Note that the second dose – 300R in air – was about as effective by itself as the first dose – 600R *in vacuo*; this is a good example for the oxygen effect on chromosome breakage.

Essentially similar results were obtained by Wolff more recently on human leukocyte cultures (9). When these were exposed to 1000R, most breaks had rejoined after 4 h. When, however, the medium contained cycloheximide – a potent inhibitor of protein synthesis in mammalian cells – all breaks were still open at this time. The results of dose rate and fractionation experiments on mouse spermatogonia were contradictory (see Auerbach and Kilbey;

Table 8.2

Effect of protein synthesis on rejoining of chromosome breaks in *Vicia faba* seeds. (After Table 1, Wolff (19))

Dose 1 (R, in vacuo)	*Interval*	*Dose 2 (R, in air)*	*Dicentrics and rings per 100 cells*	*Expected for additivity*	*Expected for interaction*
600	75 min H_2O	–	10.0 ± 1.8	–	–
600	75 min in CA	–	9.3 ± 1.8	–	–
–	75 min in H_2O	300	9.7 ± 1.9	–	–
–	75 min in CA	300	10.3 ± 1.9	–	–
600	75 min in H_2O	300	25.9 ± 2.4	19.7	39.3
600	75 min in CA	300	37.3 ± 3.6	19.7	39.3

* CA = chloramphenicol, 300 μg ml^{-1}

Bibliography); this is easily understood from the complexity of radiation effects in populations of heterogeneous, asynchronous, actively metabolizing cells. We shall return to this difficulty in the next chapter.

An interesting feature of Wolff's experiments is the dependence of repair efficiency on the dose of irradiation (10). The time taken for repair was longer after a high dose of X-rays than after a low one. It seems that radiation itself may damage the repair system. In later chapters, we shall meet with other observations that show effects of mutagenic treatment on repair systems.

The role of oxygen in repair

When the oxygen effect on rearrangement formation was first discovered, the question whether oxygen acts on breakage or rejoining was much debated. We now know that oxygen affects both processes. The situation is complicated by the fact that the effect on rejoining depends on whether oxygen is applied during or after radiation. When present *during* irradiation, oxygen not only promotes the production of breaks, it also inhibits the repair system, so that these breaks take a longer time to rejoin than breaks produced in the absence of oxygen. When present *after* irradiation, oxygen acts in the opposite way: by allowing oxidative metabolism, it speeds up the rejoining process. In plant cells, the final yield of rearrangements does not seem to depend on the length of the rejoining time, presumably because even under anoxic conditions rejoining can go to its limits before the next mitosis. In *Drosophila* zygotes, breaks in paternally irradiated chromosomes rejoin only in the zygote, and they become incapable of doing so during the first 16 min after fertilization. Under these circumstances, the final yield of rearrangements does depend on speed of rejoining (11).

The role of protein synthesis in repair

Why do repair or misrepair of broken chromosomes require protein synthesis? Wolff favoured the interpretation that chromosomes consist of stretches of DNA, joined to each other by protein linkers. There is, however, hardly any evidence to support this model; on the contrary, there is good evidence for the model of a chromosome that is traversed by a continuous fibre (or several continuous fibres) of DNA (12,12a). Wolff doubts this because, (*a*) in his experiments, rejoining took place at a stage in the cell cycle when DNA is not being synthesized, (*b*) inhibitors of DNA-synthesis (hydroxyurea, fluorodeoxyuridine) did not prolong the time between breakage and rejoining, and (*c*) rearrangements were formed equally well in cells with and without the ability for 'dark repair' (13) (see Chapter 14). However, the amount of DNA synthesis that is required for repair or misrepair may be too little to be recognizable autoradiographically or may not be prevented by the inhibitors used, and the enzyme that is missing in cells without dark repair may not be the one that joins broken chromosomes together.

The alternative interpretation, also suggested by Wolff, appears more attractive. It postulates a rejoining enzyme that is not available in sufficient amounts and has to be synthesized *ad hoc*. This view tallies well with the observation that in the spermatozoa of *Drosophila* there is neither repair nor misrepair of broken chromosomes before fertilization; it seems reasonable to assume that spermatozoa lack the relevant enzyme, and that this is provided in the ovum. Recently, this interpretation has received support from experiments (14) in which irradiated *Drosophila* males were mated to females that had or had not been treated with actinomycin-D, an inhibitor of DNA-directed RNA-synthesis and, through this, of protein synthesis. Dominant lethals and translocations were scored, the former as a measure of unrestituted breaks, the latter as a measure of rejoined (misrepaired) ones (see Chapter 2). If inhibition of protein synthesis prevents rejoining of breaks that would otherwise have formed translocations, then the frequency of translocations will decrease, while that of dominant lethals will increase concomitantly. This was indeed found to be true for zygotes from eggs that had been near maturity at the time of actinomycin treatment. Zygotes from eggs treated when fully mature, showed no treatment effect, either because actinomycin did not penetrate into them or because they had already synthesized a sufficient amount of rejoining enzyme. It should be noted that in these experiments, because of the time limit on rejoining (see above), the final yield of translocations was decreased by a delay in rejoining time.

Experiments in which plants or *Drosophila* were exposed to infrared (near-red) light before X-radiation (15-17) seemed to indicate that light of this wavelength favours the formation of rearrangements from broken chromosomes. A later analysis by Wolff (18) showed that, at least for plant cells, the effect was spurious and due to mitotic delay which led to a shift in the proportion of cell stages with different sensitivities in the X-rayed population. More important, and still problematic, are the results obtained by a combination of UV-radiation and X-rays. These will be mentioned in Chapter 10.

References

1. Sax, K. (1939), 'The time factor in X-ray production of chromosome aberrations', *Proc. Nat. Acad. Sci. U. S. A.,* **25**, 225-233.
2. Haas, F.L., Dudgeon, E., Clayton, F.E. and Stone, W.S. (1964), 'Measurement and control of some direct and indirect effects of X-radiation', *Genetics,* **39**, 453-471.
3. Wolff, Sh. and Luippold, H.E. (1956), 'The production of two chemically different types of chromosomal breaks by ionizing radiation', *Proc. Nat. Acad. Sci. U.S.A.,* **42**, 510-514.
4. Haendle, J. (1971), 'Röntgeninduzierte mitotische Rekombination bei *Drosophila melanogaster, II.* Beweis der Existenz und Charakterisierung zweier von der Art des Spektrums abhängiger Reaktionen', *Molec. Gen. Genetics* **113**, 132-149.
5. Gilot-Delhalle, J., Thakare, R. and Moutschen, J. (1973), 'Fast rejoining processes in *Nigella damascena* chromosomes revealed by fractionated ^{60}Cobalt γ-ray exposures', *Radiation Bot.* **13**, 229-242.

6. Wolff, S. and Luippold, H.E. (1958), 'Modification of chromosomal aberration yield by postirradiation treatment', *Genetics* **43**, 493-501.

7. Wolff, S. and Luippold, H.E. (1955), 'Metabolism and Chromosome-Break rejoining', *Science,* **122**, 231-232.

8a. Kihlman, B.A., Odmark, G., Norlen, K. and Karlsson, M.B. (1971), 'Caffeine, caffeine derivatives and chromosomal aberrations, I. The relationship between ATP concentration and the frequency of 8-ethoxycaffeine-induced chromosomal changes in *Vicia faba', Hereditas* **68**, 291-304.

8b. Bempong, M.A. and Newsome, Y.L. (1972), 'The effect of adenosine triphosphate on mitomycin C-induced aberration yield in *Vicia faba', Canad. J. Genet. Cytol.,* **14**, 655-660.

9. Wolff, S. (1972), 'The repair of X-ray induced chromosome aberrations in stimulated and unstimulated human lymphocytes', *Mutation Res.* **15**, 435-444.

10. Wolff, S. and Atwood, K.C. (1954), 'Independent X-ray effects on breakage and reunion', *Proc. Nat. Acad. Sci. U. S. A.,* **40**, 187-192.

11. Würgler, F.E. (1971), 'Radiation-induced translocations in inseminated eggs of *Drosophila melanogaster', Mutation Res.* **13**, 353-359.

12. MacGregor, H.C. and Callan, H.G. (1962), 'The action of enzymes on lampbrush chromosomes', *Quart. J. Microscop. Sci.* **103**, 173-203.

12a. Kavenoff, R. and Zimim, B.H. (1973), 'Chromosome-sized DNA molecules from *Drosophila', Chromosoma* **41**, 1-27.

13. Wolff, S. and Scott, D. (1969), 'Repair of radiation-induced damage to chromosomes. Independence of known DNA dark repair mechanisms', *Exptl. Cell. Res.* **55**, 9-16.

14. Proust, J.P., Sankaranarayanan, K. and Sobels, F.H. (1972), 'The effects of treating *Drosophila* females with actinomycin-D on the yields of dominant lethals, translocations and recessive lethals recovered from X-irradiated spermatozoa', *Mutation Res.,* **16**, 65-76.

15. Kaufmann, B.P., Hollaender, A. and Gay, H. (1946), 'Modification of the frequency of chromosomal rearrangements induced by X-rays, in *Drosophila. I.* Use of near infrared radiation', *Genetics,* **31**, 349-367.

16. Kaufmann, B.P. and Gay, H. (1947), 'The influence of X-rays

and near infrared on recessive lethals in *Drosophila melanogaster'*, *Proc. Nat. Acad. Sci. U. S. A.*, **33**, 366-372.

17. Swanson, C.P. and Hollaender, A. (1946), 'The frequency of X-ray-induced breaks in *Tradescantia* as modified by near infrared radiation', *Proc. Nat. Acad. Sci. (U.S.A)* **32**, 295 - 302.
18. Wolff, S. and Luippold, H.E. (1965), 'Mitotic delay and the apparent synergism of far-red radiation and X-rays in the production of chromosomal aberrations', *Photochemistry and Photobiology* **4**, 439-445.
19. Wolff, S. (1960), 'Radiation studies on the nature of chromosome breakage', *Am. Nat.* **94**, 85-93.

CHAPTER NINE

Repair of X-ray damage II: Repair of mutagenic lesions. Molecular action of X-rays. Sensitivity differences

Already in 1953, Thoday and Swanson had independently suggested that X-ray-induced chromosome breaks may remain latent for some time because during condensed stages of the chromosomes the separation of broken ends and, through this, the formation of rearrangements would be mechanically inhibited (1,2). A number of observations could be explained by this assumption, e.g. the fact that breaks produced in plant chromosomes at metaphase I of meiosis form rearrangements only during the second meiotic division.

The present concept of repair from genetical damage goes much beyond this. It visualizes changes in the genetic material that may or may not give rise to a mutation or chromosome break, and cellular repair systems that are responsible for repair. It should be realized that the type of potential damage envisaged here is different from the very short-lived organic radicals which have been postulated as intermediates in radiation damage (Chapter 7); the lifetime of reparable premutational or pre-breakage lesions is very much longer. I also want to stress that the existence of premutational changes in no way invalidates the old saying – so important in discussions of genetic hazards from radiation and other mutagens – that 'there is no genetic recovery from mutation'. Once a mutation has been 'fixed', i.e. once it has passed the reparable stage, which usually takes no more than one or a few cell cycles, it cannot be reversed except by the exceedingly rare event of a reverse mutation.

Differential repair of potential lesions is one of the causes of sensitivity differences between stages in the cell cycle and between different kinds of germ cell. It is also an important factor in the modification of dose-effect curves away from target-expectations.

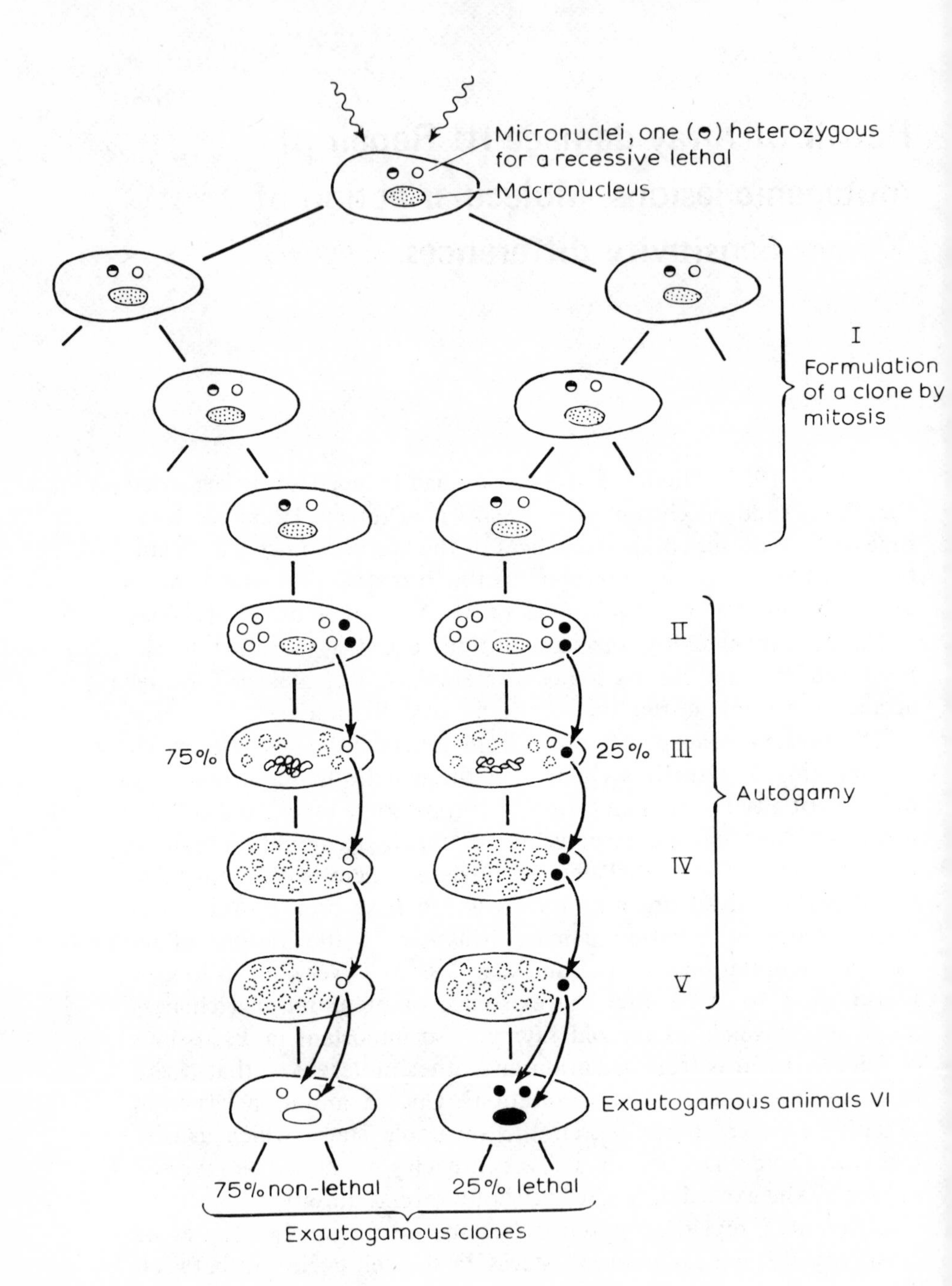

Micronuclei, one (◓) heterozygous for a recessive lethal
Macronucleus
I
Formulation of a clone by mitosis
II
75%
25%
III
Autogamy
IV
V
Exautogamous animals VI
75% non-lethal
25% lethal
Exautogamous clones

I shall give examples for both, selecting them from organisms that display clearly one or the other effect.

Paramecium. Differences in sensitivity between stages of the cell cycle

The first data that pointed to a repair of potential X-ray mutations were obtained by Kimball and his collaborators on *Paramecium.* They scored recessive lethals by letting irradiated and control animals go through autogamy and counting the frequency of inviable exautogamous clones (3). The method is shown in Fig. 9.1.

A *Paramecium* carries its transmissible genetic information in two diploid micronuclei. At meiosis, these give rise to eight haploid nuclei, only one of which forms an exautogamous clone. If one of the micronuclei is heterozygous for a newly arisen recessive lethal, two of the eight meiotic products carry it, and there is a chance of 25% that the exautogamous clone will be inviable. A special advantage of *Paramecium* is that, by starting with individuals that have just undergone autogamy, one can work with a synchronous population, which remains synchronized during the first cell cycle. One can therefore compare mutation frequencies in the different parts of the cell cycle: G1 (presynthetic stage), S (DNA synthesis), G2 (post-synthetic stage), M (meta-anaphase of mitosis). (The terminology derives from earlier autoradiographic work on plant cells in which the two interphase periods without uptake of labelled precursors into DNA were called Gap 1 and Gap 2).

Using this system, Kimball found that the frequency of lethals produced by a given dose of X-rays depends on (*a*) cell stage during irradiation and (*b*) conditions after irradiation.

Fig. 9.1. Test for recessive lethals in *Paramecium.* I. During clone formation, the micronuclei divide by mitosis. An animal that carries a recessive lethal in one of its micronuclei forms a clone in which every animal has a similar heterozygous micronucleus. II. At the beginning of autogamy, both micronuclei undergo meiosis, yielding eight haploid nuclei. In the present case, two of them carry a lethal. III. The macronucleus and seven micronuclei degenerate. Since it is a matter of chance which of the eight micronuclei survives, the lethal-bearing nuclei have a 25% chance of surviving. IV. The surviving micronucleus divides mitotically. V. The two resulting nuclei fuse again. VI. The fusion nucleus gives rise to the new macronucleus and the two new micronuclei. Exautogamous animals are homozygous for all genes and give rise to homozygous clones. In the present case, 25% of the exautogamous animals are homozygous for the lethal and cannot form viable clones. (Fig. 8, Auerbach, 1962 (bibl.))

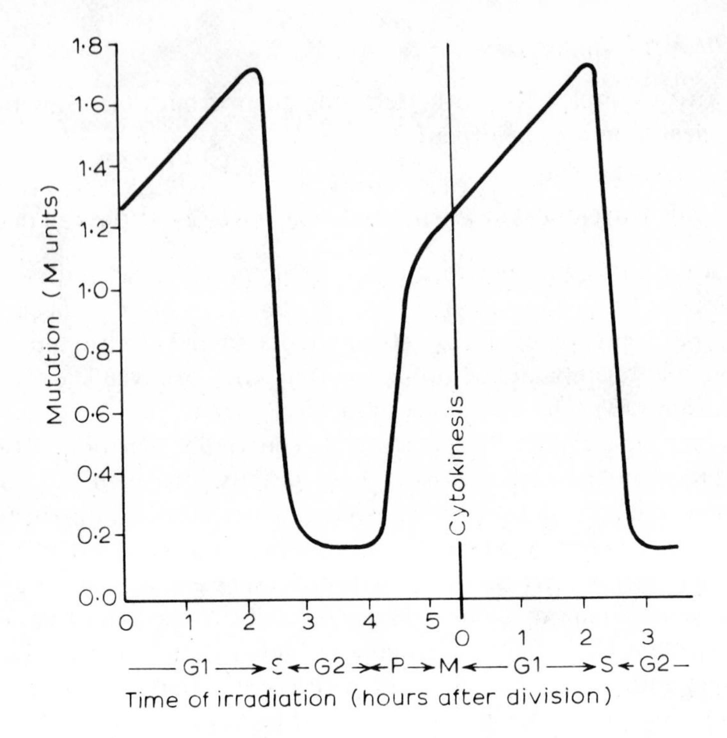

Fig. 9.2. *Caption opposite.*

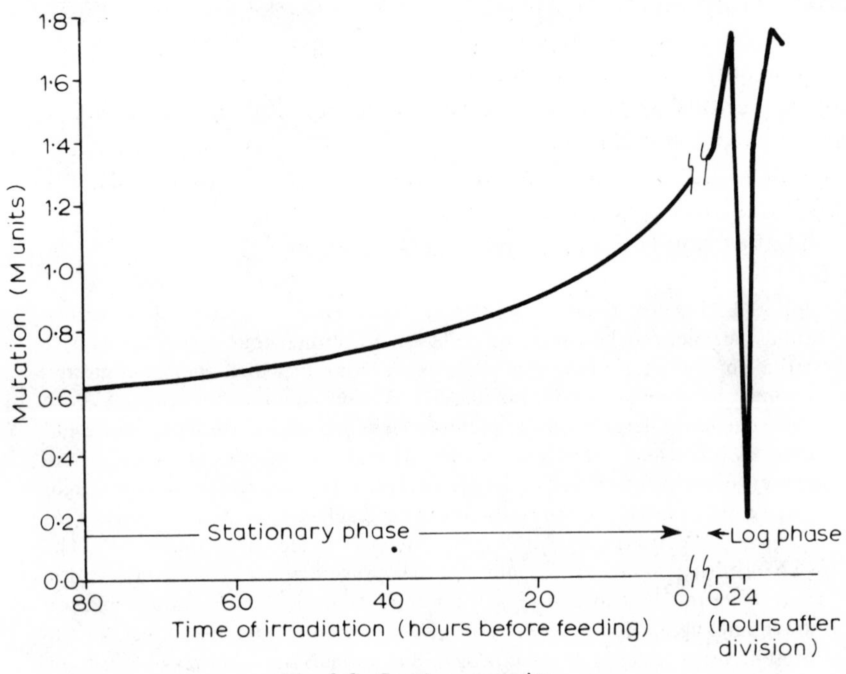

Fig. 9.3. *Caption opposite.*

(*a*) Fig. 9.2 shows the dependence of mutation frequency on stage in the cell cycle. Mutation frequency increases from early to late G1, drops drastically in S, remains very low in G2 and, during mitosis, rises again to the value characteristic for early G1. Kimball speculated that these variations could be due to repair processes that take place between the time of irradiation and the time of replication, when premutational damage is irreversibly fixed. Experiments in stationary phase (arrested in G1) supported the hypothesis. Fig. 9.3 shows that mutation frequency decreased continuously with increasing time between irradiation and feeding, i.e. beginning of growth. The finding that mutation frequency was independent of dose rate and dose fractionation made it unlikely that the lesions to be repaired were broken chromosomes; instead, some kind of premutational change in DNA was postulated.

(*b*) Experiments in which post-irradiation conditions were varied also agreed with the repair hypothesis (4). All conditions that slow down growth – low temperature, starvation, dinitrophenol, caffeine, streptomycin, chloramphenicol – decreased mutation frequency, presumably by allowing more time for repair. Complications arose for agents like streptomycin that act via inhibition of protein synthesis; for these not only prolong the *time* for repair, they also slow down the *rate* of repair. The final result is therefore the resultant of two opposite effects: decreased repair, resulting in more mutations, and delay of growth, resulting in fewer. Such situations are frequent in biological experiments. Wherever possible, it is of course desirable to isolate individual components of a complex interaction experimentally for further analysis. But where this is not yet possible, complications should not be pushed aside for the sake of simplification; this can only lead to unbiological interpretations.

Fig. 9.2. Sensitivity to the mutagenic action of X-rays during the log phase cell cycle of *Paramecium*. M units = numerical value that is proportional to the frequency of recessive lethals. (Fig. 1, Kimball (3). Courtesy Pergamon Press)

Fig. 9.3. Increase in the frequency of recessive lethals in *Paramecium* with decreased interval between irradiation in stationary phase and feeding. Diagram for log phase cell cycle is appended to the same scale for comparison. (Fig. 2, Kimball (3). Courtesy Pergamon Press)

Drosophila. Sensitivity differences between germ cells

A finding that is of the utmost importance for estimates of genetic radiation hazards to mankind is the difference in response that a given dose of X-rays elicits in different types of germ cell. Much work has been carried out to determine the underlying causes of these differences. Repair processes certainly play a role, but they are far from being alone responsible. Many other factors are involved, such as degree of contraction of the chromosomes, relative lengths of the stages in the nuclear cycle, degree of oxygenation of the cells, their susceptibility to killing and mitotic delay by radiation, amounts of radioprotective substances in the cytoplasm. It is well to keep in mind Kimball's timely warning (5): 'Repair is a valuable concept, but it should be used with restraint and not as a general explanation for any and all variations in mutation yield'.

Very thorough analyses have been carried out on *Drosophila* (6-8). Sobels and his group studied male germ cells; German and Swiss workers studied germ cells in females and embryos. As an example of the complex background of sensitivity differences, I will review some of these results.

Sobels scored sex-linked lethals and translocations in successive broods from irradiated males (see p. 76). Use of a ring-X ensured that the lethals were all point mutations (p. 93). A typical 'brood pattern' for lethals is shown in Fig. 9.4; the brood pattern for translocations has a similar shape. It will be seen that early spermatids have the highest sensitivity, yielding between two and three times as many sex-linked lethals as do mature spermatozoa. Part of this difference is due to the better oxygenation of spermatids: when flies are irradiated in nitrogen, the difference in sensitivity between spermatids and spermatozoa is largely abolished (see Figs. 7.3 and 7.4). Yet, degree of oxygenation cannot be the sole cause of the sensitivity difference. If it were, this difference should disappear also when both cell types are fully oxygenated through irradiation in pure oxygen. This, however, was not observed. Irradiation in oxygen, while increasing mutation frequency in both stages, still yielded more mutations from spermatids than from spermatozoa. In fact, this difference persisted even when spermatids irradiated in air were compared with spermatozoa irradiated in oxygen. Other factors must determine the sensitivity pattern of the *Drosophila* testis or, at least, contribute to it.

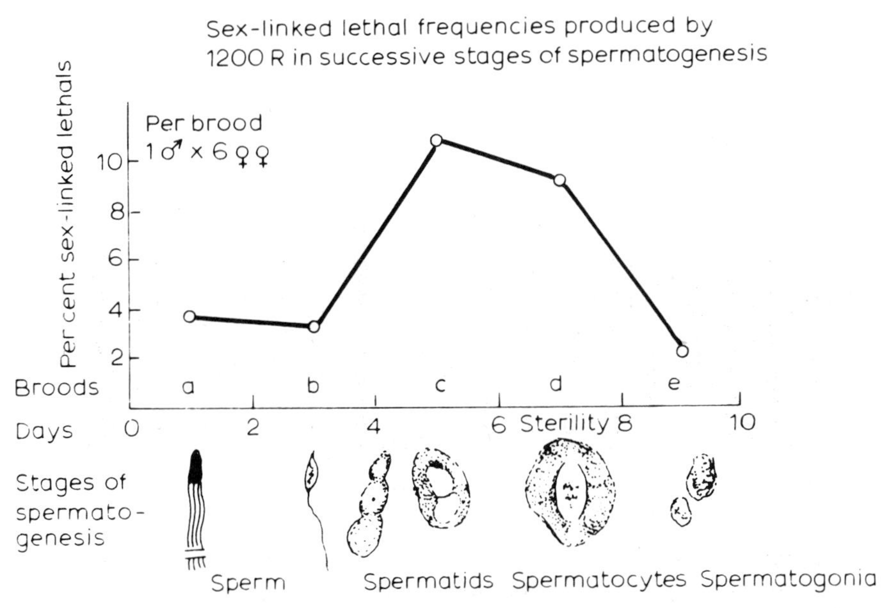

Fig. 9.4. Repair in *Drosophila* spermatozoa. Differential radiosensitivity in air for the induction of recessive sex-linked lethals in 5 successive 2-day broods, obtained by mating the irradiated males to 6 fresh females per brood. (Fig. 1, Sobels (8). Courtesy F.H. Sobels)

Sobels and his co-workers studied the role of repair by exposing irradiated flies to various post-treatments and noting the effects on the yield of sex-linked lethals and translocations. Repair was thought of as abolishing premutational or pre-breakage lesions, not as resulting in restitution of already broken chromosomes. The latter seemed unlikely because the duration of post-treatment was about 15 min, while broken chromosomes remain open for many hours in spermatids, and until after fertilization in spermatozoa. The most thoroughly studied post-treatments were oxygen *versus* nitrogen. The flies were irradiated in nitrogen so as to avoid carry-over of oxygen into the anoxic post-treatment conditions. Comparisons were carried out for all germ cell stages, but I shall review only those pertaining to spermatids and spermatozoa. Post-treatment with either nitrogen or oxygen led to exactly opposite results in the two stages (Figs. 9.5 and 9.6). In spermatozoa, the lowest yield of mutations and translocations was obtained by post-treatment in nitrogen; in spermatids, the yield was lowest after post-treatment in oxygen. Thus, recovery

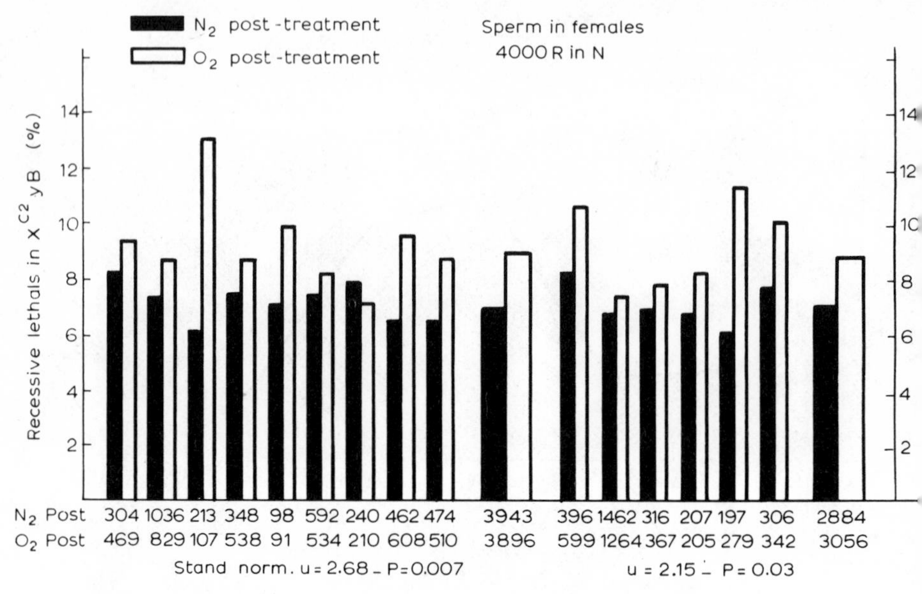

Fig. 9.5. Following X-irradiation under anoxia of spermatozoa of homogeneous radio-sensitivity in the inseminated females, post-treatment for 15 min with nitrogen results in a significant reduction of the frequencies of both recessive lethal mutations and translocations in comparison to the effect obtained with oxygen post-treatment. (Fig. 1, Sobels (58). Courtesy Academic Press)

from potential damage is favoured by nitrogen in spermatozoa, and by oxygen in spermatids. After other explanations, such as differential killing, had been excluded, Sobels drew the tentative conclusion that repair of potential damage in spermatids resembles repair of broken plant chromosomes (p. 112) in deriving its energy from oxidative metabolism, while repair in spermatids derives it from glycolysis. Experiments with metabolic inhibitors of these two pathways lent some support to the hypothesis (9).

These results go some way towards explaining the sensitivity differences between spermatozoa and spermatids, but they do not explain them fully. In conventional experiments, intrinsic sensitivity differences and differential repair are confounded. This is also so when irradiation in nitrogen is followed by post-treatment in air. It is true that this almost abolishes the sensitivity difference between spermatozoa and spermatids; but this equality is in part spurious because oxygen affects repair in opposite ways in the two cell types. If one minimizes repair by irradiation in oxygen, the sensitivity

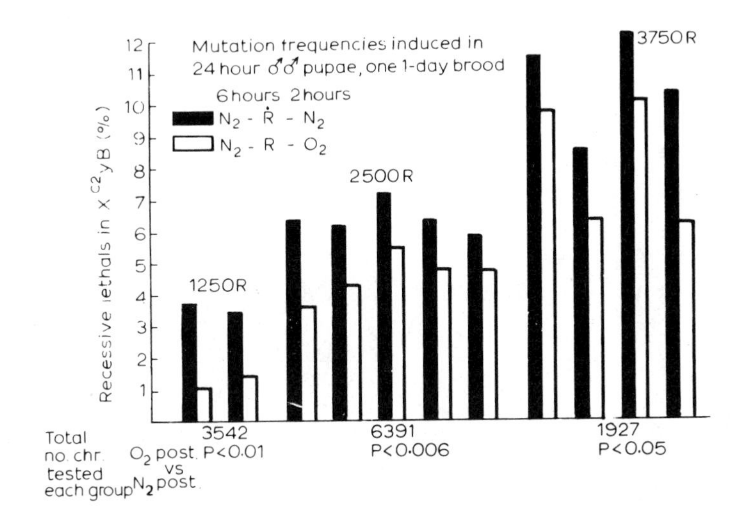

Fig. 9.6. After exposure of spermatids in 24-hour-old pupae to X-irradiation given at 46 R/sec under anoxia, post-treatment with oxygen consistently lowers the mutation frequency as compared to the effects observed with nitrogen post-treatment. Note that after exposure to 1250r, oxygen is more effective in reducing the mutation frequency than after exposure to the higher doses. (Fig. 3, Sobels (58). Courtesy Academic Press)

difference between spermatids and spermatozoa is even more pronounced than in air, because the oxygen enhancement ratio (OER, see p. 110) is about twice as high for spermatids as for spermatozoa. Why this is so, is not understood, but it is clear that the sensitivity differences between these two well-studied germ cell types have a complex basis, which ultimately must be connected with metabolic differences between them.

Among *Drosophila* oocytes, there is a strongly marked stage dependence of sensitivity. Mature oocytes are highly sensitive to the killing action of X-rays. They appear to have very little, if any, capacity for repair of radiation damage. Immature stages have efficient repair systems and can stand much higher doses than mature ones. Earlier experiments with high doses seemed to show that no reciprocal translocations can be produced in females; the only translocations obtained were the so-called 'half-translocations', i.e. detachment of attached X's accompanied by 'capping' of the broken end with a piece from another chromosome, usually the fourth or the Y. This is the only type of translocation that can be formed in immature oocytes, perhaps in part because the arrangement of the chromosomes at these stages does not provide suitable 'sites' for exchanges between two major chromosomes, and in part because any reciprocal translocations that may be formed will be eliminated through segregation at the later meiotic stages. These strictures do not apply to mature oocytes, and reciprocal translocations have now been obtained from females exposed to doses that were sufficiently low to allow survival of the mature oocytes (10). This is a good example for the complications that have to be taken into account when interpreting differences between the response of meiotic cells to mutagenic treatment.

The question whether, in *Drosophila* oocytes, repair acts on rejoining of broken chromosomes or on potential breaks, cannot be answered with certainty because it is difficult to establish whether lethality of oocytes is due to dominant lethality, i.e. chromosome breakage, or to some cytoplasmic effect. Assuming, however, that at least most of it is due to dominant lethality – and the dose-effect curves are in favour of it – the high frequency of dominant lethals in the presumed repairless mature oocytes (11) speaks for lack of rejoining rather than lack of breakage. It should be realized that this is the same argument used by Proust when studying repair in zygotes (p.129). We shall meet it again when we consider the effects of chemical mutagens (Chapter 15).

In the early zygotes, there are sensitivity differences between the maternal and paternal chromosomes, presumably linked to differences in chemistry and degree of condensation (12,13). After entry of the sperm into the egg, male and female pronucleus remain separate until the first mitosis. During this period, the male pronucleus unspirals from its highly condensed state in the spermhead to become similar to the female one; concomitantly, the arginine-rich histone is replaced by lysine-rich histone. When the frequencies of translocations were measured in zygotes irradiated with 500 R at 3 min interval up to 14.5 min, those in the maternal chromosome remained uniformly low (less than 1%), while those in the paternal X started high (over 2%) and gradually dropped to that of the maternal one. Over the same period, sex-linked lethal frequency remained uniform for the maternal chromosome; in the paternal one, a peak was reached at the time of the biological change-over from the arginine-rich histones of the spermhead to the lysine-rich histones of the male pronucleus.

Considering how many diverse factors enter into determination of germ cell sensitivities, differences in sensitivity patterns between species and higher taxonomic orders are to be expected and have, indeed, been observed. Yet they are remarkably slight even between taxonomically so widely removed species as *Drosophila* and the mouse.

Mice; silkworm. Dose rate, dose fractionation.

Although Muller had warned against genetical damage from X-rays almost from the time when he had obtained the first X-ray mutations, geneticists at large became seriously alarmed only after the introduction of nuclear fission for purposes of war and peace. Estimation of danger levels became an urgent task. It was felt that for this purpose results gained on a mammal would be more reliable than those hitherto obtained with *Drosophila,* and mutation research on mice was started on a very large scale, first at Oak Ridge, U.S.A. and, a little later, at Harwell, U.K. The method used in both places was the scoring of specific mutations in F_1. Irradiated ♂♂ were mated to tester ♀♀ that were homozygous for seven recessive markers, and the F_1 was scored for mutations or deletions of the corresponding genes in the irradiated germ cells (see p. 27). In mice, as in *Drosophila,* spermatozoa are more sensitive to the mutagenic action of X-rays than are spermatogonia; but spermatogonia, because of their long life span, are the most important germ cell at risk in a mammal. Most of the work deals therefore with spermatogonial mutations. When male mice are irradiated with moderately high doses of X-rays, utilization of postmeiotically treated germ cells in early litters is separated by a long sterile period from utilization of premeiotically treated germ cells in late litters. This makes it easy to limit sampling to spermatogonia.

Evidence for repair was first reported in 1958 (14a). It arose out of experiments in which different dose rates were used. Contrary to what had been found for *Drosophila,* there was a clear dependence of mutation frequency on dose rate. Although the dose-effect curves for both low and high dose rates could be fitted by straight lines (Fig. 9.7) (we shall return to the discrepant 1000 R point later), the frequency of mutations produced by the high dose rate of 90 R min^{-1} lay at all doses above that produced by the low dose rate of 0.009 R min^{-1}. Since X-rays had been used for the

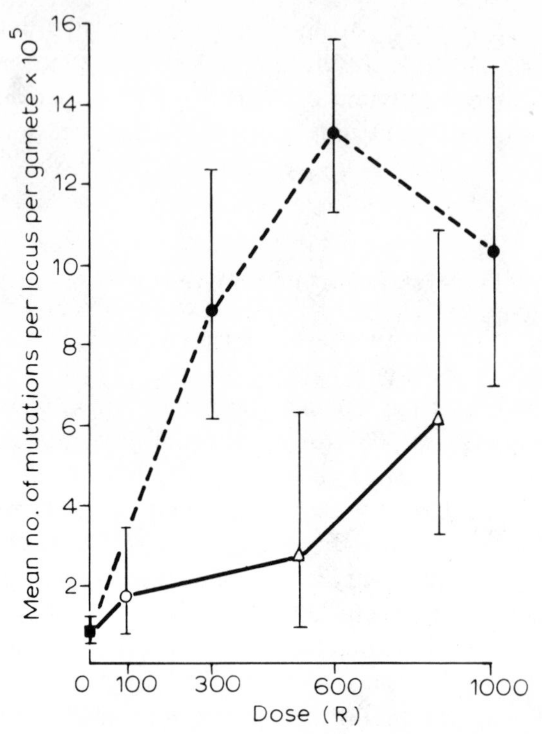

Fig. 9.7. Mutation rates at seven specific loci in spermatogonia of mice with 90% confidence intervals; effect of variation in dose rate. Solid circles: 80-90 R/min; open triangles: 90 R/week; open circle: 10 R/week. (Fig. 1, Russell (14), Courtesy Americ. Assoc. Advancement of Science)

high dose-rate, and cobalt γ rays for the low one, the effect might have been due to a difference in quality of radiation. This was excluded when the high dose-rate test was repeated using cobalt-rays. Selection among the heterogeneous population of spermatogonia might conceivably play a role. If this were true, no dose-rate effect would be expected in oocytes, the vast majority of which are in the same – so-called dictyate – stage. Yet, an even greater dose-rate effect was found for oocytes than for spermatogonia. On the other hand, there was no dose-rate effect for spermatozoa, and this removed what at first sight appeared to be a discrepancy between *Drosophila* and mice.

Russell interpreted these results as due to a repair process which takes place in the metabolically active spermatogonia or oocytes but not in the metabolically inert sperm, and which is inhibited by radiation delivered at high intensity. There has been, and still is, some debate as to the nature of the postulated repair. Is it rejoining between broken chromosomes, or is it repair of potential mutations? At the time of the first report the former process was well established; for the latter, there was then very little evidence, especially in regard to X-ray induced mutations. Because of this, and because of their bearing on estimates of radiation hazards, the results created a great amount of interest. Attempts to explain them in terms of rejoining of broken chromosome fragments have continued until the present time; but it seems much more likely that one is dealing with repair of potential genetic lesions. This has been the opinion of the Russells from the start, and they have given a good deal of evidence (14b, 15a) for their conclusion that the majority of specific locus mutations in spermatogonia and oocytes are not due to two-hit deletions, as the repair-by-rejoining hypothesis would demand. In any case, repair from X-ray induced premutational lesions is now so well established for a number of other organisms that there is no longer any reason to deny its existence in one of the earliest cases in which it had been claimed to play a role. More recently, a 'repair' or 'protective' system that is sensitive to radiation intensity has been hypothesized also for *Nicotiana* in order to explain dose rate effects on somatic mutation frequencies at two loci (15b). Dewey (15c) has shown that, in Chinese hamster cells, potential X-ray breaks are either repaired or 'confirmed' within minutes after irradiation.

When the Russells extended their experiments on mice over a wide range of dose-rates they found that there are upper and lower limits to the effect. At the upper level, there was no difference between mutation frequencies induced by 90 and 1000 R min^{-1}; at the lower level, there was none between mutation frequencies induced by 0.009 and 0.001 R min^{-1}. The existence of an upper limit is taken to mean that the repair system is completely inhibited or completely saturated already at 90 R min^{-1}. The existence of a lower limit suggests a state of optimal repair, which prevents all but a hard core of potential mutations from developing into actual ones. There are indications that, at dose rates even lower than the optimum, mutation frequency is again increased, perhaps because full repair requires stimulation by a minimum dose rate (16). Evidence for induction of repair by radiation-induced (probably also by certain chemically induced) lesions in DNA has been accumulating in recent research (17).

If it can also be assumed in man that low doses are mutagenically less effective than high ones, this will have far-reaching implications for estimates of genetic radiation hazards (Chapter 23). The same is true for observations which, although not themselves concerned with dose rate, were stimulated by the dose-rate results. The Russells argued that, if a high dose given at low intensity favours repair, then a very low dose given at high intensity may do the same. Indeed, when female mice were exposed to a dose of 50 R given at 90 R min^{-1}, the number of mutations was significantly lower than would have been expected from extrapolation of the frequencies at higher doses (18). Thus, it seems that in repair-competent cells the linear dose-effect relationship may not hold at very low doses (see p. 85). This

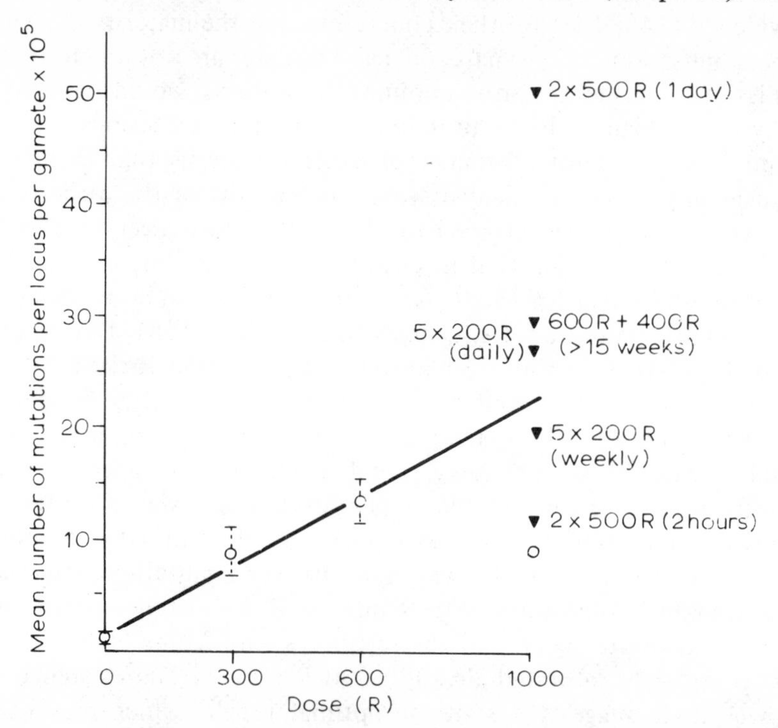

Fig. 9.8. Mutation rates at specific loci in spermatogonia of mice exposed to fractionated (▲) and single (○) doses of acute (90 R/min) X-irradiation. (Fig. 4, Russell (20). Courtesy Pergamon Press)

has been reported also for *Drosophila* (19).

Other factors that influence the yield of radiation-induced mutations in mouse spermatogonia were discovered in split-dose experiments in which a total dose of 1000 R was given in fractions separated by varied intervals (20). Fig. 9.8 shows that, depending on the pattern of fractionation, mutational yield was near the value expected from linear extrapolation, below it or above it. The lowest mutation frequency was obtained by 1000 R given in one treatment. This point is the average of the experiments recorded in Fig.9.7 and one subsequent one. While this and the other downward-deviations from linearity may be attributed to selective killing of cells that are highly sensitive to both killing and mutation, the very high mutational yield of a dose split into two fractions of 500 R separated by one day cannot be so explained. The Russells assume that the first dose affects the cell cycle in such a way that, at the time of the second dose, most spermatogonia are in a stage that is highly sensitive to mutation.

Similar results were obtained for spermatogonia of the silkworm (21). Tazima found that fractionated exposure with intervals up to 48 h increased mutational yield above that of the same dose given in one exposure. His interpretation, which has to account also for the 'reverse dose rate effect' (more mutations at low than at high dose rates) in certain germ cell stages is still tentative and cannot be discussed here.

The results from both mouse and silkworm reveal a very intricate relation between incident radiation dose and observed mutation frequency. How many of the potentially mutagenic lesions in the DNA of immature germ cells will give rise to mutations, clearly depends on a variety of factors: presence or absence of protective or sensitizing substances in the cell; degree of oxygenation; selective killing of the most sensitive cells; repair, its inhibition and, perhaps, its stimulation by the treatment itself; the proportions of cells that are in different stages of the cell cycle, and changes in these proportions under the influence of treatment. Additional factors enter in determining the yield of chromosome rearrangements. It is therefore not surprising that the dose-effect curves for genetic changes in spermatogonia do not follow strict target expectations. The possibility that the dose-response of specific-locus mutations might be fitted to a quadratic rather than a linear shape has been used in support of the hypothesis that the dose-rate effect is due to repair-by-rejoining. But since translocations, i.e. bona fide two-hit events, arise with linear kinetics (22,23), this argument cannot be used. In general, it is difficult or impossible to utilize dose effect curves obtained on populations of heterogeneous, asynchronous, actively metabolizing cells for drawing conclusions on the nature of the primary lesions in DNA. Instead, deviations from the theoretically expected shapes of dose-effect curves may be helpful in the analysis of those cellular processes that intervene between the primary lesions and the observed genetical effects.

Microorganisms. Dose effect curves

There is much evidence for repair of lethal X-ray damage in these organisms. Evidence for repair of premutational damage, on the contrary, is scarce. In part this must be attributed to a paucity of data. Ionizing radiations, because of their strong killing action on haploid cells, are far inferior to UV or certain chemicals as mutagenic agents for micro-organisms, and are therefore rarely used in mutation experiments. Perhaps more important is the difference between the roles of repair in survival and mutation. Lethality – insofar as it is due to nuclear damage – is always reduced by repair; mutation frequency, on the contrary, depends in diverse ways on a variety of repair mechanisms, which themselves may be inhibited or stimulated by repair (see Bridges, *Ann. Rev. Nuclear Science*; Bibliography). We shall come back to this presently.

These complexities are reflected in the dose-effect curves for mutations. They may be linear as expected from target theory (e.g. 24-27), but often they are not. In bacteria, linear dose effect curves have been found for X-ray induced mutations and small deletions (Fig. 9.9). In *Neurospora,* forward mutations to adenine-auxotrophy follow a linear curve (28), in contrast to the quadratic curve followed by multilocus deletions in the same region (p. 92). Exceptions, however, are frequent, as would be expected from experiments on asynchronous populations of repair-competent cells. Most frequently, the dose-effect curve is concave, suggesting a dose-exponent of more than 1 (e.g. 29,30). The usual interpretation is to attribute such curves to an admixture of two-hit, chromosomal, changes to the one-hit intragenic ones. There is, however, little ground for this assumption, especially when prototrophic reversions are scored. It seems much more likely that modification of an intrinsically linear curve is due to effects of higher doses on secondary steps in the mutation process. In *Neurospora* (31), a near-exponential dose-effect curve for adenine-reversions produced by chemical treatment has been attributed to inhibition of a repair process by the treatment (Chapter 16); it seems conceivable that a similar explanation applies to a case in bacteria (32) where the

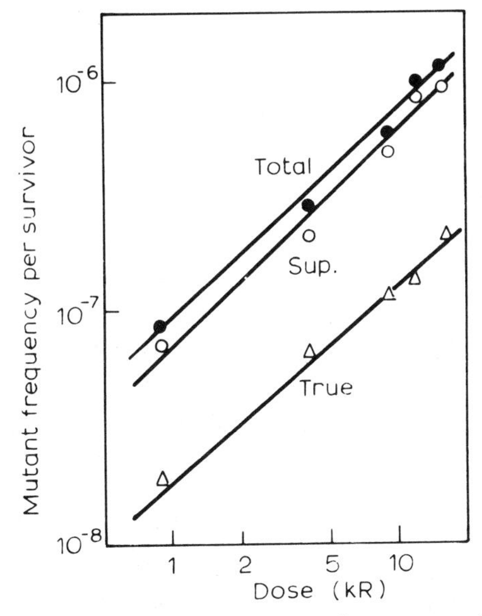

Fig. 9.9. Frequencies of true revertants and suppressors in a tryptophan-less strain of bacteria. (*E. coli WT2try*) after γ radiation in air. (Fig. 1a, Bridges (25). Courtesy Springer-Verlag)

dose-response of mutations to tryptophan independence could be fitted best by an exponential curve (i.e. a curve of the form $m = ae^{bD}$, where m = mutation frequency, D = dose, e = base of natural logarithms, a and b ≐ constants; not to be confused with the quadratic curve $m = a + bD^2$).

Dose-effect curves that are convex (i.e. bend down at higher doses) cannot be explained on pure hit-principles. Nor are they likely to arise from radiation effects on repair, although – as mentioned above (17) – there is some evidence for the *induction* of repair by radiation-induced lesions in DNA. It is more probable that convex curves are due to modification, by irradiation, of those processes that are intercalated between the 'fixation' of a mutational lesion in DNA and the appearance of a mutant clone of cells: transcription, translation, enzyme synthesis, enzyme activity, membrane formation etc. Effects of ionizing radiation on these processes have been shown to exist (33-40). In *Neurospora* (31), the dose-effect curve for X-ray induced adenine-reversions (at the *ad* 3A 38701 site) levels off very early, at a dose when that for inositol reversions (at the 37401 site) still continues to rise. This suggests that there are different radio-sensitive steps in the mutational pathways of these two reversion types. Incidentally, the difference between the shapes of the two dose-effect curves results in a change with dose of the ratio between the two reversions, often taken as a measure of the 'mutagen specificity' of the agent used. On strict target principles, such a change in relative target sizes would be very difficult to explain. We shall return to this Chapter 20.

The molecular mechanisms of X-ray mutagenesis

It is curious to realize that X-radiation, which was the first discovered mutagen, is also the one whose primary mechanism is least understood. The difficulty lies not in a shortage of possible mechanisms, but rather in their multiplicity (41-43). Chemical changes in nucleic acid bases, breakage of hydrogen bonds in the double helix, single- and double-strand breakage in DNA, cross-linking between the two strands of DNA, between different molecules of DNA and between DNA and protein – all these and more indirect ones such as the release of endonuclease from lysosomes – have been reported.

Double-strand breaks of DNA are lethal in prokaryotes (p.115); in eukaryotes they probably are the primary cause of chromosomal changes. Single-strand breaks are not lethal, or only rarely, and this makes them a likely source of intragenic mutations (44-47a). This conclusion was drawn more specifically from experiments in which mutagenesis by ionizing radiation, by ultraviolet light and by thymine starvation were shown to have some early step in common (25). Since the primary lesions of the last two agents are known to be breaks or short gaps in single strands of DNA (Chapter 14), it is likely that this is true also for ionizing radiation. Radiation-induced modification of nucleotide bases may also contribute to mutagenesis; very recent experiments suggest that they may be the main source of mutations in phage exposed to ionizing radiation (47b). Whatever the nature of the primary lesion, its final fate depends largely on the type of repair to which it is exposed. In the same way that chromosome breaks may be truly repaired by restitution or misrepaired by rearrangement, premutational lesions may be repaired by 'error-proof' or 'error-prone' mechanisms. The former restore the original conditions, thus removing both lethal and mutagenic damage. The latter also remove lethal damage but, in doing so, change the sequence of nucleotides, thus producing mutations. This interplay of mutagenic lesions and repair mechanisms has been thoroughly studied for UV-mutagenesis and will be discussed in this context (Chapter 14). It is, however, clear that repair mechanisms – some of them identical with those acting on UV lesions – enter in similar ways into X-ray mutagenesis. In particular, the functions of two repair genes (*recA* and *exrA*, p.) that are essential for UV mutagenesis also play a major role in the production of mutations by X-rays. As Bridges has formulated it (48), it may be 'the nature of the repair system that is used, and not necessarily that of the initial lesion, which determines mutability by ionizing radiation'. Probably, this statement applies not only to overall mutability but also to the type of intragenic change that radiation produces in a given type of cell. Base changes of various kinds have been reported for bacteria and yeast (49,50). Frameshifts could not be identified in the systems used for these analyses: a large admixture of frameshifts to X-ray induced mutations in *Neurospora* was inferred from reversion analysis (Chapter 17) of *ad*-3B mutants obtained from low X-ray doses (51).

Species differences in sensitivity

Target sizes for individual genes can be calculated from specific-loci tests (p. 27). Averaged over a number of genes, they yield an average target size for the species examined. It is true that the loci used for the test will affect the result. Indeed, in mouse spermatogonia, average mutation frequencies per gene per rad differ by a factor of about four between two sets of seven loci each (52). This, however, is a small degree of variation compared with the very large species differences, which range over several orders of magnitude from bacteria to barley. In *Drosophila* spermatozoa (53), the average mutation rate for eight autosomal loci was 6×10^{-8} /R per locus; in spermatogonia, it was 1.3×10^{-8} /R per locus. In the mouse, the corresponding values are at least ten times as high (52). From the point of view of genetic radiation hazards to man, this was a disquieting finding; but can it be generalized? Will it be safe to conclude that, in general, the genes of mammals, including man, offer manifold larger targets to ionizing radiation than do the genes of *Drosophila*? Recently, all available data have been used for a correlation analysis (54). This yielded a remarkably close correlation between target size and the DNA content of the haploid genome (Fig. 9.10). When the mutation rates per locus per rad were corrected for the difference

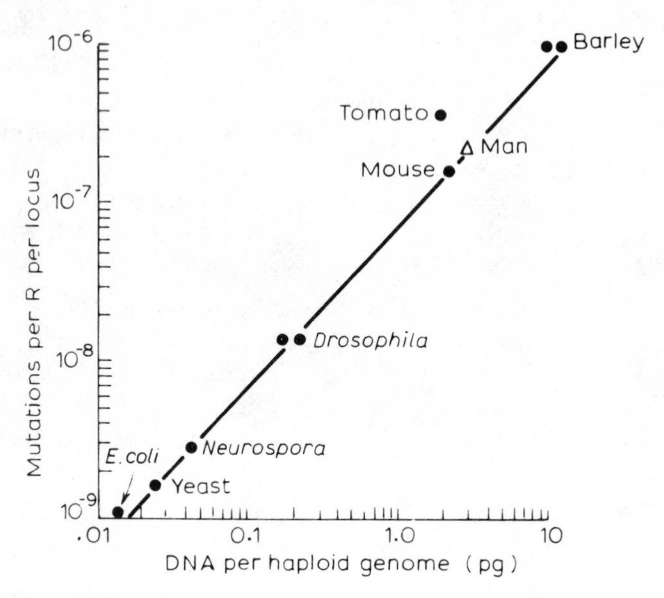

Fig. 9.10. Relation between forward mutation rate per locus per rad and the DNA content per haploid genome. Line drawn with slope one through mouse point. Point for man estimated from DNA content. (Fig. 1, Abrahamson (54). Courtesy Macmillan Journals)

in DNA content, variations between species were cut down from 1000-fold to 3-fold. For estimates of genetic radiation hazards, this is a highly relevant result. It will make it possible to extrapolate with some confidence from lower organisms to man. Judged by their DNA content, human cells should have genes that are slightly more radiosensitive than those of the mouse.

From the viewpoint of fundamental genetics, the result is not less important. It appears at a time when there is much discussion about the meaning of the species differences in amount of DNA per cell: differences which are much too large to be attributable to increased gene numbers in complex organisms or to differences in amount of satellite DNA. On the other hand, it is not possible to believe that a structural gene for, say, an enzyme should have 1000 times as many nucleotides in a mouse than in a bacterium. An attractive hypothesis is to attribute regulatory functions to the excess DNA (55). In *Drosophila,* where there is close correspondence between genes and bands in salivary gland chromosomes (56), each gene seems to have its own regulatory machinery attached to it. Moreover, the fact that no intracistronic complementation was observed within 13 adjacent genes (57) suggests that all parts of the machinery have to be undamaged for a gene to function. The results on target sizes strongly support this model.

References

1. Thoday, J.M. (1952), 'Sister-union isolocus breaks in irradiated *Vicia faba.* The target theory and physiological variation', *Heredity* **6** suppl., 299-309.
2. Swanson, C.P. (1955), 'Effect of oxygen tension on the production of chromosome breakage by ionizing radiations: an interpretation', pp. 254-261 in *Radiobiology Symposium 1954,* ed. Z.M. Bacq and P. Alexander, Butterworths Scientific Publications, London.
3. Kimball, R.F. (1963), 'The relation of repair to differential radiosensitivity in the production of mutations in *Paramecium',* pp. 167-176 in *Repair from Genetic Radiation Damage and Differential Radiosensitivity in Germ Cells,* ed. F.H. Sobels, Pergamon Press, Oxford, New York, London Paris.
4. Kimball, R.F. (1961), 'Postirradiation processes in the induction of recessive lethals by ionizing radiation', *J. Cell. Comp. Physiol.,* **58**, Suppl. 1, 163-170.
5. Kimball, R.F. (1968), 'The relation between repair of radiation damage and mutation induction', *Photochemistry and Photobiology* **8**, 515-520.
6. Glass, B. (1955), 'Differences in mutability during different stages of gametogenesis in *Drosophila', Brookhaven Symposia in Biology* **8**, 148-170.
7. Sankaranarayanan, K. and Sobels, F.H. (in press), 'Radiation Genetics', in *The Genetics and Biology of Drosophila,* ed. M. Ashburner and E. Novitski, Academic Press, New York.
8. Sobels, F.H. (1966), 'Processes underlying repair and radiosensitivity in spermatozoa and spermatids of *Drosophila',* in *Genetical Aspects of Radiosensitivity: Mechanisms of Repair.* Int. Atomic Energy Agency, Vienna, 49-65.
9. Mukherjee, R.N. and Sobels, F.H. (1968), 'The effects of sodium fluoride and iodoacetamide on mutation induction by X-irradiation in mature spermatozoa of *Drosophila', Mutation Res.* **6**, 217-225.
10. Traut, H. (1967), 'X-ray induction of 2;3 translocations in

mature and immature oocytes of *Drosophila melanogaster'*, *Genetics* **56**, 265-272.

11. King, R.C., Darrow, J.B. and Kaye, N.W. (1956), 'Studies on different classes of mutations induced by radiation of *Drosophila melanogaster* females', *Genetics* **41**, 890-900.
12. Graf, U., Piatkowska, B. and Würgler, F.E. (1969), 'X-ray-induced recessive lethals in newly inseminated eggs of *Drosophila melanogaster'*, *Mutation Res.* 7, 385-392.
13. Würgler, F.E. (1971), 'Radiation-induced translocations in inseminated eggs of *Drosophila melanogaster'*, *Mutation Res.* **13**, 353-359.

14a. Russell, W.L., Russell, L.B. and Kelly, E.M. (1958), 'Radiation dose rate and mutation frequency', *Science* **128**, 1546-1550.

14b. Russell, W.L. (1969), 'Observed mutation frequency in mice and the chain of processes affecting it', 216-228 in *Mutation as Cellular Process,* (biblio).

15a. Russell, L.B. (1971), 'Definition of functional units in a small chromosomal segment of the mouse and its use in interpreting the nature of radiation-induced mutations', *Mutation Res.* **11**, 107-123.

15b. Sand, S.A. and Smith, H.H. (1973), 'Somatic mutational transients. III Response by two genes in a clone of Nicotiana to 24 roentgens of γ radiation applied at various intensities', *Genetics,* **75**: 93-111.

15c. Dewey, W.C. (1972), 'Confirmation of lesions having the potential for forming chromosomal aberrations', *Int. J. Rad. Biol.,* **22**, 95-97.

16. Lyon, M.F., Papworth, D.G. and Phillips, R.J.S. (1972), 'Dose-rate and mutation frequency after irradiation of mouse spermatogonia', *Nature New Biology* **238**, 101-104.
17. Witkin, E.M. (1974), 'Thermal enhancement of ultraviolet mutability in a *tif-1 uvrA* derivative of *Escherichia coli* B/r: evidence that ultraviolet mutagenesis depends upon an inducible function', *Proc. Nat. Acad. Sci., U.S.A.* **71**, 1930-1934.
18. Russell, W.L. and Kelly, E.M. (1965), 'Mutation frequency in female mice exposed to a small X-ray dose at high dose rate', *Genetics,* **52**, 471.
19. Meyer, H.U. and Abrahamson, S. (1971), 'Preliminary report on mutagenic effects of low X-ray doses in immature germ cells of adult *Drosophila* females', *Genetics,* **68**, 244.

20. Russell, W.L. (1963), 'The effect of radiation dose rate and fractionation on mutation in mice', pp. 205-217 in *Repair From Genetic Radiation Damage and Differential Radiosensitivity in Germ Cells.* ed. F.H. Sobels. Pergamon Press, Oxford, London, New York, Paris.
21. Tazima, Y. (1969), 'Analysis of radiation sensitivity of silkworm spermatogonia and its implications in the study of mutation', *Proc. of Int. Seminar on Comparative cellular and species radiosensitivity in animals.* Kyoto. pp. 164-172.
22. Léonard, A. and Deknudt, W. (1968), 'Chromosome rearrangements after low X-ray doses given to spermatogonia of mice', *Can. J. Genet. Cytol.*, **10**, 119-124.
23. Gerber, G.B. and Léonard, A. (1971), 'Influence of selection, non-uniform cell population and repair on dose-effect curves of genetic effects', *Mutation Res.*, **12**, 175-182.
24. Demerec, M. (1958), 'Dose-effect relationships for X-ray induction of mutations in three genes of *Escherichia coli*', *Science,* **127**, 1059.
25. Bridges, B.A., Law, J. and Munson, R.J. (1968), 'Mutagenesis in *Escherichia coli.* II. Evidence for a common pathway for mutagenesis by ultraviolet light, ionizing radiation and thymine deprivation', *Molec. Gen. Genetics,* **103**, 266-273.
26. Munson, R.J. and Bridges, B.A. (1973), 'The LET factor in mutagenesis by ionizing radiations. I. Reversion to wild type of a bacteriophage T4 *Amber* Mutant', *Int. J. Rad. Biol.*, **24**, 257-273.
27. Ishii, Y. and Kondo, S. (1972), 'Spontaneous and radiation-induced deletion mutations in *Escherichia coli* strains with different DNA repair capacities', *Mutation Res.*, **16**, 13-25.
28. de Serres, F.J., Malling, H.V. and Webber, B.B. (1967), 'Dose-rate effects on inactivation and mutation induction in *Neurospora crassa*', *Brookhaven Symposia in Biology,* **20**, 56-76.
29. Hrishi, N. and James, A.P. (1964), 'The induction of mutation in yeast by thermal neutrons', *Can. J. Genet. Cytol.*, **6**, 357-363.
30. Deering, R.A. (1963), 'Mutation and killing of *Escherichia coli WP-2* by accelerated heavy ions and other radiation', *Radiation Res.*, **19**, 169-178.
31. Auerbach, C. and Ramsay, D. (1968), 'Analysis of a case of mutagen specificity in *Neurospora crassa. I.* Dose-response curves', *Molec. Gen. Genetics,* **103**, 72-104.

32. Bridges, B.A. (1963), 'Effect of chemical modifiers on inactivation and mutation-induction by γ radiation in *Escherichia coli'*, *J. Gen. Microbiol.*, **31**, 405-412.
33. Dale, W.M. (1966), 'Irradiation effects on Enzymes', in *Encyclopedia of Medical Radiology,* Vol. II, Part 1: 1-38. Berlin-Heidelberg-New York, Springer.
34. Pollard, E.C., Ebert, M.J., Miller, C., Kolacz, K. and Barone, T.F. (1965), 'Ionizing radiation: effect of irradiated medium on synthetic processes', *Science,* **147**, 1045-1047.
35. Yamamoto, O. (1967), 'Biochemical studies of radiation damage. I. Inactivation of the pH5 fraction in amino acyl sRNA synthesis *in vitro* and the binding of amino acids with protein and nucleic acid by γ-ray irradiation', *Int. J. Rad. Biol.,* **12**, 467-476.
36. Kirrman, J.M. (1966), 'Sur la radiosensibilité de la déhydroginase lactique d'un organe embryonnaire de poulet cultivé *in vitro', Compt. Rend. Acad. Sci.,* **263**, 426-429.
37. Harrington, H. (1964), 'Effect of X irradiation on the priming activity of DNA', *Proc. Nat. Acad. Sci., U.S.A.,* **51**, 59-66.
38. Goddard, J.P., Weiss, J.J. and Wheeler, C.M. (1969), 'Error frequency during *in vitro* transcription of poly-U is increased with γ-irradiated RNA Polymerase', *Nature,* **222**, 670-671.
39. Kučan, Z. (1966), 'Inactivation of isolated *Escherichia coli* ribosomes by γ irradiation', *Radiation Res.,* **27**, 229-236.
40. Skinner, L.G. (1968), 'Effect of X-radiation on transfer ribonucleic acid', *Int. J. Rad. Biol.,* **14**, 245-256.
41. Alexander, P. (1966), 'Changes in macromolecules produced by ionizing radiations', in *Encyclopedia of Medical Radiology.* Springer-Verlag, Berlin, Vol. 1, Part 1, 183-213.
42. Wacker, A. (1963), 'Molecular mechanisms of radiation effects', in *Progress in Nucleic Acid Research.* eds. J.N. Davidson and W.E. Cohn, Academic Press, New York, Vol. 1, 369-399.
43. Ullrich, M. and Hagen, V. (1971), 'Base liberation and concomitant reactions in irradiated DNA solutions', *Int. J. Rad. Biol.,* **19**, 507-518.
44. Freifelder, D. (1966), 'Lethal changes in bacteriophage DNA produced by X-rays', *Radiation Res.,* Suppl. **6**, 80-96.
45. Freifelder, D. (1968), 'Rate of production of single-strand breaks in DNA by X-irradiation *in situ', J. Mol. Biol.,* **35**, 303-309.
46. Schans, van der, G.P., Bleichrodt, J.F. and Blok, J. (1973),

'Contribution of various types of damage to inactivation of a biologically-active double-stranded circular DNA by γ-radiation', *Int. J. Rad. Biol.,* **23**, 133-150.

47a. Fox, B.W. and Fox, M. (1973), 'DNA single-strand rejoining in two pairs of cell-lines showing the same and different sensitivities to X-rays', *Int. J. Rad. Biol.,* **24**, 127-135.

47b. Bresler, S.E., Kalinin, V.L., Kopylova, Yu. I., Krivisky, A.S., Rybnin, V.N. and Shelegedin, V.N. (in press), 'Study of genetic effects of high energy radiation with different ionizing capacities on extracellular phages', *Mutation Res.*

48. Bridges, B.A. and Mottershead, R.P. (1972), γ ray mutagenesis in a strain of *Escherichia coli* deficient in DNA polymerase I', *Heredity,* **29**, 203-211.

49. Phillips, S.L., Person, S. and Newton, H.P. (1972), 'Characterization of genetic coding changes in bacteria produced by ionizing radiation and by the radioactive decay of incorporated ^{3}H-labelled compounds', *Int. J. Rad. Biol.,* **21**, 159-166.

50. Prakash, L. and Sherman, F. (1973), 'Mutagenic specificity: reversion of iso-1-cytochrome *c* mutants of yeast', *J. Mol. Biol.,* **79**, 65-82.

51. Malling, H.V. and de Serres, F.J. (1973), 'Genetic alterations at the molecular level in X-ray induced ad-3B mutants of *Neurospora crassa', Radiation Res.* **53**, 77-87.

52. Lyon, M.F. and Morris, T. (1966), 'Mutation rates at a new set of specific loci in the mouse', *Genet. Res. Camb.,* **7**, 12-17.

53. Alexander, M.L. (1954), 'Mutation rates at specific autosomal loci in the mature and immature germ cells of *Drosophila melanogaster', Genetics,* **39**, 409-428.

54. Abrahamson, S., Bender, M.A., Conger, A.D. and Wolff, S. (1973), 'Uniformity of radiation-induced mutation rates among different species', *Nature,* **245**, 460-462.

55. Britten, R.J. and Davidson, E.H. (1969), 'Gene regulation for higher cells: a theory', *Science,* **165**, 349-357.

56. Judd, B.H., Shen, M.W. and Kaufman, T.C. (1972), 'The anatomy and function of a segment of the X chromosome of *Drosophila melanogaster', Genetics,* **71**, 139-156.

57. Shannon, M.P., Kaufman, T.C., Shen, M.W. and Judd, B.H. (1972), 'Lethality patterns and morphology of selected lethal and semi-lethal mutations in the zeste-white region of *Drosophila melanogaster', Genetics,* **72**, 615-638.

58. Sobels, F.H. (1968), 'Genetic repair phenomena and dose-rate effects in animals', *Adv. Biol. Med. Phys.,* **12**, 341-352.

CHAPTER TEN

Mutagenesis by ultraviolet and visible light. I: Early work on macro-organisms

The problems studied

Research on the mutagenic action of UV started very soon after the discovery of X-ray mutagenesis. By the early thirties, it had been established that UV can produce mutations in *Drosophila* and flowering plants. Research was carried on with three principal aims in mind. One concerned the relationship between intragenic mutations and intergenic structural changes, which was then a very much debated topic. The X-ray data allowed no decision as to whether the two types of event have the same cause. Indeed, as mentioned in Chapter 6, some geneticists believed that all X-ray induced mutations were in reality chromosome rearrangements. Since UV, in contrast to X-rays and other high-energy radiation, produces only atomic excitations but no ionizations, the question could be asked whether perhaps structural changes required ionization for their production. The answer, as we shall see presently, was negative. All the same, the proportion of rearrangements among UV-induced mutations was generally low, and this led to the second aim, which was to analyse the nature of gene mutations by means of an agent that is less destructive than X-rays. This approach was to bear full fruit later, when the chemical nature of the genetic material had been elucidated. We shall deal with it in Chapter 14. Finally, the fact that different biological molecules have different absorption spectra for UV held out the hope that comparisons between the mutagenic efficiencies of different wavelengths would show which component of the genetic material had to absorb energy

* See review articles by Muller and by Swanson and Stadler in *Radiation Biology*, bibliography.

in order for a mutation to be produced. Since at that time most geneticists thought that the specificity of the gene resided in the protein moiety of the nucleoprotein molecule, it even did not seem impossible that chemical differences between genes might result in different mutational responses to monochromatic UV.

Methods of mutation research on maize and other flowering plants

Most of the early work on UV-mutagenesis was carried out on maize and *Antirrhinum*. This chapter seems therefore a suitable place for going into some of the experimental techniques used. A general survey of them can be found in Auerbach, *Mutation* (Bibliography); a more recent one in the article by Ehrenberg in vol. 2 of *Chemical Mutagens* (Bibliography). Special mention should be made of the small crucifer *Arabidopsis* that has recently become a favourite for mutation work on higher plants in the laboratory (1a). Maize is undoubtedly the most suitable higher plant for a sophisticated analysis of mutational effects, and it has played a prominent role in the study of mutagenesis by UV in eukaryotes. I shall therefore deal in some detail with the methods of mutation research on maize as developed by Stadler and his collaborators (1b). Familiarity with these methods will be required again in Chapter 20 for the discussion of paramutational phenomena.

The life cycle of maize

Fig. 10.1 is a diagrammatic representation of the life cycle of maize. Maize is a monoecious species, i.e male and female flowers develop on the same plant. The male inflorescence or 'tassel' forms the tip of the plant; the female inflorescence, from which at maturity the stigmata of the flowers protrude as 'silk', sits lower down the stem. The essential feature in the life cycle of all flowering plants is an alternation of generations. The diploid sporophyte generation takes up most of the life cycle; the haploid gametophyte generation forms a brief interlude between meiosis and fertilization. In male flowers, meiosis results in four haploid microspores, which develop into pollen grains. The original microspore nucleus divides into the vegetative and generative nucleus, and the latter divides once more into two. The mature pollen grain thus contains three identical haploid nuclei: one vegetative nucleus and two generative or sperm nuclei. All three migrate into the pollen tube, but only the generative

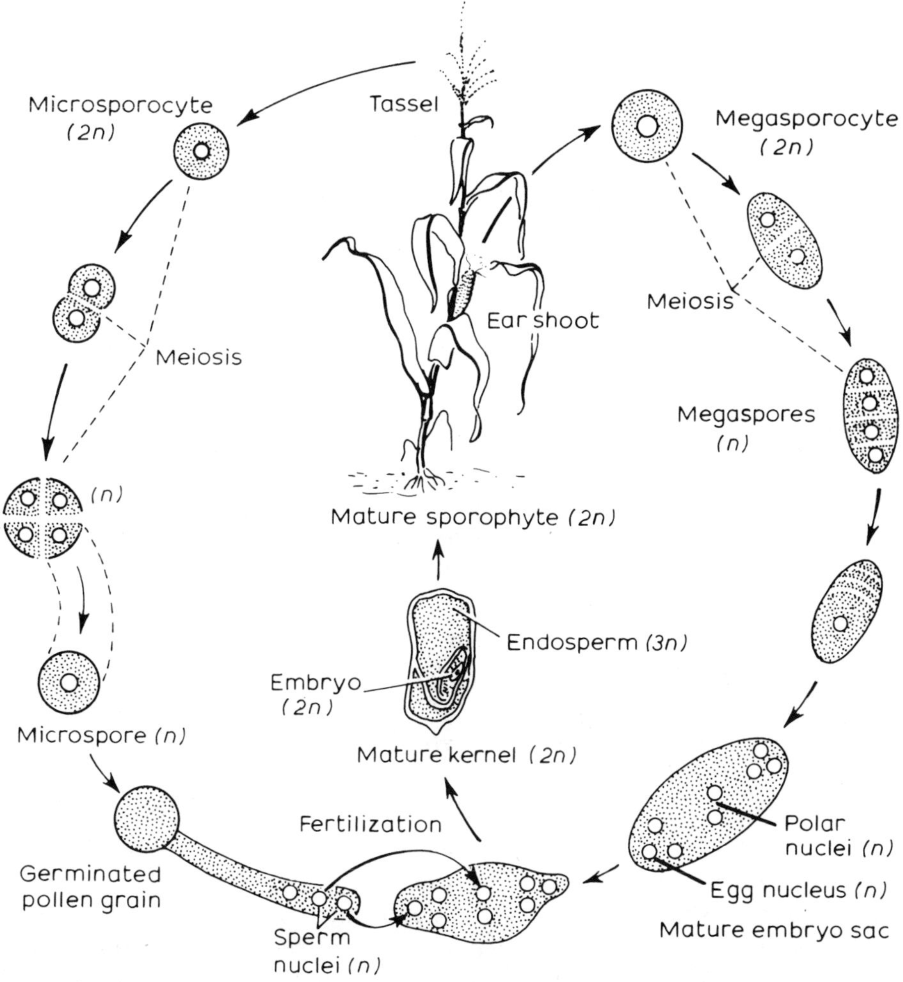

Fig. 10.1. The life cycle of corn (Zea mays). (Fig. 4.2, Srb, Owen and Edgar, 1965, (bibl.). Courtesy authors)

ones take part in fertilization. In the female flower, meiosis results in four megaspores, three of which degenerate. The remaining one undergoes three divisions, giving rise to eight identical haploid nuclei, three of which – the egg nucleus and the two polar nuclei – take part in fertilization. At fertilization, one of the two sperm nuclei fuses with the egg nucleus and forms the zygote, from which the embryo of the sporophyte develops. The other sperm nucleus fuses with both polar nuclei and forms the triploid endosperm which surrounds the embryo in the seed. The endosperm thus forms a kind of alternative F_1 for mutation studies; the fact that many genes affect the endosperm of maize and that the effects can be easily analysed in the large seed grains has proved a tremendous asset.

Genetical effects scored

After treatment of pollen grains, the following genetical effects can be scored in the progeny.

Dominant lethals

These result in zygotic lethality and are scored as embryo abortions in the seed.

Deficiencies. Visible mutations at marked loci

The vast majority of deficiencies are lethal in the haplo-phase, especially in the male gametophyte. An F_1 plant that is heterozygous for even a small deficiency segregates 50% aborted or defective pollen. This makes a distinction between gene mutations and small deletions much easier in maize than in *Drosophila.* The large number of available plant and endosperm markers makes maize a suitable material for *specific-locus* tests. It will be remembered (Chapter 2) that, in these tests, the untreated sex is homozygous for a series of recessive alleles and that mutations or deletions at the marked loci in the treated germ cells become apparent through the uncovering of one or more of the specific markers in the F_1. Fortunately, the wild-type alleles of many endosperm markers are dominant or semi-dominant over the two recessive alleles with which they are combined in the endosperm, so that endosperm as well as sporophyte can be used for the scoring of mutations at specific loci. Use of the endosperm has the advantage that large numbers of kernels can be scored in the F_1; it has the disadvantage that no genetic analysis of the mutants can be carried out. In mature pollen grains of maize, the two generative nuclei are already separate, so that the endosperm of a given kernel and the plant growing out of this kernel have been fertilized by different sperm nuclei. When treatment is given, not to pollen grains, but to microspores in which the generative nuclei are not yet separate, the plant grown from a mutant kernel will be heterozygous for the same mutation as the kernel. For UV with its low penetrating power, microspore irradiation is not a suitable technique. A distinction between UV-induced point mutations and deletions requires the use of sporophyte markers; mutated plants are grown to maturity and tested for segregation of defective pollen. This technique has been used in the experiments by Stadler and by Mott quoted in Chapter 6.

This seems the right place for mentioning a technique that was developed later and is now being much used in mutation research on maize (2a). It consists of scoring pollen grains for mutations to 'waxy', or from 'waxy' to the normal 'starchy'. The character is a property of the individual pollen grain, and large numbers can be scored by their colour reactions. Mutational sites within the locus can be determined by intercrossing mutants.

Visible mutations at unmarked loci

Dominant visible mutations are as rare in maize as they are in *Drosophila* and mice. The detection of recessive mutations requires a generation of inbreeding. An F_1 that is heterozygous for a recessive mutation will, on self-fertilization, segregate mutant and non-mutant progeny in a ratio of 1:3, or less if the mutant seedlings should be less viable than the non-mutant ones. The most frequently scored visible recessives are those affecting chlorophyll synthesis, a process in which a very large number of genes is involved. Many chlorophyll mutations act as lethals for the growing plants but can be scored in the seedlings.

Chromosome breaks and rearrangements

Translocations, inversions and sufficiently large deficiencies can be seen cytologically at meiosis. Translocations can also be detected genetically by the fact that translocation heterozygotes, like deficiency heterozygotes, produce 50% defective pollen; this corresponds to the semi-sterility of translocation heterozygotes in mice. As we have seen in Chapter 2, the consequences of chromosome breakage are different in the endosperm and in the embryo. In the latter, broken ends may 'heal', and this results in terminal deletions. In the endosperm, broken chromosomes that do not rejoin undergo a breakage-fusion-bridge cycle. The pattern formed in the kernel by the gradual uncovering of recessive colour markers may make it possible to infer the fate of the successive bridges (Fig. 2.1).

Some special problems of UV-mutagenesis

Penetration; physiological damage

Experimental work with UV presents difficulties that are not found in X-ray work. The low penetrating power of the radiation restricts the systems in which genetical tests can be performed and enters into the evaluation of the results. The often severe physiological injury sets limits to the doses that can be used, and may affect the results of mutation tests in a variety of ways.

In experiments on flowering plants, the difficulty created by the low penetrating ability of the rays could be largely overcome by using pollen grains for treatment. Even here, positional differences of the nuclei in the grains had to be taken into account. In *Drosophila,* treatment of spermatozoa *in vitro* would have been the ideal technique but, since attempts at artificial insemination of *Drosophila* never have been successful, alternative techniques had to be developed. One consisted of gently compressing the abdomen of males between quartz plates and applying radiation to the ventral surface; the use of strains in which the testis sheath lacks the wild-type yellow pigment made this technique somewhat more efficient. Even so, it has been estimated that about 99.9% of the radiation energy is absorbed on the way to the germ cells. The most useful technique turned out to be irradiation of the posterior end of the fertilized ovum during the short period when the future germ cells are close to the surface and still outside the blastoderm in the so-called polar cap. Data obtained by this technique require correction for the fact that the population of tested mature germ cells is derived from a much smaller population of treated ones, so that each mutation will develop into a cluster of mutant spermatozoa or ova.

Measurement of the genetically effective dose

When geneticists tried to compare the results obtained by UV with those obtained by ionizing radiation, they came up against a fundamental difficulty which assails all efforts to compare the effects of different mutagens. How can one compare a dose of UV with a dose of X-rays? Both can be expressed in ergs per volume, but this takes no account of the fact that a large part of the energy of UV is dispersed by interaction with molecules outside the chromosomes and that the degree to which this happens depends on wave length as well as on conditions in the cell. In such a situation – and it is one that later was to play a great role in work with chemical mutagens – the only way out is to measure dose in terms of *one* undoubtedly genetic effect and to calibrate all other genetic effects against this standard. This precludes comparisons in terms of overall utilization of energy, but makes it possible to carry out the much more important comparisons between the spectra of genetical effects produced by different mutagens. In *Drosophila,* the effective dose is usually measured in terms of sex-linked lethals; in maize, in terms of F_1 plants that segregate defective pollen; in *Tradescantia* pollen grains, in terms of cytologically detectable terminal deletions. Clearly, some amount of circularity is inherent in this procedure, because it is based on the assumption that the two mutagens to be compared act identically when producing the chosen standard effect. This may not be true, as we shall see below. Still, no better way of comparing doses has yet been discovered, and corrections for differences in the production of the standard effect may be possible. Thus, when it had been found that large rearrangements are much more frequent among X-ray-induced than among UV-induced sex-linked lethals in *Drosophila,* it became possible to estimate doses by the frequencies of lethals that were free of such rearrangements. Sometimes, a more meaningful comparison between the effects of mutagens can be obtained by the use of two different effects as standards. We shall see an example for this in Table 10.1.

Results

Types of effect and their relative frequencies

Like X-rays, UV produces point mutations, small deficiencies, chromosome breaks and chromosome rearrangements, but it does so at different relative frequencies. In *Drosophila,* translocations are very much rarer in relation to sex-linked lethals than they are after X-radiation (2b). Small deficiencies, on the other hand, appear to be as frequent among UV-induced as among X-ray induced sex-linked lethals (3). In *Tradescantia* pollen tube mitoses, terminal deletions were the most commonly found type of aberration (4). Their frequency increased linearly with dose (5), provided that all pollen grains had been irradiated at the same stage of germination. No translocations were found among over 900 chromosomes in pollen tubes in which UV had induced 10% terminal deletions per chromosome; a dose of X-rays producing the same frequency of terminal deletions yielded 4.4% translocations. Swanson drew the conclusion that UV, while capable of producing chromosome breaks, in some way interferes with their rejoining into new combinations. In agreement with this assumption, X-rays induced fewer translocations in pollen grains that had been pretreated with UV than in not so pretreated grains (6). Also in *Drosophila* spermatozoa, the frequency of X-ray induced translocations was reduced by pre-treatment with UV (7). Swanson suggested that UV might favour restitution of broken chromosomes as against misrepair, perhaps by stiffening their 'matrix'; nowadays we might prefer to think of an effect on some step in the rejoining of broken chromosomes. However, as mentioned in Chapter 9, the interaction between UV and X-rays on chromosome breakage may be synergistic as well as inhibitory (8). Probably, more than one step in the production of chromosome breaks and their rejoining is modifiable by UV. Gray (9) has pointed out that UV would be expected to have much greater effects on cellular mechanisms than X-rays because, at the doses used for obtaining comparable genetic effects, a very much larger amount of UV-energy than of X-ray energy is dissipated in the cell. In order to test the response of the rejoining mechanism to radiation, Bailey and Wolff (10) scored chromosome breaks and deletions in *Tradescantia* pollen grains in which protein synthesis after treatment was inhibited. They found that chloramphenicol in the germination medium increased the frequency of X-ray-induced aberrations (see Chapter 8), but not of UV-induced ones. They took this to mean that the required high doses of UV had already inhibited the rejoining mechanism to its limits. Alternative interpretations come to mind, but cannot be fruitfully discussed on the basis of the available data. The exciting progress in the molecular and biochemical analysis of UV-mutagenesis, to be discussed in Chapters 13 and 14, has channelled most research efforts into this area. Yet, it is unlikely that the behaviour of the chromosomes in UV-irradiated cells will be explicable in molecular terms before it has been studied more closely by cytological and genetical means.

Meanwhile, there can be no doubt that UV can cause chromosome breakage. This follows not only from the direct cytological observations just quoted, but also from the high frequency of dominant lethals in UV-treated *Drosophila* sperm (7), from the frequent losses of ring chromosomes (see p. 93) in *Drosophila* and maize (11,12), and from the ease with which breakage-fusion-bridge cycles can be induced in maize endosperm by UV-treatment of the pollen (13).

Differences between the effects of UV and X-rays

The most thorough analysis of the spectrum of UV-induced genetic changes in an eukaryote has been carried out by Stadler and his associates (1b,14). They concluded that UV differs from X-rays in four essential ways.

(1) UV yields a lower frequency of translocations in relation to mutations.

(2) The apparent point mutations include true intragenic changes.

(3) A high proportion of the mutations and deficiencies appear first as mosaics.

(4) UV-induced genetic changes are much rarer in the sporophyte than in the endosperm.

We shall consider these points in order.

(1) Table 10.1 presents the results of two experiments in which different criteria were used for comparing the doses of the two kinds of radiation. In the first, doses were measured by the frequencies of F_1 plants segregating for defective pollen; in the second, by the frequencies in which recessive alleles at two loci were uncovered in the endosperm. In both experiments, translocations were scored cytologically. In addition, the first experiment included the scoring of seed and seedling mutations in F_2; the second, the scoring of F_1 plants segregating for defective pollen. Thus, the latter effect was the independent variable in the first experiment, and a dependent one in the second.

Table 10.1. Comparison of X-ray and UV effects in maize. (Combined from Table 1, Stadler (14) and data from Stadler (16)

Treatment	*Frequency of F_1 plants segregating for defective pollen (%)*	*Frequency of interchanges (%)*	*Frequency of mutations*	
			F_2 seeds and seedlings (%)	*Uncovering of recessive markers in endosperm*
X-rays	st*	4.1	0.9	—
UV	st	0	17.0	—
X-rays	77	44	—	st*
UV	14	1	—	st

* st = effect used for standardizing dose, approximately equal in the X-ray and UV-series of the same experiment.

Both experiments agree with each other in showing a paucity of UV-induced translocations. (A peculiar type of 'deficiency-translocation' is not listed in the table but will be mentioned presently). In addition, the first experiment shows that UV produces many more mutations in relation to deficiencies than do X-rays. In fact, the evidence against the production of true gene mutations in maize by X-rays (Chapter 6) makes it likely that even the few apparent X-ray mutations were small deficiencies of a kind that allows some of the defective pollen to function and yield an F_2. In contrast, it is most unlikely that all or even most of the mutations scored after UV-irradiation were deficiencies. In the first experiment, the frequencies of deficiencies at unselected loci (F_1 plants segregating for defective pollen) was approximately the same after either type of treatment; yet UV had produced many more mutations in F_2. In the second experiment, the situation was the reverse. A dose of UV that was greatly inferior to one of X-rays in the production of deficiencies at unselected loci, yet had yielded approximately the same frequency of endosperm 'mutations', many of which probably were true gene mutations. The second experiment shows also that segregation for defective pollen does not represent identical damage after the two kinds of radiation. After X-ray treatment, many of the plants that segregate for defective pollen presumably carry translocations, and the remainder, deletions. After UV-treatment, very few of these plants are likely to carry translocations; the remainder carry either deletions or gene mutations that affect viability of the gametophyte. The fact that UV does produce this type of mutation was confirmed in a separate experiment.

(2) Stadler analysed UV-induced mutations from *A* (purple kernel) to *a* (colourless kernel) in the same way as he had analysed X-ray induced ones. While all X-ray induced mutants had been shown to carry deficiencies (p. 99), the UV-induced ones included several that gave no evidence of deficiency by the criteria used.

(3) In experiments in which endosperm mutations and deficiencies at specific loci were scored, those produced by X-rays usually affected the whole endosperm, while 75% of those produced by UV affected only part of the endosperm. The mutant area varied widely between kernels, but the varients were normally distributed about a mean of one-half.

(4) When marker genes were used that could be scored in both endosperm amd sporophyte, X-rays were found to be equally effective for both systems, as would indeed be expected from the fact that the two generative nuclei are exposed to the same dose.

UV-irradiation, because of its low penetrating ability, will often affect one nucleus more than the other but, in the population of pollen grains as a whole, these differences in exposure should average out and one would again expect approximately equal mutation frequencies in endosperm and sporophyte. This, however, was not found. While UV-induced mutations and deficiencies were very frequent in the endosperm, they were very rare in the plants grown from the same seed sample. Thus, when a sample of seeds containing 493 endosperm deficiencies or mutations to *a* was grown into plants, only five of these showed a deficiency or mutation at the *A* locus.

Stadler tentatively fitted all these observations into a model of UV-action that presupposed a delay between the chromosomal lesion and its realization as break or mutation. We have seen that, subsequently, the idea of potential damage was extended also to ionizing radiation; it seems likely, however, that the potential lesions produced by UV persist longer than those produced by X-rays. The best proof for delay of the final effect after both types of radiation is the influence of the physiological environment: in the endosperm, chromosome breaks result in breakage-fusion-bridge cycles, in the embryo in terminal deletions (Chapter 2). While this probably is due to a delay in rejoining between already open breaks, the inefficiency of UV in producing translocations paired with its efficiency in producing single breaks is more plausibly attributed to a delayed opening of potential breaks (Chapter 9). In order for an interchange to take place, breaks in two chromosomes must be open during the same short period. If one of them opens later than the other, there is a chance for the first break to have already restituted when the second one opens, or for the fragments of the first broken chromosome to have moved too far apart to take part in the same rearrangement. The latter situation may account for the curious 'deficiency-translocations', comprising a deficiency as well as a translocation, which in Stadler's experiments formed the major type of the few UV-induced translocations. Of special interest is the paucity of UV-induced point mutations in the sporophyte as compared with the endosperm. It points to delay in repair or fixation of potentially mutagenic lesions, a phenomenon that now is known to occur in UV-mutagenesis of micro-organisms. In contrast, it is no longer necessary to invoke delayed action of UV to account for the high frequency of mosaics in F_1; for this is the expected outcome of a mild mutagenic treatment of two-stranded DNA.

Wavelength dependence

In the earliest studies, pollen grains of *Antirrhinum* and maize were used to compare the mutagenic efficiencies of different wavelengths of UV. The results showed a peak of efficiency in the region of absorption by nucleic acid, but the interpretation was complicated by problems of penetration. These were overcome in experiments on fungal spores and bacteria (see next chapter) and on the liverwort *Sphaerocarpus* (15). The spermatozoids of this plant are only 0.5μ thick and represent almost naked nuclei. Lethal and visible mutations can be scored by tetrad analysis. Fig. 10.2 shows the good correspondence between mutagenic efficiency and the absorption spectrum

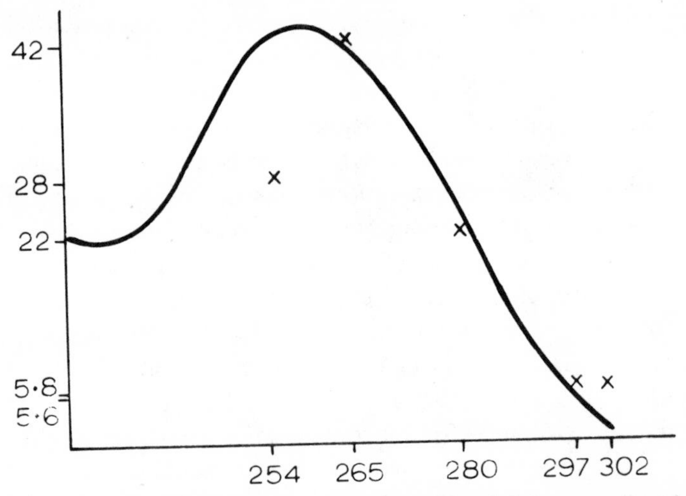

Fig. 10.2. Mutation frequency in spermatozoids of *Sphaerocarpus* plotted against wavelength of UV. Abscissa: wavelength in nm; ordinate: number of mutations per 100 tetrads from fertilization of female plants with irradiated spermatozoids. (Fig. 1, Knapp (15). Courtesy University Press Cambridge

of nucleic acid (DNA or, as it was then called, thymonucleic acid). The divergence of the first experimental point from the expected value is statistically not significant. Altogether, the values of the first three points do not differ significantly from each other, but the drop in mutation frequency at longer wavelengths is striking and real. Nevertheless, as mentioned in Chapter 1, these findings were not considered sufficient reason for implicating the nucleic acid of the gene as that part in which mutagenic changes take place, for energy transfer to the protein moiety could not be excluded. These doubts were strengthened when it was shown that visible light can produce mutations in cells that have been sensitized by fluorescent dyes, such as erythrosin or acridine orange ('photodynamic effect'). We shall come back to this in Chapter 12. Clearly, if energy transfer is involved in photodynamic action there was no reason to exclude it for the action of UV.

References

1a. McKelvie, A.D. (1963), 'Studies in the induction of mutations in *Arabidopsis Thaliana* (L) Heinh., *Radiation Bot.* **3**, 105-123.

1b. Stadler, L.J. (1939), 'Genetic studies with ultraviolet radiation', *Proc. 7th Int. Congress of Genetics.* ed. R.C. Punnett, Cambridge University Press, Cambridge, 269-276.

2a. Nelson, O.E. Jr. (1957), 'The feasibility of investigating "genetic fine structure" in higher plants', *Am. Nat.* **91**, 331-332.

2b. Mackenzie, K. (1941), 'Mutation and lethal effects of ultraviolet irradiation on *Drosophila*', *Proc. Roy. Soc. Edinburgh,* **B61**, 67-77.

3. Slizynski, B.M. (1942), 'Deficiency effects of ultraviolet light on *Drosophila melanogaster*', *Proc. Roy. Soc. Edinburgh,* **B61**, 297-315.

4. Swanson, C.P. (1940), 'A comparison of chromosomal aberrations induced by X-ray and ultraviolet radiation', *Proc. Nat. Acad. Sci., U.S.A.*, **26**, 366-373.

5. Swanson, C.P. (1942), 'The effects of ultraviolet and X-ray treatment on the pollen tube chromosomes of *Tradescantia*', *Genetics,* **27**, 491-503.

6. Swanson, C.P. (1944), 'X-ray and ultraviolet studies on pollen tube chromosomes. I. The effect of ultraviolet (2537Å) on X-ray induced chromosomal aberrations', *Genetics,* **29**, 61-68.

7. Kaufmann, B.P. and Hollaender, A. (1946), 'Modification of the frequency of chromosomal rearrangements induced by X-rays in *Drosophila. II.* Use of ultraviolet radiation', *Genetics,* **31**, 368-376.

8. Kirby-Smith, J.S. (1963), 'Effects of combined UV and

X-radiations on chromosome breakage in *Tradescantia* pollen', pp. 203-214 in *Radiation-Induced Chromosome Aberrations.* ed. S. Wolff, Columbia University Press, New York, London.

9. Gray, L.H. (1963), p. 212 in *Radiation-Induced Chromosome Aberrations.* ed. S. Wolff, Columbia University Press, New York, London.
10. Bailey, P.C. and Wolff, S. (1964), 'A comparison of X-ray and ultraviolet-induced aberrations in pollen tube chromosomes of *Tradescantia. II.* Influence of protein synthesis inhibitors', *Radiation Botany,* **4**, 121-125.
11. Fabergé, A.C. and Mohler, J.D. (1952), 'Breakage of chromosomes produced by ultraviolet radiation in *Drosophila', Nature,* **169**, 278-279.
12. Schultz, J. (1951), 'The effect of ultraviolet radiation on a ring chromosome in *Zea mays', Proc. Nat. Acad. Sci U.S.A.* **37**, 590-599.
13. Fabergé, A.C. (1936), 'Ultraviolet induced chromosome aberrations in maize', *Genetics,* **36**, 549-550.
14. Stadler, L.J. (1941), 'The comparison of ultraviolet and X-ray effects on mutation', *Cold Spring Harbor Symp. Quant. Biol.* **9**, 168-178.
15. Knapp, E. and Schreiber, H. (1939), 'Quantitative Analyse der Mutations-auslösenden Wirkung monochromatischen U-V-Lichtes in Spermatozoiden von *Sphaerocarpus', Proc. 7th Int. Cong. Genetics.* ed. R.C. Punnett, Cambridge University Press, Cambridge. 175-176.
16. Stadler, L.J. and Sprague, G.F. (1937), 'Contrasts in the genetic effects of ultraviolet radiation and X-rays', *Science,* **85**, 57-58.

CHAPTER ELEVEN

Mutagenesis by ultraviolet and visible light II. Methods of mutation research on micro-organisms and mammalian cell cultures

I. GENERAL

The most important experiments on the mutagenic effects of ultraviolet and visible light have been carried out on micro-organisms. For a critical appreciation of the results, it is necessary to understand the rationale, scope and limitations of the test methods, and the kind of conclusions that can or cannot be drawn validly from the data. I have dealt with this aspect of the methodology more fully in a previous book, in which additional references to the literature up to 1961 can also be found ('Mutation'; Bibliography). Detailed descriptions of techniques are found in 'Chemical Mutagens' Bibliography).

In the 1940s, fungi, bacteria and viruses became objects of mutation research; algae entered the field a little later. Presently, mammalian cell cultures are used in much the same way as cultures of micro-organisms, so that for the purpose of this book they will be included among micro-organisms (see books by Stent; Hayes; Fincham; Merchant and Neel; (Bibliography).

The first mutation experiments on fungi were carried out in the late 1930s (1). The mutants scored were morphological variants; this introduced an unavoidable subjective element into the scoring. This drawback was overcome when Beadle and Tatum first isolated biochemically deficient, auxotrophic mutants in *Neurospora* (2); this was soon followed by the production of auxotrophs in other micro-organisms. By scoring for reverse mutations to prototrophy, the personal error could be largely eliminated. Resistance mutations to antibiotics were first used by Demerec (3) in his studies on bacteria;

the use of resistance mutations was later extended to other micro-organisms, including mammalian cell cultures. Viruses, which lack autonomous metabolism, do not yield auxotrophs; they are also resistant to antibiotics. The most commonly used mutations in research on viruses concern the type of lesion they produce, the types of host they can affect, and their resistance to temperature.

A great advantage of mutation research on micro-organisms lies in the large numbers that can be tested. While a test for sex-linked lethals in *Drosophila* is considered large already when it is carried out on 1000 F_1 cultures, samples for mutation tests on micro-organisms may contain 10^9 or more cells or viruses. Several points must, however, be kept in mind when comparing mutation experiments on macro- and micro-organisms. The first is a statistical one (4). It is, of course, much easier to reach statistical significance with a test carried out on 10^8 cells than with one carried out on 10^3. But in mutation work, conclusions can hardly ever be drawn from a single experiment, however large its scale. Even if the difference between control and treated series, or between two differently treated series, should be statistically highly significant, this does not necessarily mean that it is due to the one controlled variable, e.g. treatment or its absence. Especially in work with micro-organisms, there are a large number of uncontrolled and often uncontrollable variables that may create differences between samples; examples will be found in the subsequent chapters. The really important variation is between replicate experiments, and the necessity for carrying out sufficient of these is not relieved by increasing the scale of the individual experiment. The second point to be remembered is that even in micro-organisms sample sizes can be very large only when 'screening tests' are used, i.e. tests that allow only mutated organisms to survive and form colonies. In techniques that do not employ screening procedures, such as tests for mutation from prototrophy to auxotrophy or from small to large plaques of bacteriophage, sample sizes range in the thousands and are, thus, not much larger than those used in tests on higher organisms. Finally, while a mutation in maize or *Drosophila* is recognized as a nuclear change by the very way in which it is detected, e.g. a 3:1 segregation in the F_2 of maize, or the absence of one type of flies in a *Drosophila* culture, mutations in micro-organisms are recognized only by phenotype. This makes it necessary to use special tests for distinguishing phenocopies and cytoplasmic changes from nuclear mutations.

II. FUNGI, ALGAE AND BACTERIA

Screening tests

Reversions from auxotrophy to prototrophy

Most of the micro-organisms that are studied in genetical laboratories can grow on a synthetic medium which contains few or no organic compounds other than a carbohydrate source of energy and, often, one or more vitamins ('minimal medium'). Auxotrophs do not grow on this medium. They form colonies only when the missing metabolite is supplied. Thus, a lysine-requirer grows only on medium supplemented with lysine, an adenine-requirer, on medium supplemented with adenine. Prototrophic cells, in which the original synthetic capacity has been restored by either true reversion or a suppressor mutation (Chapter 4) can grow on medium without the supplement. Screening for mutations from auxotrophy to prototrophy (reverse mutations in the wider sense, i.e. true revertants and suppressors) consists of plating cell samples of an auxotrophic strain on medium not containing the required metabolite, and counting the colonies that are formed. Fig. 11.1 illustrates diagrammatically a screening test for reverse biochemical mutations in *Neurospora*. Apart from technical details, the method is the same for all micro-organisms and mammalian cells.

Conidia (asexual spores) of an auxotrophic strain are washed free of traces of the supplemented medium on which the culture had been grown. They are then divided into two portions: one of these is treated, the other serves as control. With each portion, two kinds of plating are carried out: one on minimal medium for the scoring of mutations, the other on supplemented medium for the scoring of survival. On the mutation plates, only mutants will grow into colonies, and since they form a small fraction of all plated spores, plating density can be very high; usually it is between 10^7 and 10^8. On the survival plates, most untreated cells and, often, a large proportion of the treated ones will form colonies. In order not to overcrowd the plates, the samples for the survival plates have to be diluted. Colonies that appear on minimal medium are scored as mutants; this may need confirmation (p. 194). Mutation frequencies are expressed in one of two ways (usually in both): as frequencies per cells originally plated, or as frequencies per surviving cells. If viability of the controls is taken as 100%, these two values will

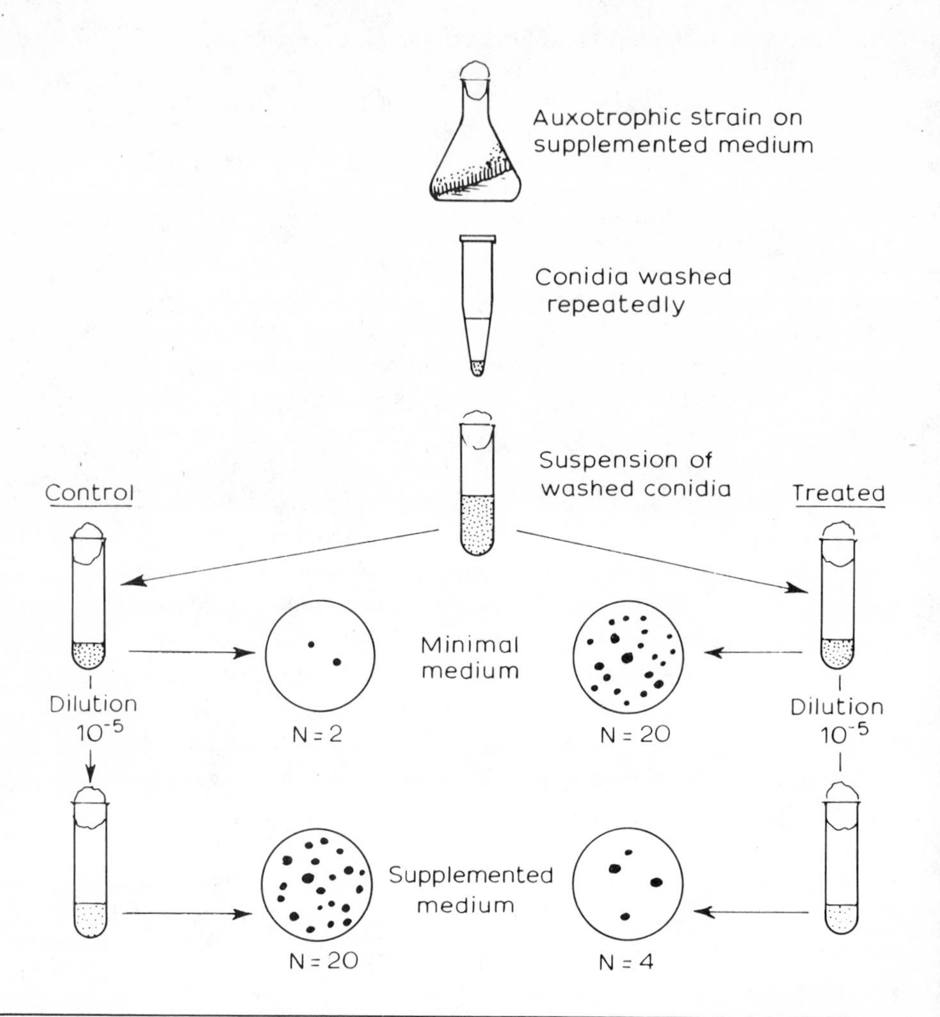

Medium		Controls	Treated
Supplemented	Survival	100%*	20%
Minimal	Number spores plated	$20 \times 10^5 = 2 \times 10^6$	2×10^6
Minimal	Number surviving spores	2×10^6	4×10^5
Minimal	Mutations/plated spores	$2/2 \times 10^6 = 1 \times 10^{-6}$	$20/2 \times 10^6 = 10 \times 10^{-6}$
Minimal	Mutations/surviving spores	1×10^{-6}	$20/4 \times 10^5 = 50 \times 10^{-6}$

* Assumed to be 100%. Actual viability, as based on haemocytometer count, will usually be less than 100%.

Fig. 11.1. Screening for reverse biochemical mutations in *Neurospora*. (Fig. 13, Auerbach, 1962, (bibl.). Courtesy author)

be identical for the controls. For the treated spores they usually diverge; the more so, the higher the killing rate of the treatment. It is customary to consider mutation frequency per plated spores as the more reliable value, because mutation frequency per survivors may be spuriously increased if the treatment favours survival of mutant spores. This is illustrated in row III of Table 11.1 for a non-mutagenic treatment which kills 99% of the non-mutant spores and none of the mutant ones. A comparison with row I shows that mutation frequency per plated spores or, expressed differently, the absolute number of mutant colonies has not changed, while mutation frequency per survivors has been raised from 10^{-5} to 10^{-3}. With this possibility in view, it is often considered that an increase in the absolute number of mutant colonies is the only clear proof for mutagenicity. This conclusion requires qualification. Even an effective mutagen may fail to increase the absolute number of mutant colonies if it acts only at highly toxic doses (row IV) or kills mutant in preference to non-mutant cells (row V). Thus, an increased mutation frequency per survivors not paralleled by an increased mutation frequency per plated cells, may be due to various causes: selection of mutant cells by a non-mutagen (row III), masking of a true mutagenic effect by toxicity (row IV), or selection against mutant cells by a mutagen (row V). Tests exist for distinguishing between these possibilities. The most important among them is the so-called reconstruction test, in which mixtures of auxotrophic and prototrophic cells are exposed to treatment in order to see whether both types of cell survive equally well.

Care must also be taken with spore density on the mutation plates. If it is too high, one of two opposite things may happen. If the cells of the original strain die quickly on minimal medium, nutrients or metabolites released by the majority may feed a small minority into spurious prototrophs. Recently, this has been shown to play a major role in mutation tests on mammalian cells (5). If, on the other hand, the original strain is 'leaky', i.e. if the majority of its cells can grow a little without their required nutrilite, they will compete with the mutant auxotrophs for resources of the minimal medium and may prevent them from growing into colonies. It is easy to guard against either contingency by using a range of dilutions on the mutation plates. Even if all these precautions have been taken, mutants may be missed because the mutated cell may require some growth before it can express its newly acquired biochemical

Table 11.1

The effects of unselective and selective killing on observed mutation frequencies in screening tests

Series	*Numbers*		*Survival %*		*Numbers surviving*		*Mut/plated*	*Mut/surviv.*
	Plated	*Mutated*	*Non-Mutants*	*Mutants*	*Total*	*Mutants*		
I Controls	10^6	10	100	100	10^5	10	$10/10^6 = 10^{-5}$	$10/10^6 = 10^{-5}$
II Treated	10^6	1000	10	10	10^5	100	$100/10^6 = 10^{-4}$	$100/10^5 = 10^{-3}$
III Treated	10^6	10	1	100	10^4	10	$10/10^6 = 10^{-5}$	$10/10^4 = 10^{-3}$
IV Treated	10^6	1000	1	1	10^4	10	$10/10^6 = 10^{-5}$	$10/10^4 = 10^{-3}$
V Treated	10^6	1000	10	1	10^5	10	$10/10^5 = 10^{-5}$	$10/10^5 = 10^{-4}$

III. Selective killing of non-mutants results in a spurious increase of mut/survivors.

IV. Strong unselective killing prevents an increase of mut/plated.

V. Selective killing of mutants masks mutagenic ability in mut/plated and reduces it in mut/surviv.

activity. This 'phenomic delay' is a very serious but often neglected source of errors in experiments with auxotrophs; mutagenic effects may be completely missed or grossly underestimated when treated auxotrophs are plated directly onto the challenging minimal medium. This can be avoided by adding to the mutation plates sufficient supplement to allow all cells a few divisions, so that also late starting prototrophs will get their chance of expression.

Additional sources of error arise from the fact that survival and mutation frequencies are not measured under identical conditions; neither are mutation frequencies at different loci. Since post-treatment conditions are well known to influence both survival and recovery of mutants, these differences may bias the results. Two kinds of data are affected by this difficulty: (1) estimates of mutation frequencies per survivors, and (2) comparisons between reverse mutations at different loci.

(1) The legend to Fig. 11.1 shows that the frequency of mutations per survivors is obtained as the ratio between the number of mutant colonies divided by that of the surviving ones times the dilution factor. It is, however, not at all sure that survival on the densely populated, unsupplemented mutation plates will always be the same as survival on the thinly populated, supplemented survival plates. If more cells should survive on the survival plates than on the mutation plates, the number of live cells tested for mutations will be overestimated and the frequency of mutations per survivors will be correspondingly underestimated. Conversely, if survival should be better on the mutation plates than on the survival plates (and this may well be the case if a delay in growth allows better recovery of damaged cells) then mutation frequencies per survivors will be overestimated. It should be emphasized that the problem cannot be solved by reconstruction experiments, for it concerns differences in survival between the *original* cells under different conditions, while reconstruction experiments test for selection for or against *mutant* cells. Tests for this source of error have rarely been carried out (6). In bacteria, an expression medium that allows a tenfold increase of auxotrophic cells can often be used to test for survival as well as for mutation. While this removes part of the difference between the scoring conditions, the difference in plating density remains and cannot be avoided. A striking example of how this can bias the result has been described by Witkin (7). She compared UV-induced mutation frequencies in two strains of *E.coli*: B10 and B10r. While

the curves relating the absolute numbers of mutations (mutations per plated cells) ran parallel for the two strains, that relating mutation frequency per survivors to dose was much steeper for B10 than for B10r. When it was found that B10 but not B10r survives UV-treatment better on crowded than on diluted seeded plates, it was assumed that in B10 mutation frequency per survivors was spuriously boosted through an underestimate of survival on the crowded mutation plates. Recently, the correctness of this assumption has been proved for a similarly reacting strain of bacteria.

(2) In tests for reverse mutations at two loci, different plating media must be used for the two kinds of reversion; the probability of obtaining independent reversions of both auxotrophs in the same cell is so low that it need not be considered. Thus, in a strain that is diauxotrophic for adenine and methionine, adenine-reversions are scored on methionine-supplemented medium, while methionine-reversions are scored on adenine-supplemented medium, and this may create different conditions for expression. A striking example of this source of error is described in Chapter 20. In many instances, it can be avoided by incubating all treated cells on medium that contains sufficient amounts of both supplements for a few divisions and then adding the full amount of one of them to half the plates, the full amount of the other to the other half.

In contrast to true reversions, supersuppressors may act simultaneously on several auxotrophic loci in the same cell. Theoretically, it should make no difference whether the screening medium for supersuppressors contains one, several, or none of the simultaneous requirements of the suppressible gene. In practice, the kind of medium used may affect mutational yield also in this case (Chapter 20).

Resistance mutations

Screening for resistance to radiation, chemicals, extremes of temperature or, in the case of bacteria, to a bacteriophage is done in essentially the same way as screening for reverse biochemical mutations. The cells are plated in the presence of the inhibitory agent, so that only mutant cells can form colonies. The possible sources of error are the same as those in experiments on reverse biochemical mutations: selection for or against mutants by the treatment; inhibition or promotion of the growth of mutant cells by a high background of non-mutant ones; influences of the plating medium. Sufficient time for expression under non-inhibitory conditions is an absolute requirement. In addition, while reverse biochemical mutations usually are sufficiently dominant to be expressed in heterokaryons or diploids, resistance mutations often are recessive and do not show their effect in diploid or multinucleate cells. In multinucleate bacteria, this 'segregational delay', while essentially different from the 'phenomic delay' of prototrophic mutations (p. 179), is taken care of in a similar way by allowing the treated cells a few divisions before challenging them with the screening conditions.

Tests that do not utilize screening techniques

Auxotrophs arise by 'forward' mutations, i.e. by mutations that inactivate or change a normal gene. Except in rare cases, they cannot be scored by screening techniques. They share this disadvantage with mutations from resistance to sensitivity, and with recessive lethal mutations. Some of the techniques for the scoring of such mutations will be briefly outlined.

Recessive lethals

As mentioned before, samples for the scoring of forward mutations are necessarily much smaller than samples for screening tests. The higher the number of loci covered by the experimental procedure, the more information can be gained from small-scale experiments. In *Neurospora,* Atwood – and others after him – have developed techniques for detecting recessive lethals, i.e. mutations (or deficiencies) that kill homokaryotic cells but can be carried indefinitely in a heterokaryon (8,9a). Essentially, these techniques are similar to those for the scoring of autosomal recessive lethals in *Drosophila*; but while the *Drosophila* worker can rely on the automatic segregation of homozygotes from intercrossed heterozygotes, the *Neurospora* worker has to enforce homokaryosity by special tricks. The scheme of Atwood's technique is shown in Fig. 11.2. Since the whole genome is being scanned for lethals, their frequency is high, even without treatment. This makes it a useful system for tests of weak mutagens or of spontaneous mutation rates.

The importance of temperature-sensitive (ts) lethals for the study of vital biological processes has been pointed out before (p. 34). Hong and Ames (9b) have described a technique by which ts lethals can be induced in selected small regions of the bacterial chromosome. The principle consists of producing mutations in a transducing phage

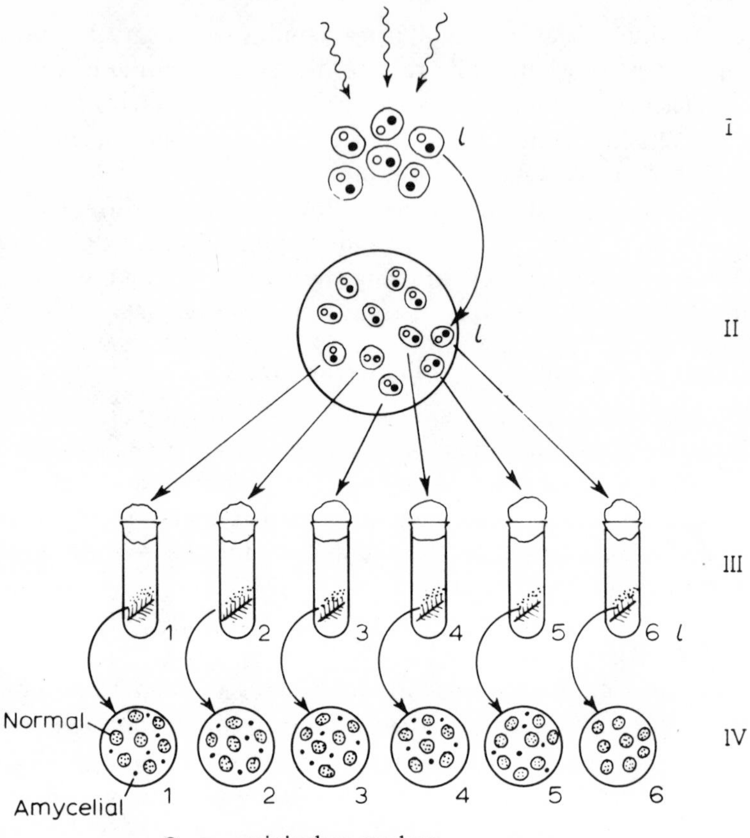

○ = arginineless nucleus
● = methionineless amycelial nucleus
l = recessive lethal

I. Heterokaryotic conidia are treated. In one, a lethal mutation has occurred in the methionineless nucleus.

II. On minimal medium, each conidium grows into a heterokaryotic colony; one carries a lethal.

III. Individual colonies have been grown into cultures on minimal medium.

IV. On methionine medium, conidia from these cultures form two types of colony; normal heterokaryotic, and amycelial methionineless. Absence of the latter from plate 6 denotes presence of a lethal.

Fig. 11.2. Atwood's test for recessive lethals, in *Neurospora*. (Fig. 17, Auerbach, 1962, (bibl.). Courtesy author)

which carries a normal piece of DNA into a selected auxotroph. Lethals scored in the phage are closely linked to the selected marker.

Auxotrophs at unselected loci

These are scored by plating treated (and control) cells on complete medium and transferring the colonies, either singly or by replica plating (p.188), to minimal medium. Those that grow on the former but not on the latter are presumed to be auxotrophs; they can be confirmed as such and classified in regard to the required nutrilite by tests for growth on a range of variously supplemented media. Most auxotrophic mutations are recessive and do not manifest themselves in diploids or heterokaryotes. This has to be considered in choosing the type of cell for treatment (preferably haploid and uninucleate), and the technique for testing (time for segregational delay in bacteria; opportunity for meiotic segregation in *Neurospora,* etc.). The number of loci able to yield auxotrophs is large enough to yield mutation frequencies of up to several per cent. If quantitative accuracy is not essential, various tricks can be used for enriching mutant cells in the tested sample. Common to all of them is preincubation of treated cells under conditions in which auxotrophs cannot grow, followed or accompanied by removal or killing of the growing non-mutant cells. In filamentous fungi, the simplest way of removing growing spores is to pass the mixture of hyphae and non-growing auxotrophic spores through a suitable filter which removes the former. Other techniques create conditions under which growing cells commit suicide, while non-growing ones survive: the best known example is the penicillin method for the enrichment of auxotrophs among mutagenically treated bacteria. It rests on the observation that penicillin kills only growing cells.

All these tricks introduce sources of inaccuracy into the quantitative estimates of mutation frequencies. Qualitatively, too, they may bias the spectrum of auxotrophs. If, for example, a certain kind of auxotroph tends to be more 'leaky' than another, it is more likely to escape detection. This accounts for the striking preponderance of nonsense mutations among penicillin-screened bacterial auxotrophs. Actually, there is no technique that yields a completely unbiased spectrum of auxotrophs. Even the least biased one, that of transferring unselected colonies from complete to minimal medium, has its pitfalls, due to biochemical interactions between certain types of auxotroph and constituents of the medium. Thus, fewer lysineless mutants are recovered from medium containing arginine than from arginine-free medium, and vice versa, probably because of competition for a permease. In various fungi, the growth of guanine auxotrophs is inhibited by adenine in the plating medium. In some cases, such an inhibition may be magnified by the mutagen used, and this leads to a spurious appearance of mutagen specificity (Chapter 20). The most reliable, but very laborious, method of determining the spectrum of auxotrophs would be to plate samples of treated cells separately on media that are supplemented singly with every one of the nutrilites that may be required by an auxotrophic mutant.

Auxotrophs at selected loci

The scoring of auxotrophs at selected loci provides mutants for several important problems: analysis of the structure of an individual locus, and analysis of the types of change produced in it by mutagenic treatment or by spontaneous mutations. The frequencies of this kind of mutation are low in any particular experiment, but may sometimes be increased by special procedures. Thus, certain auxotrophs are resistant to conditions that kill prototrophs; they can then be obtained from large samples of cells by means of a screening test. In *Neurospora,* lysine- and arginine- auxotrophs are resistant to parafluorophenylalanine and can be screened on plates containing this amino acid analogue (10a). Where screening tests cannot be applied, the detection of auxotrophs of a particular type is sometimes facilitated by a coincident colour change in the mutant as compared with the original colony. In *Aspergillus,* Alderson and Scazzocchio (10b) have developed a system for scoring forward mutations in at least eight loci; it is based on the observation that certain mutants retain the ability to form green conidia on a medium on which wild-type conidia turn yellow. In yeast and several other fungi, blockage of one of the last steps in adenine-synthesis results in the accumulation of purple pigment, and this can be used for the detection of auxotrophs at these loci (11) ('white-to-red' technique). Mutations to adenine-dependence at loci that occur earlier in the pathway prevent the formation of purple pigment and can be recognized as white colonies in an otherwise purple-adenineless strain (12,13) ('red-to-white' technique). In bacteria, a colour technique for scoring loss of the ability to ferment lactose makes use of an indicator medium (EMB-agar), on which fermenting colonies stain purple, while non-fermenters are white (14). Colour markers have proved of special value in scoring mosaic mutations (Chapter 19).

Mutations in growing cultures

The fluctuation test. Chemostat and turbidostat

When treatment is applied to suspensions of cells, mutation frequency can be estimated directly from the number of mutants obtained. This is not so for mutations that arise during growth of a culture. Here, the mutant cells form clones the size of which is determined by the time when the mutation occurred and, possibly, by competition between mutant and non-mutant cells. Strictly speaking, this applies also to the controls in experiments on cell suspensions; for a chance mutation early during growth of the parent culture will result in a high 'background' mutation frequency. This difficulty is, however, easily overcome by using the same suspension for treatment and controls and by subtracting the number of pre-existing mutants from the number found on the mutation plates.

For studies aimed at the analysis of mutagenesis in growing cultures, clone formation becomes a crucial problem. One frequently used method for dealing with it has been developed by Luria and Delbrück (15). It is applicable to all organisms which divide by fission, but is only of limited use for organisms with filamentous growth. It consists of scoring the number of mutants present in a series of parallel cultures, which have been started simultaneously with small inocula from the same parental culture. If, during growth, mutations arise at a constant low rate, their frequencies in the individual cultures will form a Poisson distribution. The average mutation frequency per culture can therefore be calculated from the number of cultures that do *not* contain any mutants, for this number forms the first term of the Poisson distribution $p_0 = e^{-m}$, where m is the mean. Knowing the initial and final number of cells per culture, one can then calculate the mutation rate, which for bacteria is usually expressed as the probability of mutation per bacterium per generation. Alternatively, mutation rate can be calculated from the average number of mutants in the parallel cultures; but this procedure requires some additional assumptions, and its accuracy depends much more on selection during growth.

Obviously, the most direct way of calculating mutation rate in a growing culture of bacteria or similarly growing unicellular organisms would be to determine both the number of cells and the number of mutants at given intervals of time. This, however, is not possible when the cells are grown in the usual containers; for here the limited supply of nutrients leads to rapid changes of growth rate with time. An initial lag phase is followed by a period of exponential growth, during which population size doubles at regular intervals; as the food supply diminishes, growth slows down and finally stops altogether. If growth could be kept permanently at the exponential stage, mutation rate could be determined directly from the number of mutants in samples taken at different times. The chemostat was designed for achieving this (16). It is an apparatus in which growth is kept exponential by continually adding limited amounts of an essential nutrilite, e.g. glucose or phosphate, or an amino acid required by a strain auxotrophic for it. Concomitantly, the excess bacteria are swept out, so that population size remains constant. A great advantage of the chemostat is the possibility of altering generation time through alterations in the amount of limiting nutrilite; this makes it feasible to measure mutation rates in units of either chronological or physiological time. It is, however, very difficult to avoid the building-up of selection pressures in the chemostat, and Munson and Jeffrey (17) have pointed to the advantages of the turbidostat (18) over the chemostat for mutation studies.

Adaptation versus mutation

The fluctuation test can also be used to decide a question which, because of its Lamarckian implications, was much discussed during the time of the Lysenko-controversy, but which has retained relevance even now. It is well known that bacteria can adapt to adverse environmental circumstances such as exposure to penicillin or the replacement of glucose in the medium by lactose, and that such adaptations, once established, may persist indefinitely. The question was asked whether certain screening agents, e.g. antibiotics, do not *produce* resistance mutations rather than only *detect* them. The question cannot be answered once and for all; in every doubtful case, it will be necessary to distinguish adaptation from mutations during previous growth with or without mutagenic treatment. One of the means for doing this is the Luria-Delbrück test (15). The relevant statistic for the present purpose is the variance between the number of mutants on plates inoculated with the contents of the parallel cultures. Note that we now are considering *not the number of mutations, but the number of mutants* to which they give rise. For mutations that arise on the plates, these two numbers will be the same. The number of mutants, like those of the mutations from which they arose, will form a Poisson distribution for which the variance is small and equal to the mean. This is not true for mutants that are found in the parallel growing cultures; for here early occurring mutations will yield large clones of identical mutants, while late-occurring mutations may yield only one or two mutants. As a result, the variance of the number of mutants in the individual cultures will be much larger than the mean. This criterion can be used to decide whether, in a particular type of experiment, mutants arise through adaptation on the screening plates or have arisen previously during growth of the parallel cultures. Table 11.2 shows data on bacterial mutations to phage resistance. It will be seen that the variance between the parallel cultures is much larger than the mean; the conclusion was drawn that the mutations had taken place prior to plating. As a control, the table shows the number of mutants in replicate samples taken from the same bulk culture; in this case, variation between plates is due only to sampling error, the numbers of mutants on the plates follows a Poisson distribution, and the variance is equal to the mean.

Another technique for attacking the same problem, the Newcombe respreading test (19), cannot be discussed here. The most elegant and decisive technique is Lederberg's replica plating test (20). In its essentials, it consists of selection for resistance among cells that themselves had not come into contact with the screening agent and cannot, therefore, have been induced to mutate under its influence. Fig. 11.3 shows the essentials of the technique as applied to phage resistance of bacteria. Large numbers of bacteria from a sensitive strain are plated on phage-free medium and grown into microcolonies, which soon cover the 'master plate'. By means of a sterile velvet pad, replicas of this plate are transferred to two or more agar plates impregnated with phage. Colonies that develop on these plates consist of resistant bacteria. By marking the position of the pad in relation to the plates, corresponding regions on master- and replica plates can be identified. A small region of the master plate surrounding one of the resistant colonies is cut out and used for seeding the first transfer plate, which likewise is free of phage. When replicas of this plate are made to plates containing phage, the proportion of resistant colonies has increased. In general, several

Table 11.2

The fluctuation test of the spontaneous origin of T1 phage resistant *E.coli* mutants

Individual cultures		*Samples from bulk culture*	
Culture no.	*Resistant bacteria*	*Sample no.*	*Resistant bacteria*
1	1	1	14
2	0	2	15
3	3	3	13
4	0	4	21
5	0	5	15
6	5	6	14
7	0	7	26
8	5	8	16
9	0	9	20
10	6	10	13
11	107		
12	0		
13	0		
14	0		
15	1		
16	0		
17	0		
18	64		
19	0		
20	35		
Mean	11.3		16.7
Variance	694		15

Combined from Tables 1 and 2 in Luria and Delbrück (15) Table 6.1 Stent 1971 (bibl.). Courtesy Freeman & Co.

cycles of replica plating are sufficient to establish a strain of bacteria that is wholly resistant to the drug. Since during the whole process of selection the strain has never been in contact with the drug, such a result is clear proof that drug resistance in this case was due to mutation and not to adaptation. The replica plating technique can be modified for use with any organism of colonial growth. It has many other uses than that discussed just now; one of them – the selection of auxotrophic mutants – has been mentioned before (p. 183).

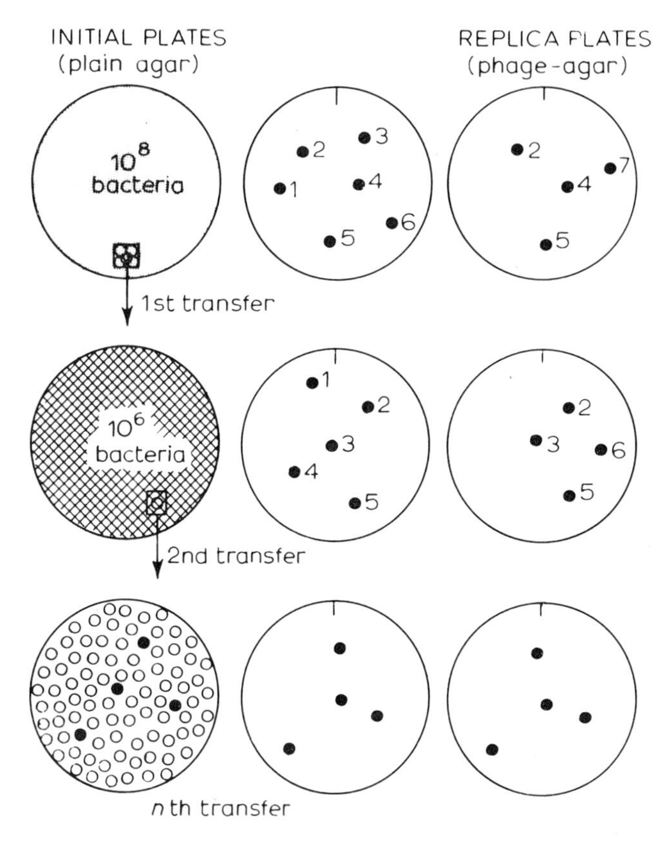

Fig. 11.3. Illustrating how the technique of replica plating can be used for the isolation of mutant clones by indirect selection. (Fig. 52, Hayes, 1968, (bibl.). Courtesy Blackwell Scientif. Public.)

III. BACTERIOPHAGES AND OTHER VIRUSES

Screening tests for reverse mutations

The host-parasite relationship between bacteria and bacteriophages is highly specific. A strain of phages can parasitize only a limited number of bacterial strains, and a bacterial strain can be parasitized by only a limited number of phage strains. Mutations that enable a phage to grow in a previously immune bacterial strain are called host-range mutations. When phage is added to a plate that is coated with sensitive bacteria, each particle, by lysing a bacterium and liberating more infectious particles, produces a clear 'plaque' on the opaque background of bacterial growth. On plates that are coated with bacteria outside the normal host range of a given phage, only mutants form plaques, and their number is used for calculating the frequencies of host range mutations.

When the host-range of a phage strain has been narrowed by a previous mutation, screening tests can be employed for the detection of reverse mutations. This method has been employed with great success for the mutational analysis of a special region of the T4 bacteriophage chromosome. Wild-type T4 forms plaques on *Escherichia coli* cells of strains B and K(λ); the latter is a substrain of K12, carrying phage λ. A certain class of mutants, rII (from 'rapid lysis'), is characterized by changed plaque morphology on B and by loss of the ability to grow on K12 (λ) or on other bacterial strains that carry λ. Reverse mutations are screened for by plating rII particles on K12 (λ), while the number of viable particles is assayed by plating on B. Similar techniques can be used for other bacteriophages and viruses, e.g. the poliomelitis virus (21).

Tests for forward mutations

Changes in plaque morphology of bacteriophage can be scored in the usual way. Mutations of phage T4 to rII are particularly useful because of the ease with which revertants can be scored. This possibility makes the rII region an ideal system for the study of forward and reverse mutations at different sites within one locus. Special importance attaches to temperature-sensitive (ts) mutants. Like lethals in eukaryotes, these can occur anywhere in the genome and are therefore much more frequent than mutations affecting only one or a few loci. Large numbers of them have been detected in bacteriophage by their inability to form plaques at high temperature (42°), while able to do so at low temperature (25°). Their significance lies in the fact that they make available for analysis mutants of genes coding for such essential processes as DNA-synthesis.

In experiments with tobacco-mosaic virus, differences in visible symptoms on infected plants have been used as indicators of mutation. In particular, the difference between necrotic and non-necrotic lesions on a given host strain can be used in the same way as differences between plaque-types formed by wild-type and mutant bacteriophages (22). The fact that the RNA of the virus can, by itself, infect plant cells, carrying with it the whole of the genetic information, has made it possible to apply mutagenic treatment successfully to naked RNA. This has opened up completely new vistas for studying mutation at the primary level. RNA, however, is exceptional as a carrier of genetic information. Moreover, methods for the genetical analysis of presumed mutations in plant viruses are not yet available. It is, therefore, of the utmost importance that mutations can now be studied in transforming principle, which consists of naked bacterial DNA and forms part of a genetical system that, in some bacterial species, is extremely well analysed.

IV. TRANSFORMING PRINCIPLE

In principle, all types of mutation that can be scored in a bacterium can be scored also in transforming DNA after its incorporation into the host chromosome. The main difficulty lies in obtaining estimates of 'survival', i.e. of the number of effectively incorporated genes from treated DNA. This difficulty is overcome by the application of screening techniques that allow only transformed bacteria to survive. Thus, when mutagenically treated DNA from streptomycin-resistant cells is used for the transformation of streptomycin-sensitive ones, only cells that can grow on streptomycin are transformants, and mutation frequencies per transformants can be calculated in the same way as mutation frequencies per survivors are calculated in cells or viruses (23). A particularly sensitive technique consists of scoring mutations on the same piece of DNA that carries the selective marker (24a). Fig. 11.4 shows the operon responsible for tryptophan synthesis in *Bacillus subtilis,* and some closely linked genes. When the host requires tryptophan, while the donor is prototrophic for it, only transformants can grow on tryptophan. Mutations in the other genes of the piece of transforming DNA can be scored in the usual way. Often, scoring is restricted to the fluorescent colonies that arise from mutation in a neighbouring gene. A similar technique has been devised for scoring mutations induced *in vitro* in the DNA of bacteriophage T4 (24b). Mutation experiments with transforming principle are not only the best means for analysing the primary mutagenic changes in DNA. Over and above this, they provide opportunities – so far hardly utilized or even realized – for studying in isolation the roles of these primary changes and of the cellular steps in mutagenesis. In particular, they should make it possible to separate the effects of a mutagenic treatment on DNA and on the remaining part of the mutation process.

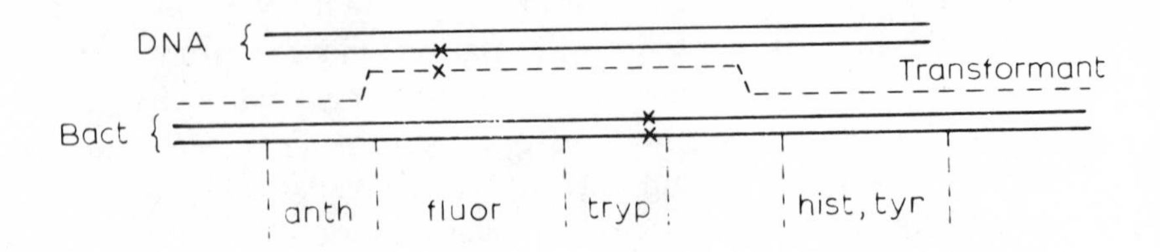

Fig. 11.4. Linked mutations in the tryptophan region of *Bacillus subtilis*. x = mutation that has converted a functionally active into a functionally inactive DNA region. bact = tryptophan dependent bacterium. DNA = prototroph transforming DNA, treated by a mutagen. Mutants in the region: *anth* grows on anthranilate, *fluor* accumulates a blue fluorescent compound when grown on indole, *tryp* grows on tryptophan, *hist* requires histidine, *tyr* requires tyrosine. The dashed lines show which information a transformant has to pick up in order to grow without tryptophan and produce a fluorescent mutant colony on limiting indole. (Fig. 1, Freese (24)). Courtesy Nat. Acad. Sciences, U.S.A.)

V. MAMMALIAN CELL CULTURES

As in micro-organisms, resistance mutations and reverse mutations from auxotrophy to prototrophy can be detected by screening techniques, while forward mutations to auxotrophy become amenable to screening only by means of special tricks. A neat trick, working on the principle of prototrophic suicide (p.183), has been developed by Kao and Puck (25). It is based on the fact that growing cells easily incorporate the thymidine analogue 5-bromodeoxyuridine (5-BUdR) into their DNA. We shall see in Chapter 17 that such cells have become mutation-prone. In the present context, this is irrelevant; what matters is that they also have become very sensitive to killing by visible or ultraviolet light. Fig. 11.5 shows how this property can be used to obtain a population of auxotrophic colonies with very low admixtures (about 10^{-4}) of prototrophic ones.

Among resistance mutations, azaguanine-resistance is much used for testing the effect of mutagens or potential mutagens (26). Its mechanism is well understood, and it has the added advantage that forward and reverse mutations can both be detected by screening tests. Moreover, the relevant gene is sex-linked, and this gets over the difficulty of detecting recessive mutations in diploid cells. The gene that is being tested for mutations codes for the enzyme HGPRT (hypoxanthine-guanine-phosphoribosyl-transferase). Active HGPRT is required for the utilization of exogenously offered purine

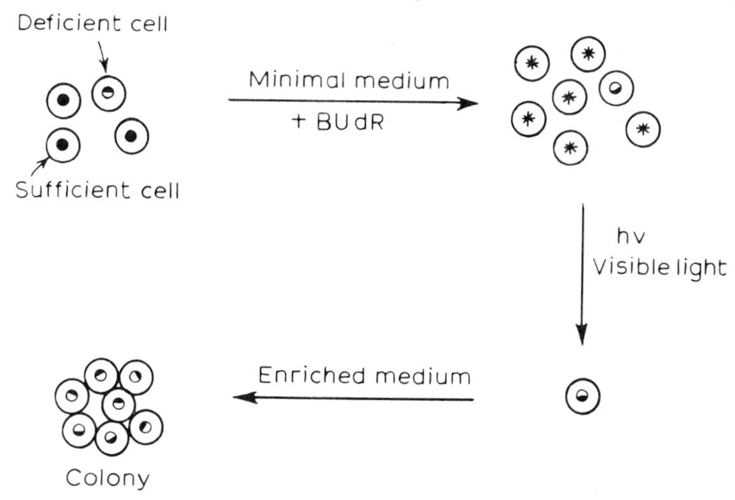

Fig. 11.5. Schematic representation of the BUdR-visible light technique for isolation of nutritionally deficient mutant clones of mammalian cells. It is assumed that the population contains a few auxotrophs that have originated through spontaneous mutation or have been produced by a mutagen. The mixed population is exposed to BUdR in a deficient medium in which only prototrophs can grow and can incorporate BUdR; these are killed by subsequent exposure to a standard fluorescent lamp. On transfer to a medium lacking BUdR and enriched with various nutrilites, only the auxotrophs grow into colonies. The very rare surviving prototrophs can be identified by testing on deficient medium. (Fig. 1, from Kao (25). Courtesy Nat. Acad. Sciences, U.S.A.)

and pyrimidine bases in DNA synthesis. If the gene is inactive, these bases are no longer utilized. Neither is there utilization of the toxic guanine analogue 8-azaguanine (8-AG), and cells that have lost HGPRT activity survive on medium with 8-AG, on which normal cells are killed. They also survive on medium without 8-AG, since they are capable of *de novo* synthesis of the necessary purine – and pyrimidine – bases. If, however, this endogenous pathway is blocked by the folic acid analogue aminopterin, the HGPRT$^-$ cells die of starvation, unless reverse mutation to HGPRT$^+$ re-opens the exogenous pathway. Thus, forward mutations can be screened on medium that is supplemented with 8-AG, while reverse mutations can be screened on medium that lacks 8-AG but is supplemented with aminopterin. Another system in which both forward and reverse mutations have been screened by specific resistance properties are mutations at the thymidine kinase locus in mouse lymphoma cells (28).

VI. VERIFICATION AND ANALYSIS OF MUTATIONS

By whatever method presumed mutants have been obtained, it is necessary to confirm that they are, indeed, nuclear mutants. Herein lies one of the shortcomings of mutation research on micro-organisms and cell cultures in contrast to, e.g. *Drosophila* or maize. In theory, every presumed mutant ought to be submitted to genetical tests; in practice, such tests are carried out on random samples. Once it has been established that a certain type of variant produced by a given type of treatment is due to nuclear mutation, it may no longer be necessary to test similar variants produced by the same treatment. But for every new type of experiment, tests should again be carried out. Phenotypically similar variants may arise as nuclear mutants under one set of conditions and as cytoplasmic changes under another; or as suppressors in one case, as true revertants in another. The problem is particularly difficult for presumed mutants in somatic cell cultures, for which no genetic tests are as yet available. Indeed, doubts have been raised as to the nature of some of the presumed mutants because mutation frequencies did not show the expected dependence on the degree of ploidy (29). In the thymidine kinase system (28), however, enzyme activities and mutation frequencies agreed with expectations for presumed homozygotes and heterozygotes at the relevant locus. Among azaguanine-resistant mutants of a mouse cell line, there were two types with altered enzyme properties (kinetic constants in one, temperature tolerance in the other) (30), while among those of a Chinese hamster line there were several that produced CRM (p. 36) (31); these variants are likely to be due to mis-sense mutations. In micro-organisms, genetic methods are available for establishing the nuclear nature of presumed mutants, for placing bona fide mutants on the chromosome map, and for testing new mutants for allelism with already known ones. These methods are described in textbooks on the particular group of micro-organisms concerned (see bibliography).

Methods for analysing new mutants *at the molecular level* are based in large part on the use of chemical mutagens and will be discussed more fully in this context. Here I shall only give a brief rationale of the most important tests. The much used 'reversion-analysis' is based on the finding that different chemicals produce different lesions in DNA. The mutants to be analysed are subjected to a battery of mutagens, some known to produce frameshifts, others known to produce base substitutions, and reverse mutations are scored by the usual screening techniques. Since frameshifts are reversed only by frameshifts, and base substitutions only by base substitutions (32), mutants can be classified into frameshifts and base substitutions according to the type of mutagen by which they can be reversed. Within the base substitution group, further subdivision is possible. Some chemicals act preferentially on GC, others, on AT. Mutants that are reversed mainly by the first kind are supposed to carry a GC pair at the mutant site; mutants responding mainly to the latter kind of mutagen, an AT pair. An obvious source of error is that suppressors may be mistaken for true reversions. For extracistronic suppressors, this can be ruled out by genetic analysis; for intracistronic suppressors, this is not possible except when normal and mutant gene products can be sequenced (Chapter 4). Another weakness of the method lies in the fact that few mutagens are sufficiently specific to produce only one type of effect; we shall come back to this in Chapter 17. Nevertheless, reversion analysis has given much useful information, especially when applied to mutants obtained *in vitro*, e.g. by treatment of transforming DNA or viruses outside the cell. When genes are treated inside cells, the primary effects of the mutagens on DNA may be overlaid and distorted by cellular processes acting before or after the reaction between DNA and mutagen.

The safest means for inferring the molecular nature of a mutation would be to sequence the DNA of the original strain and of the mutant one. This cannot yet be done; but a good approximation to this goal is possible in cases where the amino acids in the gene product can be sequenced, as in the lysozyme of phage T4 (33), the tryptophan-synthetase of *E.coli* (34) and the *iso*-1 cytochrome-c of *Saccharomyces cerevisiae* (35). The nature of frameshift mutations can be analysed in the polypeptide formed by double mutants in which the reading frame that had been shifted by a first frameshift has been restored by a second; we have discussed an example of this in Chapter 4. In mis-sense mutations that form CRM (p. 36) the base change can often be inferred from a comparison between the polypeptide chain in normal and CRM protein. Nonsense mutations do not usually produce CRM, but they have the advantage that their codon is unambiguously known so that the mutational change can be inferred from the amino acid that, in reverted or suppressed mutants, is inserted at the site of the chain termination. For true revertants, this change must have taken place in the nonsense codon itself

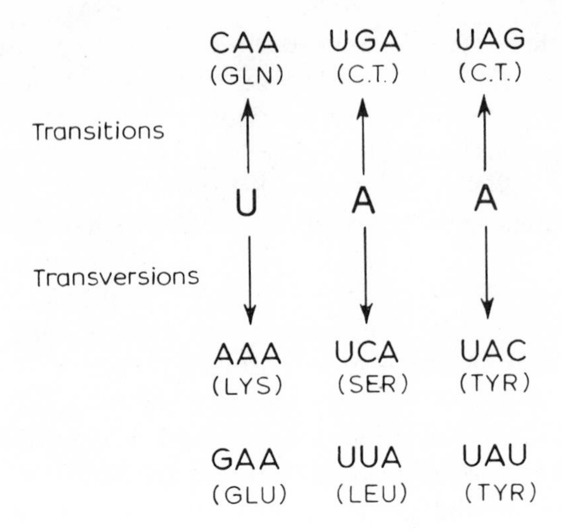

Fig. 11.6. The m-RNA conversions corresponding to the possible transitions (upper row) and transversions (lower row) which could occur at each base in the UAA codon. C.T. = chain terminating (nonsense) codon. (Fig. 1, Ripley and Drake (37). Courtesy Springer Verlag)

(e.g. from UAA to GAA); for revertants by means of supersuppressors, the base change often – although not always – has occurred in the anticodon of the t-RNA that inserts the mutant amino acid (Chapter 4). Clear results can be obtained by interconversion of amber mutants, which respond to amber suppressors and are recalcitrant to ochre suppressors, into ochre mutants, which show the opposite reactions to the two types of suppressor (36). This mutation, from UAG to UAA, unambiguously can be classified as a transition from GC to AT, while the opposite conversion from ochre into amber mutants requires the opposite transition from AT into GC. The ochre codon has also been used to test for transversions among rII mutants of phage T4 (37); Fig. 11.6 shows the scheme employed. Its usefulness depends on the possibility of distinguishing phenotypically between the true revertants (CAA) and the false revertants arising from transversions.

An indirect method for using nonsense mutations in the identification of base changes in *E.coli* DNA has been developed by Person and his collaborators (38). They started from the finding that the efficiency with which a given supersuppressor acts on a given nonsense mutation depends in part on the amino acid it inserts (Chapter 4). Bacterial suppressors with known amino acid insertions produce characteristic 'suppression patterns' when acting on an array of nonsense mutations in phage T4. A newly arisen suppressor that produces the same pattern as one of the standard ones is presumed to insert the same amino acid; this often makes it possible to infer the change in the anticodon of the mutant t-RNA. Thus, a particular suppressor is known to insert glutamine (CAG) instead of amber (UAG) by means of an anticodon change from GC to AT. A new suppressor mutant that produces the same suppression pattern is classified as due to a transition from GC to AT.

In *Neurospora,* de Serres and his group (39,40) have used complementation analysis in heterokaryons for gaining insight into the molecular nature of mutations at the purple adenine 3B locus. With the aid of a large number of already existing mutant strains, they divided the complementation map of the locus into 17 units or 'complons'. New mutants were tested for the complementation pattern which they produced in this map. In dimeric or polymeric proteins, complementation occurs at the polypeptide level by the mutual covering-up of localized errors in the amino acid sequences (Chapter 3). This leads to definite expectations for the type of complementation patterns yielded by mis-sense mutations on the one hand, nonsense mutations and frameshifts on the other. The former have only one incorrect amino acid in their polypeptide chain; they will be complemented by any mutant that carries the correct sequence in this region. Thus, all strains complementing a mis-sense mutation will have to complement only one particular part of the map; but apart from this, no pattern is expected. Nonsense mutations produce polypeptide chains that are broken off at the site of the chain-terminating codon. They can be complemented only by polypeptides that carry the correct sequence of amino acids from a little in front of the nonsense codon to the end of the gene. The complementation pattern will be 'polarized' (not to be confused with the 'polarity' that refers to effects of mutations beyond the confines of one cistron, (p. 47). Frameshifts, by giving rise to nonsense codons, may also yield polarized complementation patterns; otherwise, they are likely to be non-complementing. So, presumably, are nonsense mutations near the beginning of the gene. Fig. 11.7 shows an example of the kind of complementation analysis carried out on chemically induced adenine-3B mutants. The non-polarized complementing mutants are attributed to mis-sense base substitutions, the polarized complementing ones

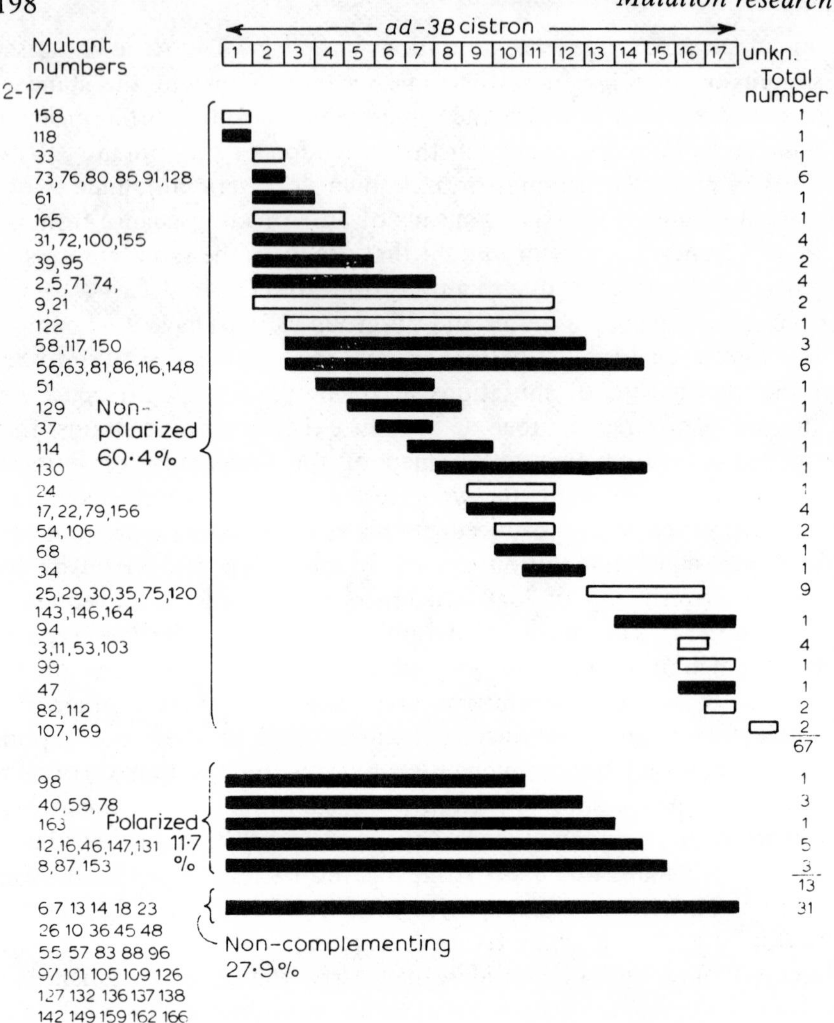

Fig. 11.7. Complementation pattern of a random sample of nitrous acid induced *ad*-3B mutants in *Neurospora*. Solid bars: non-leaky mutants, open bars: leaky mutants. (Fig. 3, de Serres (41). Courtesy Elsevier Scientific Publ. Cy.)

and the non-complementing ones, to nonsense base substitutions or frameshifts. This scheme, together with the possibility of detecting multi-gene deletions in the *ad*-3 region of *Neurospora* has made this region into one of the mutationally best analysed ones, at least as regards the broader molecular classification. Attempts to establish the mutant base pair in mis-sense mutations have been made by reversion analysis; for reasons mentioned above and more fully discussed in Chapter 17, these conclusions always contain an element of doubt.

References

1. Hollaender, A. and Emmons, C.W. (1941), 'Wavelength dependence of mutation production in the ultraviolet with special emphasis on fungi', *Cold Spring Harbor Symp. Quant. Biol.*, **9**, 179-186.
2. Beadle, G.W. and Tatum, E.L. (1945), '*Neurospora*. II. Methods of producing and detecting mutations concerned with nutritional requirements', *Am. J. Bot.* **32**, 678.
3. Demerec, M. (1951), 'Studies of the streptomycin-resistance system of mutations in *E.coli*', *Genetics*, **36**, 585-597.
4. Auerbach, C. (1970), 'Remark on the 'Tables for determining the statistical significance of mutation frequencies', *Mutation Res.*, **10**, 256.
5. Bürk, R.R., Pitts, J.D. and Subak-Sharpe, J.H. (1968), 'Exchange between hamster cells in culture', *Exp. Cell Res.*, **53**, 297-301.
6. Auerbach, C. and Ramsay, D. (1971), 'The problem of viability estimates in tests for reverse mutations', *Mutation Res.*, **11**, 353-360.
7. Witkin, E.M. (1967), 'Mutation-proof and mutation-prone modes of survival in derivatives of *Escherichia coli B* differing in sensitivity to ultraviolet light', *Brookhaven Symposia in Biology*, **20**, 17-55.
8. Atwood, K.C., Mukai, F. and Pittenger, T.H. (1958), ' "Punch tube" and "squirting" methods for *Neurospora*', *Microbial Genetics Bull.*, **16**, 34-35.

9a. De Serres, F.J. (1968), 'Genetic analysis of the extent and type of functional inactivation in irreparable recessive lethal mutations in the *ad*-3 region of *Neurospora crassa*', *Genetics*, **58**, 69-77.

9b. Hong, J.S. and Ames, B.N. (1971), 'Localized mutagenesis of any specific small region of the bacterial chromosome', *Proc. Nat. Acad. Sci. U.S.A.* **68**, 3158-3162.

10a. Kinsey, J.A. and Stadler, D.R. (1969), 'Interaction between analogue resistance and amino acid auxotrophy in *Neurospora*', *J. Bacter.*, **97**, 1114-1117.

10b. Alderson, T. and Scazzocchio, C. (1967), 'A system for the study of interlocus specificity for both forward and reverse mutation in at least eight gene loci in *Aspergillus nidulans'*, *Mutation Res.* **4**, 567-577.

11. De Serres, F.J. and Kφlmark, H.G. (1958), 'A direct method for determination of forward mutation rates in *Neurospora crassa', Nature,* **182**,1249-1250.

12. Roman, H. (1956), 'A system selective for mutations affecting the synthesis of adenine in yeast', *C.R. Labor. Carlsberg,* **26**, 299-314.

13. Leupold, L.S. (1955), 'Versuche zur Genetischen Klassifizierung Adeninn-Abhängiger Mutanten von *Schizosaccharomyces pombe', Arch. Julius Klaus-Stift. Vererbungsforsch. Sozialanthropol. Rassenhyg.,* **30**, 506-516.

14. Lederberg, J. (1950), 'Isolation and characterisation of biochemical mutants of bacteria', *Methods in Medical Research,* **3**, 5-22.

15. Luria, S.E. and Delbrück, M. (1943), 'Mutations of bacteria from virus sensitivity to virus resistance', *Genetics,* **28**, 491-511.

16. Novick, A. and Szilard, L. (1950), 'Description of the Chemostat', *Science,* **112**, 715-716.

17. Munson, R.J. and Jeffrey, A. (1964), 'Reversion rate in continuous cultures of an *Escherichia coli* auxotroph exposed to γ rays', *J. Gen. Microbiol.,* **35**, 191-203.

18. Bryson, V. (1952), 'Microbial selection. Part II: The Turbidostatic Selector – a device for automatic isolation of bacterial variants', *Science,* **116**,48-51.

19. Newcombe, H.B. (1949), 'Origin of bacterial variants', *Nature,* **164**, 150.

20. Lederberg, J. and Lederberg, E.M. (1952), 'Replica plating and indirect selection of bacterial mutants', *J. Bacter.,* **63**, 399.

21. Vogt, M., Dulbecco, R. and Wenner, H.A. (1957), 'Mutants of poliomyelitis viruses with reduced efficiency of plating in acid medium and reduced neuropathogenicity', *Virology*, **4**, 141-155.

22. Mundry, K.W. and Gierer, A. (1958), 'Die Erzeugung von Mutationen des Tabakmosaikvirus durch chemische Behandlung seiner Nukleinsäure *in vitro*', *Zeitsch. Vererb. Lehre,* **89**, 614-630.

23. Litman, R. and Ephrussi-Taylor, H. (1959), 'Inactivation et mutation des facteurs génétiques de l'acide desoxyribonucléique du pneumocoque par l'ultraviolet et par l'acide nitreux', *C.R. Acad. Sc.*, **249**,838-840.
24a. Freese, E. and Strack, H.B. (1962), 'Induction of mutation in transforming DNA by hydroxylamine', *Proc. Nat. Acad. Sci. U.S.A.*, **48**, 1796-1803
24b. Baltz, R.H. and Drake, J.W. (1972), 'Bacteriophage T4 transformation: an assay for mutations induced *in vitro*', *Virology*, **49**, 462-474.
25. Kao,F.T.and Puck,T.T.(1968),'Genetics of somatic mammalian cells. VII Induction and isolation of nutritional mutants in Chinese hamster cells', *Proc. Nat. Acad. Sci., U.S.A.* **60**, 1275-1281.
26. Szybalski, W., Ragni, G. and Cohn, N.J. (1964), 'Mutagenic response of human somatic cell lines', *Symp. Int. Soc. Cell Biol.*, **3**, 209-221.
27. Albertini, R.J. and de Mars, R. (1970), 'Diploid azaguanine-resistant mutants of cultured human fibroblasts', *Science*, **169**, 482-485.
28. Clive, D., Flamm, W.G., Macheska, M.R. and Bernheim, N.J. (1972), 'A mutational assay system using the thymidine kinase locus in mouse lymphoma cells', *Mutation Res.*, **16**, 77-87.
29. Harris, M. (1971), 'Mutation rates in cells at different ploidy levels', *J. Cell. Physiol.*, **78**, 177-184.
30. Sharp, J.D., Capecchi, N.E. and Capecchi, M.R. (1973), 'Altered enzymes in drug-resistant variants of mammalian tissue culture cells', *Proc. Nat. Acad. Sci. U.S.A.* **70**, 3145-3149.
31. Beaudet, A.L., Roufa, D.J. and Caskey, C.T. (1973), 'Mutation affecting the structure of hypoxanthine: guanine phosphoribosyltransferase in cultured Chinese hamster cells', *Proc. Nat. Acad. Sci., U.S.A.*, **70**, 320-324.
32. Orgel, A. and Brenner, S. (1961), 'Mutagenesis of bacteriophage T4 by acridines', *J. Mol. Biol.*, **3**, 762-768.
33. Streisinger, G., Okada, Y., Emrich, J., Newton, J., Tsugita, A., Terzaghi, E. and Inouye, M. (1966), 'Frameshift mutations and the Genetic Code', *Cold Spring Harbor Symp. Quant. Biol.*, **31**, 77-84.
34. Yanofsky, C., Ito, J. and Horn, V. (1966), 'Amino acid replacements and the Genetic Code', *Cold Spring Harbor Symp. Quant. Biol.*, **31**, 151-162.

35. Prakash, L. and Sherman, F. (1973), 'Mutagenic specificity: reversion of *iso*-1-cytochrome-c mutants of yeast', *J. Mol. Biol.,* **79**, 65-82.
36. Person, S. and Osborn, M. (1968), 'The conversion of amber suppressors to ochre suppressors', *Proc. Nat. Acad. Sci., U.S.A.,* **60,** 1030-1037.
37. Ripley, L.S. and Drake, J.W. (1972), 'A genetic assay for transversion mutations in bacteriophage T4', *Mol. Gen. Genet.,* **118,** 1-10.
38. Osborn, M., Person, S., Phillips, S. and Funk, F. (1967), 'A determination of mutagen specificity in bacteria using nonsense mutants of bacteriophage T4', *J. Mol. Biol.,* **26**, 437-447.
39. De Serres, F.J. (1964), 'Mutagenesis and chromosome structure', *J. Cell. Comp. Physiol.,* **64**, Suppl. 1, 33-42.
40. De Serres, F.J. and Malling, H.V. (1969), 'Identification of the genetic alterations in specific locus mutants at the molecular level', *Jap. J. Genetics,* **44**, Suppl. 1., 106-113.
41. De Serres, F.J., Brockman, H.E., Barnett, W.E. and Kølmark, H.G. (1967), 'Allelic complementation among nitrous acid-induced *ad*-3B mutants of *Neurospora crassa', Mutation Res.,* **4**, 415-424.

CHAPTER TWELVE

Mutagensis by ultraviolet and visible light III: Early work on micro-organisms

In what I have called the third period of mutation research (Chapter I), a number of new phenomena were discovered in experiments on micro-organisms exposed to ultraviolet light (see article by Pomper and Atwood in 'Radiation Biology', vol. 2, Bibliography). How fundamental these discoveries were to our understanding of the whole process of mutagenesis became apparent only in the next period, when knowledge of the chemical nature of the gene made it possible to formulate and test molecular models. In the present chapter I shall deal only with the early results and the conclusions that were drawn from them at the time. Some later findings will be included where they round off the picture.

Wavelength

Ultraviolet light proved to be an excellent mutagen for the easily penetrated cells of micro-organisms. Studies on the effective wavelengths in the UV range were carried out on fungal spores (1) and gave essentially the same results as those obtained with *Sphaerocarpus* (Chapter 10). Experiments on the genetical effect of visible light, with or without previous 'photosensitization' of the cells, have been reviewed by Dubinin and by Eisenstark (Bibliography).

Dose-effect curves

The complexities which, as we have seen in Chapter 9, lead to deviations of dose-effect curves from target expectations, are even more pronounced for UV than for ionizing radiations. This is to be expected from the fact that so much more of the energy of UV is dissipated in the cell before reaching DNA (p.164). Recent studies have provided many examples for the action of UV on those macromolecules and cellular processes that are intercalated between a mutational change in DNA and the appearance of a mutant clone (2-10). Moreover, the repair processes that play such an important role in UV-mutagenesis (Chapter 14) are in part themselves influenced by UV. It is therefore not surprising that dose-effect curves for UV-induced mutations very rarely are linear and often cannot be fitted by any equation that is based on pure target principles.

At high doses, mutation frequency often declines following a maximum at an intermediate dose. In the early years of research on UV-mutagenesis, this gave rise to much discussion. Various possibilities, not mutually exclusive, were considered. One is the existence, in the treated sample, of a small subpopulation with high resistance to both killing and mutation induction. At high doses, the majority of survivors will belong to this subpopulation and exhibit its characteristic low mutation frequency. While this could explain some cases, it clearly did not apply to others, in which the survivors could be shown to be exactly as UV-sensitive as the parent strain. An alternative explanation, which has a very modern ring about it, was suggested by Muller in 1954 (Chapter 8 in 'Radiation Biology', vol. 1, Bibliography). He postulated a repair process that 'acting against mutagenesis and perhaps initiated, in part at least, even by the radiation of mutagenic wavelength, might play an important part in the decline in mutagenic efficiency of ultraviolet with increasing dose'.

At levels below the maximum, the dose-effect curves for mutations are rarely linear. An exception is formed by extracellularly irradiated bacteriophage (11); this supports the notion that cellular processes are responsible for the deviations from linearity in cellular systems, including phage irradiated inside the bacterium (12). In *E.coli*, mutations to phage resistance arose with linear kinetics when treatment was given chronically in the chemostat (p.185), but with a dose exponent between 2 and 3 when the bacterial suspension was exposed to acute irradiation (13). This is reminiscent of the dose rate effect on X-ray induced mutations in spermatogonia of mice (Chapter 9) and may similarly be due to inhibition of a repair system by radiation delivered at a high dose rate. Dose-effect curves with dose square kinetics are frequently found in bacteria and have been attributed speculatively to various types of two-hit interaction, where one of the hits acts on a cellular target outside the mutagenic lesion itself (14-16). One should also consider the possibility that moderate and high doses of UV inhibit repair of the lesions that they themselves have produced and thus lend an upward bend to the dose-effect curve. There is evidence that 'near-UV' of 365 nm does inhibit repair of the lesions that have been produced by the mutagenically active 'far-UV' of 254 nm (17). In *Neurospora,* forward mutations at the *ad*-3 locus arise with two-hit kinetics, independent of whether they are due to intragenic point mutations or to deletions covering several genes (18). It will be remembered that, after X-radiation, the former type arises with linear kinetics, as expected from target theory (Chapter 9). Reverse mutations in a di-auxotrophic strain of *Neurospora* yielded dose-effect curves that saturated at different doses for the two loci; this suggests that processes involved in expression are differently affected by high doses of UV (19). Two strains of *Aspergillus nidulans* that differed in UV-sensitivity, differed also in the dose-response curves for auxotrophic reversions at two loci (20). Of special interest is the finding that this effect of the residual genotype on kinetics differed between the two loci, resulting in a reversal of the 'mutagen specificity' of UV. We shall return to this case in Chapter 20.

While most of the examples referred to in the last paragraph come from later work, an involvement of cellular processes in the dose effect curves for UV had suggested itself already to some of the early research workers in the field; Muller's quotation on p. 204 is an example for this. So is Stadler's idea of a delay between the primary lesion in chromosome or gene and its expression as chromosome break or mutation (p. 169). This pointed to the existence of a 'premutational' stage in UV mutagenesis. In the early post-war years, the reality of this concept was shown by experiments which will be reviewed in the remainder of this chapter and in the two following ones.

Photoreactivation

This important phenomenon owes its discovery to an accidental observation which was followed up in spite of the lack of any plausible interpretation. Like the discovery of penicillin, it exemplifies the value of a scientist's willingness to follow observational leads even if they cannot yet be fitted into any known theory. In 1948, Kelner (21) in Cold Spring Harbor was analysing a well-known phenomenon: the ability of irradiated and seemingly dead bacteria to recover colony-forming capacity during post-irradiation storage. The degree of recovery varied widely, even between experiments that had been carried out under what appeared to be identical conditions. A possible source of this variation was temperature, and Kelner carried out a careful study of the effects of post-irradiation temperature. Variability persisted, even between series exposed to the same temperature. In replicates of the same experiment, recovery rates might vary from zero to 100 000-fold improved survival. The highest degree of recovery was found in a bacterial suspension that had been kept on an open shelf exposed to light from the window. This suggested to Kelner that the responsible factor for recovery was visible light. Experiments carried out to prove this hypothesis brought spectacular confirmation. Data from one of them are shown in Table 12.1.

Since then, a great variety of cells have been tested for susceptibility to photoreactivation (22,23). The majority have been shown to be capable of it although to varying degrees (for exceptions see Chapter 13). There is always a hard core of damage that cannot be photoreactivated; its magnitude varies between cell types and types of damage studied. The effective wavelengths comprise the shortwave part of the visible spectrum, up to about 480 nm; experiments in which photoreactivation has to be avoided must be carried out in red or yellow light. Stored in the dark, irradiated cells or viruses may retain their response to visible light for differing times, depending on the type of cell and, especially, on its metabolic state. Bacteria in the stationary phase lose response to photorepair in a few hours, while bacteria kept at $-10°$ or extracellularly stored viruses retain it over days or weeks. The phenomenon has been called by many terms: photorestoration, photorecovery, photoreversal, photoreactivation, photorepair. Nowadays, the last two are mainly in use.

Many types of biological damage are subject to photorepair. Mutation is one of them. This discovery in the late forties and early fifties (24-27) provided clear evidence for the existence of a transient, premutational stage in UV-mutagenesis. The nature of this stage could be conceived of in two essentially different ways, depending on whether it was supposed to occur before or after the

Table 12.1

Effect of duration of visible light illumination on recovery of *Streptomyces* from ultraviolet irradiation injury. (Table 2, Kelner (21)). Courtesy Nat. Acad. Sciences U.S.A.

Illumination time (min)	*Viable cells per ml. of suspension*	*Relative increase in survival rate*
0	2.5*	
10	2.5×10^3	1 000 fold
20	9.2×10^3	3 700 fold
30	1.3×10^5	52 000 fold
40	1.6×10^5	64 000 fold
50	2.0×10^5	80 000 fold
60	5.3×10^5	210 000 fold
145	5.5×10^5	220 000 fold
173	7.7×10^5	310 000 fold
240	8.0×10^5	320 000 fold

* The count of the non-ultraviolet-irradiated suspension was 4.2×10^6, so that the survival rate at time zero was 6.0×10^{-7}.

production of lesions in the gene. On the first model, UV would produce mutations indirectly by means of a chemical that acts as the actual mutagen and that is partially destroyed by visible light: a 'photosensitive mutagen poison' (28). On the second model, UV would be the primary mutagen, but the lesions that it produces in the gene would remain reparable for a considerable length of time. We shall see in the next chapter that the second model is, in fact, correct. At the time of the early experiments, however, there was no reason for preferring one model to the other, and most geneticists adopted the first. Experiments of quite a different nature, carried out during the same period, were taken to support this view.

The production of mutations by irradiation of medium

In 1947, a group of workers at the University of Texas reported the production of mutations in *Staphylococcus aureus* that had been grown for several hours in UV-irradiated medium (29). Table 12.2 shows the results of an experiment in which resistance mutations to antibiotics were scored. The left part compares the numbers of resistant bacteria among those that had been grown in irradiated broth

Table 12.2

Increase in the rate of mutation to streptomycin and penicillin resistance after growth of *Staphylococcus aureus* in UV-irradiated medium (number of live bacteria). (Table 3, Stone (29). Courtesy Nat. Acad. Sciences U.S.A.)

	Nutrient broth		*Synthetic medium*	
Inhibitor units/ml	*control*	*irradiated*	*control*	*amino acids irradiated*
0	300×10^6	260×10^6	1250×10^6	900×10^6
Penicillin				
0.04	13 000	120 000	12 000	55 000
0.07	10	310	30	1 520
Streptomycin				
1.0	42 000	140 000	30 000	168 000
3.0	5 000	33 000	2 700	23 000

with the control value obtained from bacteria that had been grown in unirradiated broth. Special tests were carried out to exclude various possibilities of selection. Thus, induction of a general tolerance to antibiotics by growth in irradiated medium was excluded by the fact that the streptomycin-resistant variants had remained sensitive to penicillin and vice versa. Selection of pre-existing mutants by growth in irradiated medium was already made unlikely by the high frequencies of mutants. It was definitely excluded for biochemical mutations involving the ability to utilize the sugar mannitol (30). Wild-type *Staphylococcus* can utilize this sugar. Growth in irradiated broth produced mutants that had lost this ability; when these in turn were grown on irradiated medium, reverse mutations to mannitol-utilization were produced. Independent of whether these had arisen as true reversions, as suppressors, or through cytoplasmic changes, the fact that *the same treatment* produced changes in both directions proves that it did not act by selection. Whether the changes had occurred in nuclear genes could not be tested at a time when no sexual or parasexual processes had yet been discovered in bacteria. To my knowledge, such tests have never been carried out; but from what we now know about bacterial mutations, it seems likely that the variants produced by irradiated broth were indeed true mutations. Mutations were equally well produced when the irradiated medium was not broth but a synthetic medium containing mineral salts, glucose, and a number of amino acids and vitamins. The right part of Table 12.2 shows that the full mutagenic effect was obtained when only the amino acids were irradiated; irradiation of the mineral salts was ineffective. Exposure of the glucose to the usual dose made the medium too toxic for use; exposure to lower doses was mutagenically ineffective. The conclusion was drawn that the active mutagen in irradiated medium is a radiochemical derivative of amino acids. Its nature was inferred from a number of additional observations (31,32). (1) Broth could be rendered mutagenic by pretreatment with hydrogen peroxide instead of irradiation with UV. (2) It retained its mutagenicity over many hours, although at the time of inoculation with bacteria no free hydrogen peroxide was detectable. (3) In synthetic medium, pretreatment of the amino acids with hydrogen peroxide had the same effects as irradiation of the amino acids with UV. (4) Whether mutagenicity of the medium was due to irradiation or hydrogen peroxide, it was destroyed by catalase. (5) Treatment of the bacteria with hydrogen peroxide produced no mutations. Taken together, these observations led to the conclusion that the active mutagen in these experiments was an organic peroxide of amino acids. Subsequently, the mutagenic capacity of organic peroxides has been proved by a more direct approach (Chapter 18).

In the absence of any model for the action of UV on the genetic material, the possibility that its 'direct' action on the genes was also mediated by a chemical radiation product was not far-fetched. It is true that the wavelength-dependence for mutagenesis by direct exposure to UV differed from that of mutagenesis via irradiated medium (33), shorter wavelengths being effective in the latter case. Moreover, in terms of energy expenditure, direct irradiation was much more effective than irradiation of the medium, for which much larger doses were required. These discrepancies were, however, not sufficient to establish an essential difference between the two modes of mutagenic action of UV. It remained conceivable that both its direct action on the cell and its indirect action via irradiated medium were due to radiochemically produced mutagens, and that direct exposure was more effective only because it produced these compounds inside the cell and close to the gene. More recent experiments (34), in which mutation frequency in unirradiated chromosomes of male bacteria was increased in crosses with UV-irradiated female bacteria, have been taken to show the production of a mutagenic substance in irradiated cytoplasm. It is interesting, and links up with X-ray experiments on higher organisms (Chapter 7), that no such effect was found when the females had been exposed to ionizing radiation.

Mutation fixation (MF) and mutation frequency decline (MFD)

In early experiments on the production of bacterial mutations by UV, it was often observed that mutations were rare when plating was done shortly after irradiation, but increased in frequency with a delay in plating. It may be recalled that Stadler had already postulated a delayed mutagenic effect of UV for the interpretation of his data in maize (p.169); but the delay in the appearance of bacterial mutations was much more pronounced. In many cases, maximal yield of mutations required conditions that permitted 10 to 12 bacterial generations between exposure and screening. This delayed effect of UV gave rise to much speculation, especially as delayed mutagenic effects had meanwhile been found to be a characteristic feature of chemically induced mutagenesis (Chapter 15). A variety of explanations were offered, such as phenomic lag in expression of the mutants (p.178), segregation of single mutant nuclei from multinucleate cells or of single mutant chromosome strands from multistranded chromosomes, specific retardation in the division of mutant cells, etc. Probably, some or all of these may play a role in delayed mutagenesis of certain systems. None of them was found to account for it in the best-analysed system: the production of prototrophic mutations in auxotrophic bacterial strains. Analysis of this situation led to the discovery of two related phenomena: mutation fixation and mutation frequency decline. These turned out to be the first pointers to the existence of dark repair from UV-induced damage as distinct from photorepair. Because of their significance in this regard, they will be discussed in some detail.

Demerec (35) made a special study of the delayed appearance of prototrophic mutants among UV-treated auxotrophic bacteria. He regulated the number of residual divisions on the screening plates by adding to the minimal medium varying small amounts of broth, each of which allowed the auxotrophic inoculum to reach a characteristic final population size. In all experiments, the number of mutants – recognized as large colonies on a lawn of residual growth – increased with the number of residual divisions and reached a maximum after a division number that varied between strains.

Witkin (36) took this situation as her starting point for an analysis of delayed UV-mutations. In turn, she tested all explanations that had been offered and, in turn, rejected every one of them. The most plausible explanation, that of a long phenomic delay, was tested in experiments in which information for prototrophy was introduced into tryptophanless cells in two different ways: by transduction from a prototrophic strain, or by UV-induced mutations. As regards expression, the situation is the same in both cases; in both, the new information has to be translated into an enzyme before the cell can grow on medium without tryptophan. Yet, while one division was sufficient for expression of the transduced information, six or more divisions were required for full mutational yield. Thus, phenomic delay cannot account for the delayed appearance of the mutations. Segregational delay (p.181) was another possibility to be considered. It was tested for by an ingenious technique which it would take us too far to describe here. The outcome was that segregational delay, too, is much too short to account for mutational delay. The dilemma that finally confronted Witkin was that, on the one hand, there was no doubt of the positive correlation between number of residual divisions and induced mutation frequency and, on the other hand, all plausible interpretations of this correlation had been ruled out by experiment. The solution of the dilemma is of general interest. It lies in a possibility that is easily overlooked: the possibility that a correlation between two events may not be due to one being the cause of the other; it may equally well be due to both events being positively correlated with a third one. In this particular case, the common cause was the amount of broth supplementation which determined, on the one hand, the number of residual divisions and, on the other hand, the frequency of UV-induced mutations. The two consequences of broth enrichment could be dissociated when it had been found that the full effect of enrichment or of its lack is obtained during the first hour after irradiation. Table 12.3 shows that irradiated bacteria which had spent the first hour on enriched medium yielded a high mutation frequency, independent of whether they were subsequently grown on rich or poor medium. Conversely, bacteria that had spent the first hour on poor medium yielded few mutations independent of the medium to which they were subsequently transferred. The last column shows that the number of

Table 12.3

Transfer of irradiated W92 (*E.coli* try$^-$) from high to low and from low to high levels of enrichment after one hour of incubation. (4.1 x 10^7 survivors plated on initial enrichment level, incubated one hour at 37^{0}C, washed from surface, concentrated by centrifugation, replated on final enrichment level, incubated 48 h.) (Table 7, Witkin (36). Courtesy Cold Spring Harbor Labor.)

Enrichment level		*No. viable cells replated*	*No. induced prototrophs per plate*	*Final No. residual divisions*
Initial	*Final*			
High	High	3.6 x 10^7	736; 673	7.1
High	Low	3.5 x 10^7	585; 619	1.6
Low	Low	3.8 x 10^7	55; 64	1.1
Low	High	3.4 x 10^7	72; 58	6.8

Low = Minimal Medium (MM). High = MM + 2.5% broth.

residual divisions was determined by the nature of the second medium and was wholly irrelevant to the frequency of induced mutations.

The loss of potential mutations through amino acid deprivation is a gradual process, which Doudney and Haas (37) have called *mutation frequency decline* (MFD). Its course is shown in Fig. 12.1. The curve labelled MFD was obtained by plating irradiated cells either directly on semi-enriched medium (SEM = MM + 5% broth) or transferring them to it after various periods on MM. It will be seen that after 15 min on MM more than half the potential mutations to prototrophy had been lost irretrievably. With subsequent transfers, mutation frequency dropped more slowly; after about 30 min, it had reached the low value typical for cells plated directly on MM. A number of observations showed that MFD is due to lack of protein synthesis during a short critical period after UV-irradiation. Thus, prevention of MFD requires a full pool of amino acids, not only the particular amino acid for which the strain is auxotrophic. Moreover, even in the presence of broth or an amino acid pool MFD occurs when protein synthesis is inhibited by chloramphenicol.

The second curve in Fig. 12.1 describes what Doudney and Haas have called *mutation fixation* (MF). It shows the period after irradiation during which the cells remain sensitive to MFD. The procedure for constructing the MF curve was exactly the reverse of that used for constructing the MFD curve. Irradiated tyr$^-$ cells

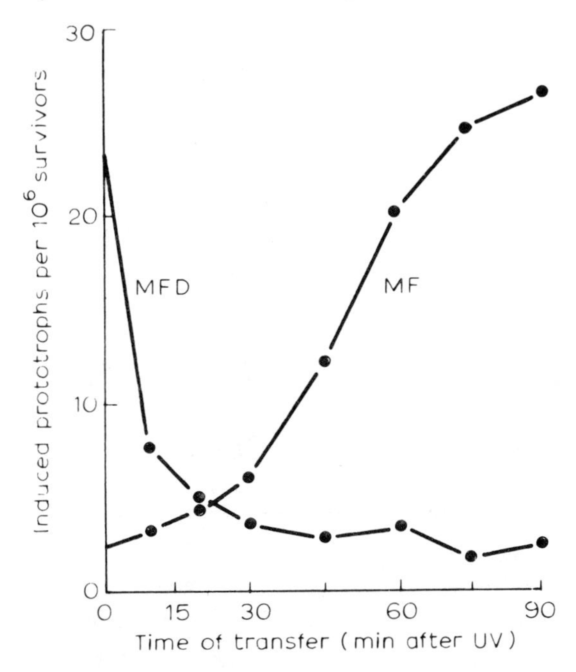

Fig. 12.1. Mutation frequency decline (MFD) and mutation fixation (MF) after UV-irradiation of a tyrosine-requiring strain of *Escherichia coli.* The mutations were to tyrosine independence. Transfer between medium with or without 5% broth, both media contained 1μg tyrosine/ml. (Fig. 1. Witkin (38). Courtesy the Wistar Inst. of Anatomy and Biology.)

were plated either directly on MM, or were transferred to it after various periods on SEM. The first point shows the low mutational yield on MM. With increasing periods of time on SEM, fewer and fewer potential mutations were lost after transfer, until after about one hour on SEM, none were lost and the mutational yield was that typical for cells plated immediately on SEM and not further transferred. During this hour, there had been a gradual conversion of potential mutations, liable to MFD, into fixed mutations, no longer sensitive to amino acid deprivation.

Witkin adduced evidence that the duration of the sensitive period is independent of the rate of protein synthesis. The terminal event leading to mutation fixation occurs about one third through the first post-irradiation cell cycle, the time of gene replication. Conditions that prolong the time between irradiation and DNA synthesis prolong the sensitive period. Most probably, the fixation of potential mutations occurs at the time of gene replication.

Two points have to be kept in mind when attempting to interpret MFD and MF. First, these results refer only to mutation, not to survival; the latter was the same in these experiments in the presence or absence of amino acids. Post-irradiation conditions that affect survival as well as mutation may act in different or more complex ways. Second, the loss of potential mutations through MFD should not be confused with loss of mutations through lack of expression. Where the system required it, Witkin added traces of the missing nutrilite in order to allow full expression of the induced prototrophs (see legend to Fig. 12.1).

Witkin's experiments and similar ones by others, in particular Doudney and Haas, agreed with those on photoreactivation in showing the existence of a premutational stage between UV-irradiation and the production of stable gene changes. Again, the question arose whether this stage was due to the production by UV of an unstable chemical mutagen, whose synthesis or stabilization is favoured by protein synthesis, or to the existence of premutational genetic lesions which, in the absence of protein synthesis, are lost through repair or lethality. Witkin favoured the latter assumption, but decisive proof for it could be produced only when the molecular action of UV on DNA had been elucidated. With this we shall deal in the next two chapters.

References

1. Hollaender, A. (1941), 'Wavelength dependence of the production of mutations in fungus spores by monochromatic ultraviolet radiation', *Proc. 7th Int. Cong. Genetics,* 153-154.
2. Swenson, P.A. and Setlow, R.B. (1964), 'β-galactosidase: inactivation of its messenger RNA by ultraviolet irradiation', *Science,* **146**, 791-793.
3. Sauerbier, W., Millette, R.L. and Hackett, P.B. Jr., (1970), 'The effects of ultraviolet irradiation on the transcription of T4 DNA', *Biochim. Biophys. Acta.,* **209**, 368-386.
4. Swenson, P.A. and Nishimura, S. (1964), 'Inactivation of S-RNA by ultraviolet radiation', *Photochem. Photobiol.,* **3**, 85-90.
5. Harriman, P.D. and Zachau, H.G. (1966), 'Ultraviolet inactivation of transfer ribonucleic acid functions', *J. Mol. Biol.,* **16**, 387-403.

6. Aoki, I., Ikemura, T., Fukutome, H. and Kawade, Y. (1969), 'Ultraviolet inactivation of the functions of phenylalanine and lysine transfer RNA's of *Escherichia coli*', *Biochim. Biophys. Acta,* **179**, 308-315.
7. Ono, J., Wilson, R.G. and Grossman, L. (1965), 'Effects of ultraviolet light on the template properties of polycytidylic acid', *J. Mol. Biol.,* **11**, 600-612.
8. Logan, D.M. and Whitmore, G.F. (1966), 'Effects of ultraviolet light on polyuridylic acid assayed *in vitro*', *J. Mol. Biol.,* **21**, 1-12.
9. Kagawa, H., Fukutome, H. and Kawade, Y. (1967), 'Inactivation of *Escherichia coli* ribosomes by ultraviolet irradiation. I. Activity of poly U-directed polyphenylalanine synthesis', *J. Mol. Biol.,* **26**, 249-265.
10. Resnick, M.A. and Holliday, R. (1971), 'Genetic repair and the synthesis of nitrate reductase in *Ustilago maydis* after UV irradiation', *Molec. Gen. Genetics,* **111**, 171-184.
11. Drake, J.W. (1966), 'Ultraviolet mutagenesis in bacteriophage T4. I. Irradiation of extracellular phage particles', *J. Bacter.* **91**, 1775-1780.
12. Drake, J.W. (1963), 'Properties of ultraviolet-induced rII mutants of bacteriophage T4', *J. Mol. Biol.,* **6**, 268-283.
13. Novick, A. (1955), 'Mutagens and antimutagens', *Brookhaven Symposia in Biology. "Mutation",* **8**, 201-215.
14. Bridges, B.A. (1966), 'A note on the mechanism of UV mutagenesis in *Escherichia coli*', *Mutation Res.,* **3**, 273-279.
15. Mennigmann, H.D. (1972), 'Pyrimidine dimers as premutational lesions in *Escherichia coli* WP2 Hcr$^-$', *Molec. Gen. Genetics,* **117**, 167-186.
16. Witkin, E.M. and George, D.L. (1973), 'Ultraviolet mutagenesis in *polA* and *uvrA polA* derivatives of *Escherichia coli* B/r: evidence for an inducible error-prone repair system', *Genetics Suppl.,* **73**, 91-108.
17. Tyrrell, R.M. and Webb, R.B. (1973), 'Reduced dimer excision in bacteria following near ultraviolet (365 nm) radiation', *Mutation Res.,* **19**, 361-364.
18. De Serres, F.J. and Kilbey, B.J. (1971), 'Differential photoreversibility of ultraviolet-induced premutational lesions in *Neurospora*', *Mutation Res.,* **12**, 221-234.
19. Auerbach, C. and Ramsay, D. (1968), 'Analysis of a case of mutagen specificity in *Neurospora crassa*', *Molec. Gen. Genetics,* **103**, 72-104.

20. Chang, L.T., Lennox, J.E. and Tuveson, R.W. (1968), 'Induced mutation in UV-sensitive mutants of *Aspergillus nidulans* and *Neurospora crassa'*, *Mutation Res.*, **5**, 217-224.
21. Kelner, A. (1949), 'Effect of visible light on the recovery of *Streptomyces griseus* conidia from ultraviolet irradiation injury', *Proc. Nat. Acad. Sci., U.S.A.*, **35**, 73-79.
22. Dulbecco, R. (1955), 'Photoreactivation' in *Radiation Biology, Vol. II: Ultraviolet and related radiations*, ed. A. Hollaender, McGraw-Hill Book Co. Inc., New York: p. 455-486.
23. Jagger, J. (1960), 'Photoreactivation' in *Radiation Protection and Recovery*, ed. A. Hollaender, Pergamon Press, New York, p. 352-377.
24. Novick, A. and Szilard, L. (1949), 'Experiments on light-reactivation of ultraviolet-inactivated bacteria', *Proc. Nat. Acad. Sci., U.S.A.*, **35**, 591-600.
25. Kimball, R.F. and Gaither, N.T. (1950), 'Photorecovery of the effects of ultraviolet radiation on *Paramecium aurelia'*, *Genetics*, **35**, 118.
26. Goodgal, S.H. (1950), 'The effect of photoreactivation on the frequency of ultraviolet induced morphological mutations in the microconidial strain of *Neurospora crassa'*, *Genetics*, **35**, 667.
27. Meyer, H.U. (1951), 'Photoreactivation of ultraviolet mutagenesis in the polar cap of *Drosophila'*, *Genetics*, **36**, 565.
28. Newcombe, H.B. and Whitehead, H.A. (1951), 'Photoreversal of ultraviolet-induced mutagenic and lethal effects in *Escherichia coli'*, *J. Bacter.*, **61**, 243.
29. Stone, W.S., Wyss, O. and Haas, F. (1947), 'The production of mutations in *Staphylococcus aureus* by irradiation of the substrate', *Proc. Nat. Acad. Sci., U.S.A.*, **33**, 59-66.
30. Stone, W.S., Haas, F., Clark, J.B. and Wyss, O. (1948), 'The role of mutation and of selection in the frequency of mutants among micro-organisms grown on irradiated substrate', *Proc. Nat. Acad. Sci., U.S.A.*, **34**, 142-149.
31. Wyss, O., Stone, W.S. and Clark, J.B. (1947), 'The production of mutations in *Staphylococcus aureus* by chemical treatment of the substrate', *J. Bacter.*, **54**, 767-772.
32. Wyss, O., Clark, J.B. Haas, F. and Stone, W.S. (1948), 'The rôle of peroxide in the biological effects of irradiated broth', *J. Bacter.*, **56**, 51-57.

33. Haas, F., Clark, J.B., Wyss, O. and Stone, W.S. (1950), 'Mutations and mutagenic agents in bacteria', *Am. Nat.*, **84**, 261-274.
34. Kada, T. and Marcovich, H. (1963), 'Sur le siège initial de l'action mutagène des rayons X et des ultraviolets chez *Escherichia coli K12', Ann. Inst. Past.*, **105**, 989.
35. Demerec, M. and Cahn, E. (1953), 'Studies of mutability in nutritionally deficient strains of *Escherichia coli', J. Bacter.*, **65**, 27-36.
36. Witkin, E.M. (1956), 'Time, temperature and protein synthesis: a study of ultraviolet-induced mutation in bacteria', *Cold Spring Harbor Symp. Quant. Biol.*, **21**, 123-140.
37. Doudney, C.O. and Haas, F.L. (1958), 'Modification of ultraviolet-induced mutation frequency and survival in bacteria by post-irradiation treatment', *Proc. Nat. Acad. Sci., U.S.A.*, **44**, 390-401.
38. Witkin, E.M. (1961), 'Modification of mutagenesis initiated by ultraviolet light through post-treatment of bacteria with basic dyes', *J. Cell. Comp. Physiol.*, **58** (Suppl. 1), 135-144.

CHAPTER THIRTEEN

The molecular action of ultraviolet and visible light. I: Photo-products. Photorepair

Already in the 1940s, some research had been carried out on the photoproducts formed in UV-irradiated nucleic acid and its components. With the realization that DNA is the essential part of the gene, these investigations gained tremendously in importance and were pursued in a number of laboratories. A variety of photoproducts were detected *in vitro* and *in vivo.* Of special value was the analysis of UV-irradiated transforming DNA, which can be treated both *in vitro* and *in vivo* and can be tested for biological effects in living cells.

Photoproducts in DNA (1-4)

Among the nucleic acid bases, pyrimidines turned out to be about 10 times as sensitive to UV as purines with hydrated cytosine the most frequent product. In native DNA, the formation of cytosine hydrate is probably restricted to regions in which the complementary strands have separated. The main photoproducts in double-stranded DNA are dimers formed from neighbouring pyrimidines on the same strand. They are of the cyclobutane-type, in which the double bonds between the 5 and 6 positions of each ring have been opened to give rise to a four-carbon ring that ties the two pyrimidines together. Although all possible kinds of pyrimidine dimer are formed ($\widehat{TT}$; $\widehat{CC}$; $\widehat{CT}$), those between two thymines are the most frequent and most stable ones. Fig. 13.1 shows the particular isomer of $\widehat{TT}$ (out of 6 possible ones) that is formed between thymines on the same strand of DNA. To a smaller extent, UV also produces cross-links between complementary strands of DNA and between

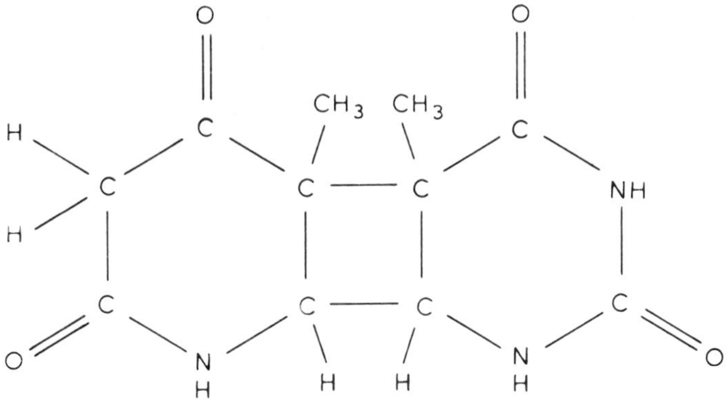

Fig. 13.1. Cyclobutane type of thymine dimer.

DNA and protein. Single-chain breaks are infrequent at biologically tolerated doses (5). Table 13.1 lists the proportion of the main photoproducts in phage and two strains of bacteria that differ widely in resistance to UV. The table shows that the more resistant strain, *B/r*, tolerates about 100 times as many pyrimidine dimers as does the more sensitive one, B_{s-1}. We shall see (Chapter 14) that this is due to the ability of *B/r*, but not of B_{s-1}, to get rid of most of the dimers by dark repair.

Even for the same strain, the proportion of pyrimidine dimers among UV-induced lesions is not the same under all circumstances. When bacteria were irradiated at $-79°$ (6), roughly seven times as many mutations were produced as in the unfrozen state, but they did not seem to be due to the ordinary kind of dimer, and this was confirmed in studies on transforming DNA (7).

Effects of dyes

A number of dyes interfere in various ways and by various mechanisms with the reaction of DNA to visible light and UV. Acridines, by intercalating between the bases of DNA and by deflecting energy from DNA, reduce the extent of dimerization by a subsequent dose of UV (8). Thus, when given *before* UV, they act as protecting agents against killing and mutation (9,10). We shall see later that,

Table 13.1

Estimates of the numbers of photoproducts in DNA per mean lethal dose* for irradiation at 254 nm in aqueous solution. (Table 1, Setlow (5). Courtesy Pergamon Press.)

Product	*T2 phage*	*E. coli* resistant (*B/r*)	*E. coli* sensitive (B_{s-1})
Interstrand cross-links	$<10^{-2}$	2	$<10^{-2}$
Chain breaks	$<10^{-2}$	2	$<10^{-2}$
DNA-protein links †	–	3	10^{-2}
Cyclobutane dimers	10	10^3	10
Hydrates	~1	10^2	~1

Note that the molecular weight of *E.coli* DNA is about 20 times that of phage DNA.

* Dose that yields on the average one 'lethal hit' per particle or cell. In an exponentially decaying population, survival is represented by the first term of the Poisson distribution e^{-m}, where m is the mean number of lethal hits. When $m = 1$, survival is $e^{-1} = 37\%$.

† No good estimate of the absolute numbers can be made.

when given *after* UV, they have the opposite effect (p.239).Acridines are also among those dyes that sensitize cells to the lethal and mutagenic action of *visible light* by means of their so-called 'photodynamic action' (11-13a).

The mechanism of photodynamic action is not yet properly understood. The suggestion has been made that the release of deoxyribonuclease from lysosomes may play a role (13b). It seems clear that it must involve some kind of energy transfer for which the dyes act as photoreceptors. There have also been reports of mutagenesis by visible or near-visible light without added dyes (14,15a); probably, the cells themselves contained an effective chromophore. The molecular nature of the lesions that are produced by the photodynamic action of dyes is not known; it probably includes a variety of lesions, but not thymine dimers. The main primary effect appears to be the destruction of guanine residues (15b). On the contrary, thymine dimers are the main product of short-wave visible light (310 nm or more) acting on DNA that has been sensitized by acetone, acetophenone, or acetophenone-derivatives (16,17). This has led to a convenient method for studying the effects of thymine dimers in isolation and for estimating their contribution to the genetic effects of UV (p. 225).

Still a different kind of photosensitization is produced by treatment with psoralen and its derivatives. These are chemicals that sensitize skin to the short-wave range of the visible spectrum and are used as tanning agents. In bacteria and phage, they produce lethal as well as mutational damage (18,19). Their main action in producing mutations is to form cyclobutane addition products with pyrimidines (20,21) and these can react further to produce cross-links with pyrimidine bases in the complementary strand. It is the latter reaction that leads to killing, while the former appears to be the basis for mutation (22).

The role of pyrimidine dimers in UV-damage

The high proportion of pyrimidine dimers among UV-induced lesions does not necessarily mean that these photoproducts are the main causes of the lethal and mutagenic effects of UV. Indeed, we have already seen that this is not true for bacteria irradiated in the frozen state at very low temperature (6,7). However, for ordinary conditions of treatment, the correlation between the proportion of dimers formed and the amount of killing and mutation induction is very close. This was established by varying the proportion of dimers after a given dose of UV in three ways: (1) by varying the wavelength of UV, (2) by photorepair, (3) by dark repair.

Wavelength dependence of dimer formation

UV does not only *form* dimers; it also *breaks* them. Each wavelength produces a characteristic balance between formation and breakage, and this determines which proportion of neighbouring pyrimidines is present as dimers. Using polythymidylic acid as a model system, Deering and Setlow (23) found that the steady-state fraction of dimers was approximately 70% for 280 nm and approximately 15% for 240 nm irradiation. A part of the dimers that had formed after exposure to 280 nm could be resolved by subsequent exposure to a shorter wavelength. The degree of biological damage was shown to be parallel with the proportion of dimerized thymine pairs. Fig. 13.2 shows data from two experiments on transforming DNA of *Haemophilus influenzae.* Splitting of dimers was measured by the increase

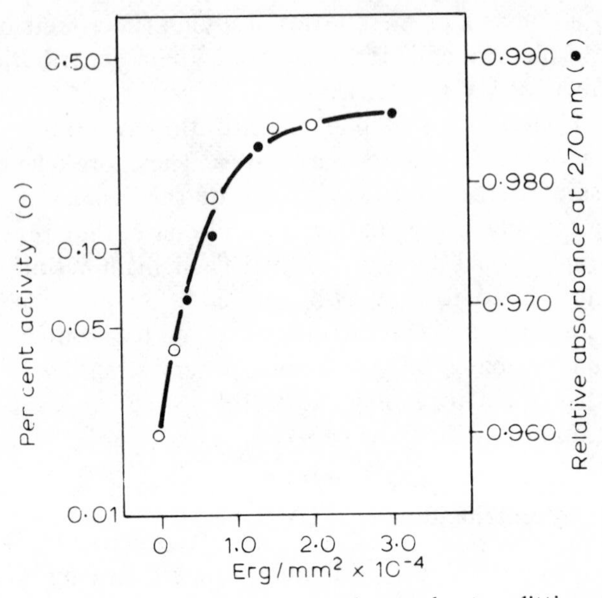

Fig. 13.2. Correlation of increase in absorbance, due to splitting of dimers with increase in biological activity after irradiation of UV-ed transforming DNA with different amounts of visible light. The solid and open circles represent data from two experiments. (Fig. 7, Setlow (48). Courtesy Nat. Acad. Sciences, U.S.A.)

in absorbance at 270 nm; biological activity was measured by the ability to transform recipient bacteria to antibiotic resistance. The correlation between the two effects is striking.

Photorepair (see Chapter 12)

The first step towards an understanding of the mechanism of photorepair was made when Goodgal, Rupert and Herriott (24) found that UV-irradiated *Haemophilus influenzae* did not respond to visible light. Neither did its transforming DNA after inactivation by UV. This suggested to the investigators that *E.coli* produces an enzyme which is required for photorepair and which is not produced by *Haemophilus.* Experiments in which UV-irradiated *Haemophilus* DNA was exposed to visible light in the presence of an extract from *E.coli* proved that this hypothesis was correct. Table 13.2 shows data from one of the experiments.

The transforming ability of unirradiated *Haemophilus* DNA was not affected by *E.coli* extract (II). That of UV-treated DNA was restored from about 4% (III) to about 30% (V) provided visible light was

Table 13.2

Photoreactivation of ultraviolet-inactivated *H. influenzae* transforming factor (TP) for streptomycin resistance by an extract of *E.coli B.* (After Table 1, Goodgal (24))

Mixture incubated 20 min at 37°C with Mg^{++} and ATP	*Conditions*	*Titer of TP*
I TP, buffer	Dark	1.5×10^4
II. TP, *E. coli* extract	Dark or light	1.9×10^4
III. UV'ed TP, buffer	Dark	6.6×10^2
IV. UV'ed TP, *E. coli* extract	Dark	4.6×10^2
V. UV'ed TP, *E. coli* extract	Light	4.4×10^3

applied at the same time. In the dark, the *E.coli* extract was ineffective (IV). During the following years, Rupert (25-27) extracted photoreactivating enzyme from baker's yeast and analysed its action. Many, but not all organisms, have been shown to possess photoreactivating enzyme. Its production is under genetic control, and mutants lacking it have been found in several micro-organisms that normally possess it (e.g. 28,29).

Haemophilus influenzae is among those organisms that lack photoreactivating ability. This makes it into an excellent tool for detecting the enzyme in other cells by means of tests like the one in Table 13.2. Such tests proved, for example, the presence of photoreactivating enzyme in *Neurospora crassa* (30,31). They failed to detect it in extracts from eggs, larvae and pupae of *Drosophila melanogaster* (32), although *Drosophila* had been one of the first organims for which the restorative effect of visible light on UV-induced cell-killing and mutation had been demonstrated (p. 206). However, not too much weight can be given in biological research to purely negative results. Indeed, newer experiments in which pyrimidine dimers were monitored in embryonic cells have shown that in *Drosophila* also visible light resolves dimers (33). In mammals, with the exception of marsupials (34), all earlier tests gave negative results. Recently, however, photoreactivating enzyme has been found in human leukocytes (35).

The molecular mechanism of photoreactivation is the enzymatic splitting of pyrimidine dimers into monomers (36). There is an essential difference between this means of resolving dimers and their resolution by short-wave UV, described above. The latter is a purely physical phenomenon, due to the different wavelength dependence of the dimerizing and monomerizing abilities of UV. Photorepair is an enzymatic reaction, for which the energy is provided by short-wave visible light. The whole process can be represented by the formula

$$E + S \underset{k_2}{\overset{k_1}{\rightleftharpoons}} ES \xrightarrow[\text{light}]{k_3} E + P,$$

where E is the photoreactivating enzyme, S its substrate (a UV-lesion in DNA), ES an enzyme-substrate complex, and P the product (a repaired lesion). Only the last step depends on the presence of visible light (26,27). Caffeine, by interfering with the formation of the ES complex, inhibits photorepair; but this can be observed only in sufficiently UV-sensitive cells or phage (37).

Photorepair is highly specific for pyrimidine dimers, whether these occur in DNA or in synthetic polynucleotides. An elegant way of showing this is to measure photorepair of transforming DNA in the presence of other polymers able to form pyrimidine dimers; the more dimers the added polymer contains, the more it reduces photo-repair of DNA by competition for the photoreactivating enzyme. Fig. 13.3 shows the results of an experiment in which irradiated transforming DNA was photoreactivated in the presence of poly dA:dT that had been subjected to different regimes of UV-irradiation and photoreactivation. In the absence of the polymer, there was the usual rapid recovery of transforming ability during the photo-reactivation period (■). Addition of unirradiated polymer made little difference (▲). When, however, the added polymer had been exposed to a dose of UV that had dimerized 2.9% of the thymines, photorepair of DNA was strongly reduced (●). The remaining curves (□, △) show the reduction of photorepair in the presence of irradiated polymer that had been given increasing periods of photoreactivation before addition of the DNA. In parallel with the loss of thymine dimers, there was a loss of competing ability. None was left after 40 min photoreactivation when all dimers had been resolved (○). Poly dA, which contains no pyrimidine, and poly d(AT):d(AT), which contains no neighbouring pairs of them, had no competing ability. This provided further proof for pyrimidine dimers being the substrate of the photoreactivating enzyme.

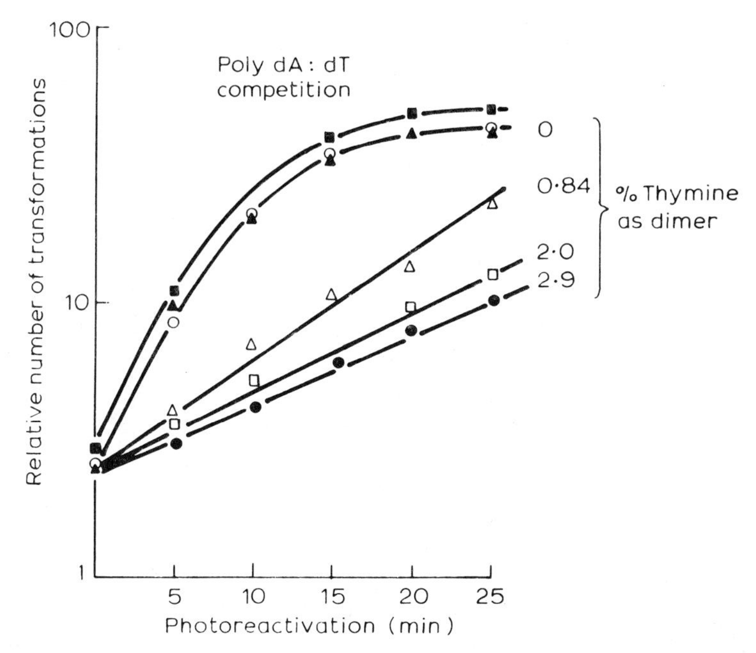

Fig. 13.3. Competition for photoreactivating enzyme between UV-irradiated transforming DNA and variously treated poly dA : dT.
■ No poly dA:dT present.
▲ Unirradiated poly dA:dT present.
● UV-irradiated poly dA:dT present, not photoreactivated.
□, △, ○ UV-irradiated poly dA:dT, photoreactivated for 10,20,40 min.
(Fig. 2. Setlow (49). Courtesy Nat. Acad. Sciences, U.S.A.)

Photorepair of UV-induced mutational damage is never complete; but it may lead to the loss – or, better, prevention – of very substantial proportions of mutations. On the contrary, photorepair is almost 100% efficient when applied to the damage caused by short-wave visible light ('black light') in acetophenone-sensitized cells (p. 221) (38,39). A comparison between the degree of photorepair of acetophenone-sensitized damage with UV-damage in the same system provides a means for estimating the contribution of pyrimidine dimers to UV-induced damage. A different way of doing this consists of comparing UV-mutagenesis in related strains that differ through the possession or lack of photoreactivating enzyme. This has led to the conclusion that most of the killing and mutagenic action of UV on *E.coli* is due to pyrimidine dimers (40). For suppressor mutations, the same conclusion was drawn from the

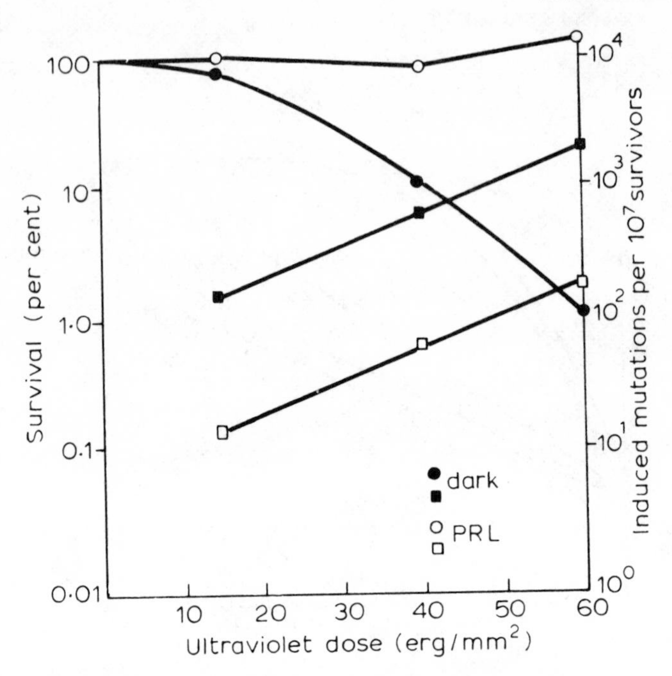

Fig. 13.4. Reversal, by visible light, of potentially lethal damage and induced mutations that result in independence of tryptophan in strain WP2 of *Escherichia coli.* Circles: survival. Squares: mutation frequency. PRL = 10 min exposure to photoreactivating light immediately after ultraviolet irradiation. (Fig. 5, Witkin (40). Courtesy Amer. Assoc. Advancement of Science.)

degree of protection against dimer formation by previous exposure to acriflavin (41). A critique of these conclusions will be considered in Chapter 14 (p. 244).

The magnitude of photorepair may vary between different types of damage scored in the same cells. Thus, in *E.coli*, mutations to streptomycin-resistance and to prototrophy show a high degree of photoreactivability (Fig. 13.4); mutations that result in the inability to ferment lactose are not photoreactivable at all (42). In a wild-type strain of *Saccharomyces cerevisiae,* lethals and mutations to prototrophy were photorepaired to the same extent; in a UV-sensitive derivative of this strain, lethal damage was removed more efficiently than mutational damage (43). In *Neurospora,* with one exception to be discussed presently, lethal and mutagenic lesions were found to be repaired to the same extent (44,45). Because of this parallelism, photorepair and mutagenesis can be measured by

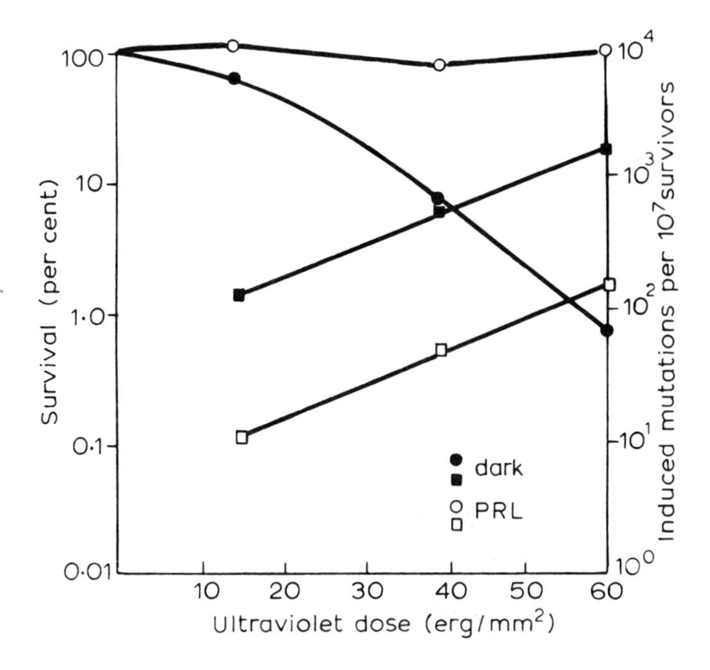

Fig. 13.4 (*continued*)

the same dose-reduction factor (DRF): after photorepair, a given dose of UV yields the same genetical damage as does a lower dose without photoreactivation. This is visualized best when mutation frequency per survivor is plotted against survival as in Fig. 13.5. It shows the results of an experiment in which a doubly auxotrophic strain, requiring adenine and inositol, was exposed to UV followed by visible light. Let us first look at the adenine-reversions (left panel). The black triangles represent conidia plated in the dark. Each has its counterpart in a hollow triangle, representing conidia that had been exposed to the same dose followed by photoreactivation. With increasing dose (left to right), survival decreased and mutation frequency increased. The important point to notice is that all points – black and white – fall *on the same curve*; but the dose required to yield a given amount of damage in the dark is smaller than the dose required to yield the same amount of damage in the light. The ratio between these doses is the DRF; 1-DRF is the 'photoreactivable sector'. In the experiment shown here, the photoreactivable sector was about 0.5, i.e. about half the lethal and mutagenic damage of UV could be repaired by visible light.

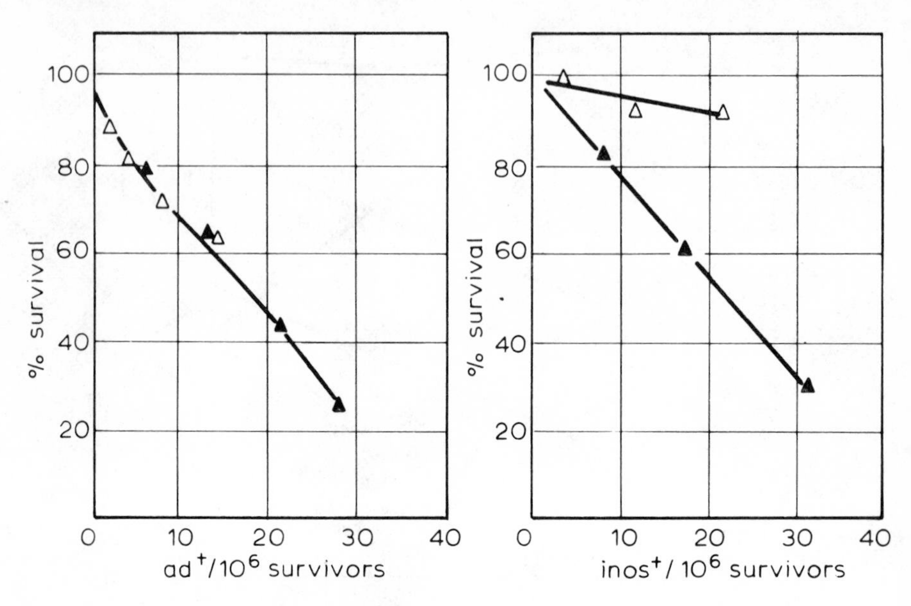

Fig. 13.5. Comparison between the photoreactivation of lethal damage and of mutations leading to the reversion of the alleles *ad*38701 and *inos* 37401 in *Neurospora.* Open triangles: with photoreactivation. Filled triangles: without photoreactivation. (Fig. 2, Kilbey (45). Courtesy Springer-Verlag.)

The exception mentioned above was found in the same experiment and is shown in the right panel of Fig. 13.5 which represents the response of the inositol-allele 37401 to UV and photoreactivation. Here, the black and white triangles do not fall on the same curve. For instance, the last black triangle has its white counterpart well above the line i.e., at a mutation frequency that corresponds to a much lower survival than that actually found after repair. Curiously enough, this particular inositol-allele is the only one that shows an abnormally weak response to photorepair. Whether this means that a photoproduct other than a pyrimidine dimer of the cyclobutane type is required for reversion of 37401, or whether a dimer at this site is in some way protected from the photoreactivating enzyme is not known.

When the forward mutations at the *ad*-3 loci were grouped into molecular types (p.197) the interesting observation was made that the proportions of frameshifts and base changes was not changed by photoreactivation (44). The same was found to be true for rII mutations in phage T4 (46). Obviously, pyrimidine dimers serve as source of both kinds of mutational change. Photoproducts that – like the inositol-allele 37401 – are recalcitrant to photorepair must contribute to the same extent to both molecular classes. On the contrary, they appear to contribute less to deletions in the *ad*-3 region that span more than one gene; for these are more photoreactivable than point mutations in this region (47).

The importance of these observations for speculations on the mechanism of UV-mutagenesis is obvious. Before, however, going on to this topic, we have to discuss repair in the absence of visible light.

References

1. Wacker, R.A., Dellweg, H., Träger, L., Kornhauser, A., Lodemann, E., Türck, G., Selzer, R., Chandra, P. and Ishimoto, M. (1964), 'Organic photochemistry of nucleic acids', *Photochem. Photobiol.*, **3**, 369-394.
2. Smith, K.C. (1966), 'Physical and chemical changes induced in nucleic acids by ultraviolet light', *Radiation Res. suppl.*, **6**, 54-79.
3. Setlow, R.B. and Setlow, J.K. (1972), 'Effects of radiation on polynucleotides', *Ann. Rev. Biophy. Bioengin.* **1**, 293-346.
4. Deering, R.A. (1962), 'Ultraviolet radiation and nucleic acid', *Scient. Am.*, **207**, 135-144.
5. Setlow, R.B. (1968), 'Photoproducts in DNA irradiated *in vivo*', *Photochem. Photobiol.*, **7**, 643-649.
6. Ashwood-Smith, M.J. and Bridges, B.A. (1966), 'Ultraviolet mutagenesis in *Escherichia coli* at low temperatures', *Mutation Res.*, **3**, 135-144.
7. Rahn, R.O., Setlow, J.K. and Hosszu, J.L. (1969), 'Ultraviolet inactivation and photoproducts of transforming DNA irradiated at low temperatures', *Biophys. J.*, **9**, 510-517.
8. Sutherland, B.M. and Sutherland, J.C. (1969), 'Mechanisms of inhibition of pyrimidine dimer formation in deoxyribonucleic acid by acridine dyes', *Biophys. J.*, **9**, 292.
9. Alper, T., Forage, A.J. and Hodgkins, B. (1972), 'Protection of normal, lysogenic and pyocinogenic strains against ultraviolet radiation by bound acriflavine', *J. Bacter.*, **110**, 823-830.

10. Alper, T. and Forage, A.J. (1974), 'The primary UV lesions for reversion to prototrophy in auxotrophic *Escherichia coli*: inferences from studies of protection by acriflavine', *Molec. Gen. Genetics,* **128**, 147-155.
11. Kaplan, R.W. (1949), 'Mutations by photodynamic action in *Bacterium prodigiosum', Nature,* **163**, 573-574.
12. Calberg-Bacq, C.M., 'Delmelle, M. and Duchesne, J. (1968), 'Inactivation and mutagenesis due to the photodynamic action of acridines and related dyes on extracellular bacteriophage T4B', *Mutation Res.,* **6**, 15-24.
13a. Ritchie, D.A. (1965), 'Mutagenesis with light and proflavine in phage T4. II. Properties of the mutants', *Genet. Res., Camb.,* **6**, 474-478.
13b. Allison, A.C. and Paton, G.R. (1965), 'Chromosome damage in diploid cells following activation of lysosomal enzymes', *Nature,* **207**, 1170-1173.
14. Kubitschek, H.E. (1967), 'Mutagenesis by near-visible light', *Science,* **155**, 1545-1546.
15a. Webb, R.B. and Malina, M.M. (1967), 'Mutagenesis in *Escherichia coli* by visible light', *Science,* **156**, 1104-1105.
15b. Simon, M.I., Grossman, L. and Van Vulakis, H. (1965), 'Photosensitized reaction of polynucleotides. I. Effect on their susceptibility to enzyme digestion and their ability to act as synthetic messengers', *J. Mol. Biol.,* **12**, 50-59.
16. Lamola, A.A. and Yamane, T. (1967), 'Sensitized photodimerization of thymine in DNA', *Proc. Nat. Acad. Sci., U.S.A.,* **58**, 443-446.
17. Meistrich, M.L. and Drake, J.W. (1972), 'Mutagenic effects of thymine dimers in bacteriophage T4', *J. Mol. Biol.,* **66**, 107-114.
18. Igali, S., Bridges, B.A., Ashwood-Smith, M.J. and Scott, B.R. (1970), 'Mutagenesis in *E.coli.* IV. Photosensitisation to near-ultraviolet light by 8-methoxypsoralen', *Mutation Res.,* **9**, 21-30.
19. Drake, J.W. and McGuire, J. (1967), 'Properties of *r* mutants of bacteriophage T4 photodynamically induced in the presence of thiopyronin and psoralen', *J. Virology,* **1**, 260-267.
20. Musajo, L., Bordin, F. and Bevilacqua, R. (1967), 'Photoreactions at 3655Å linking the 3-4 double bond of fucocumarins with pyrimidine bases', *Photochem. Photobiol.,* **6**, 927-931.

21. Musajo, L., Bordin, F., Caporale, G., Marciani, S. and Rigatti, G. (1967), 'Photoreactions at 3655 Å between pyrimidine bases and skin-photosensitizing fucocumarins', *Photochem. Photobiol.*, **6**, 711-719.
22. Ben-Hur, E. and Elkind, M.M. (1973), 'Psoralen plus near ultraviolet light inactivation of cultured chinese hamster cells and its relation to DNA cross-links', *Mutation Res.*, **18**, 315-324.
23. Deering, R.A. and Setlow, R.B. (1963), 'Effects of ultraviolet light on thymidine dinucleotide and polynucleotide', *Biochem. Biophys. Acta.*, **68**, 526-534.
24. Goodgal, S.H., Rupert, C.S. and Herriot, R.M. (1957), 'Photoreactivation of *Haemophilus influenzae* transforming factor for streptomycin resistance by an extract of *E.coli B'*, in *The chemical basis of Heredity.* ed. W.D. McElroy and B. Glass; John Hopkins Press, Baltimore. 341-343.
25. Rupert, C.S. (1960), 'Photoreactivation of transforming DNA by an enzyme from bakers' yeast', *J. Gen. Physiol.*, **43**, 573-595.
26. Rupert, C.S. (1962), 'Photoenzymatic repair of ultraviolet damage in DNA. I. Kinetics of the reaction', *J. Gen. Physiol.*, **45**, 703-724.
27. Rupert, C.S. (1962), 'Photoenzymatic repair of ultraviolet damage in DNA. II. Formation of an enzyme-substrate complex', *J. Gen. Physiol.*, **45**, 725-741.
28. Harm, W. and Hillebrandt, B. (1962), 'A non-photoreactivable mutant of *E.coli B'*, *Photochem. Photobiol.*, **1**, 271-272.
29. Resnick, M.A. (1969), 'A photoreactivationless mutant of *Saccharomyces cerevisiae'*, *Photochem. Photobiol.*, **9**, 307-312.
30. Terry, C.E., Kilbey, B.J. and Howe, H.B. (1967), 'The nature of photoreactivation in *Neurospora crassa'*, *Radiation Res.*, **30**, 739-747.
31. Terry, C.E. and Setlow, J.K. (1967), 'Photoreactivating enzyme from *Neurospora crassa'*, *Photochem. Photobiol.*, **6**, 799-803.
32. Muhammed, A. and Tresko, J. (1967), 'A search for a photoreactivating enzyme in *Drosophila melanogaster* extracts', *Genetics,* **56**, 578.
33. Trosko, Y.E. and Wilder, K. (1973), 'Repair of UV-induced pyrimidine dimers in *Drosophila melanogaster* cells *in vitro'*, *Genetics,* **73**, 297-302.

34. Cook, J.S. and Regan, J.D. (1969), 'Photoreactivation and photoreactivating enzyme activity in an order of mammals (Marsupialia)', *Nature,* **223**, 1066-1067.
35. Sutherland, B.M. (1974), 'Photoreactivating enzyme from human leukocytes', *Nature*, **248**, 109-112.
36. Setlow, J.K. (1966), 'Photoreactivation', *Radiation Res. Suppl.*, **6**, 141-155.
37. Harm, W. (1970), 'Analysis of photoenzymatic repair of UV lesions in DNA by single light flashes. VIII Inhibition of photoenzymatic repair of UV lesions in *E.coli* DNA by caffeine', *Mutation Res.*, **10**, 319-333.
38. Mennigmann, H.D. (1972), 'Pyrimidine dimers in pre-mutational lesions in *Escherichia coli* WP2 Hcr^{-}', *Molec. Gen. Genetics,* **117**, 167-186.
39. Meistrich, M.L. and Drake, J.W. (1972), 'Mutagenic effects of thymine dimers in bacteriophage T4', *J. Mol. Biol.*, **66**, 107-114.
40. Witkin, E.M. (1969), 'The role of DNA repair and recombination in mutagenesis', *Proc. 12th Int. Cong. Genetics,* **3**, 225-245.
41. Alper, T. and Forage, A.J. (1974), 'The primary UV lesions for reversion to prototrophy in auxotrophic *Escherichia coli*: inferences from studies of protection by acriflavine', *Molec. Gen. Genetics,* **128**, 147-155.
42. Witkin, E.M. (1966), 'Radiation-induced mutations and their repair', *Science,* **152**, 1345-1353.
43. Resnick, M.A. (1969), 'Induction of mutations in *Saccharomyces cerevisiae* by ultraviolet light', *Mutation Res.*, **7**, 315-332.
44. Kilbey, B.J. and de Serres, F.J. (1969), 'Quantitative and qualitative aspects of photoreactivation of premutational ultraviolet damage at the *ad*-3 loci of *Neurospora crassa*', *Mutation Res.*, **4**, 21-29.
45. Kilbey, B.J. (1967), 'Specificity in the photoreactivation of premutational damage induced in *Neurospora crassa* by ultraviolet', *Molec. Gen. Genetics,* **100**, 159-165.
46. Drake, J.W. (1966), 'Ultraviolet mutagenesis in bacteriophage T4. II. Photoreversal of mutational lesions', *J. Bacter.*, **92**, 144-147.
47. De Serres, F.J. and Kilbey, B.J. (1971), 'Differential photo-

reversibility of ultraviolet-induced premutational lesions in *Neurospora*', *Mutation Res.*, **12**, 221-234.

48. Setlow, R.B. and Setlow, J.K. (1962), 'Evidence that ultraviolet-induced thymine dimers in DNA cause biological damage', *Proc. Nat. Acad. Sci., U.S.A.* **48**, 1250-1257.
49. Setlow, J.K., Boling, M.E. and Bollum, F.J. (1965), 'The chemical nature of photoreactivable lesions in DNA', *Proc. Nat. Acad. Sci., U.S.A.* **53**, 1430-1436.

CHAPTER FOURTEEN

The molecular action of ultraviolet and visible light. II: Dark repair. Mechanisms of mutation induction

At the end of Chapter 10, two related phenomena in UV-mutagenesis of bacteria have been described: mutation frequency decline (MFD) and mutation fixation (MF). It will be recalled that MFD is the irreversible loss of potential mutations to prototrophy by inhibition of protein synthesis and that MF is the event, presumably DNA synthesis, that terminates the period during which MFD can occur. In 1961 (1) Witkin and, independently, Lieb (2) suggested that MFD might be due to an enzymatic repair process that did not require visible light. This was the first suggestion that there might be dark repair of UV damage. At present, several dark repair systems are known in a variety of organisms, prokaryotes as well as eukaryotes, and their study forms a major area of research on radiation damage (see Bibliography: *Radiation Res.* Suppl. 1966; *Cold Spring Harbor Symp.* 1968; Witkin, 1969).

We have seen that the study of photorepair profited from the fact that *Haemophilus influenzae* lacks photoreactivating enzyme. In a similar way, the study of dark repair profited from the discovery of UV-sensitive strains that lack certain dark repair systems. The first highly UV-sensitive strain of *E.coli*, B_{s-1}, was discovered in 1958 (3) and still is much used in experiments on dark repair. We have met it already in Table 13.1, where it was contrasted with the resistant strain *B/r*. Even more resistant strains are found among other bacterial species. The most resistant one has been called *Micrococcus radiodurans*; it can tolerate without loss of viability a dose of UV that reduces the survival of *B/r* to about 10^{-3} (4). When pyrimidine dimers are assayed in the DNA of irradiated bacteria, they are found to be of approximately equal frequency in sensitive and resistant strains. Thus, the DNA of resistant strains is not less susceptible to UV-damage than is the DNA of sensitive ones. Rather, resistance is due to the capacity for dealing with pyrimidine dimers either before or after replication of DNA. We shall deal first with pre-replication repair.

Pre-replication repair

Excision repair

This type of repair was discovered simultaneously and independently by two groups of workers in the U.S.A., using essentially the same procedure (5,6). The DNA of two closely related bacterial strains – one resistant, the other sensitive to UV – was labelled with ^{3}H-thymidine and the fate of the UV-induced pyrimidine dimers was followed in extracts from treated cells during one or two hours of post-treatment growth. In both strains, the total frequency of pyrimidine dimers remained unchanged. In the sensitive strains, there was also no change in the distribution of the dimers among the acid-insoluble fraction of the extract (DNA) and the acid-soluble fraction (free polynucleotides). In the resistant strains, the picture was different. The frequency of dimers in DNA decreased gradually and, concomitantly, their frequency in the soluble fraction increased. Fig. 14.1 shows the results of an experiment in which the disappearance of dimers from irradiated DNA was compared in B/r and B_{s-1}. When isolated from the soluble fraction, the dimers are found in short oligonucleotides, but this may in part be due to degradation of longer pieces by a cytoplasmic nuclease.

Both groups of workers drew the conclusion that dark repair acts by excising pyrimidine dimers from UV-irradiated DNA and that the gaps left after excision are filled in again by nucleotides, with the complementary strand of DNA serving as template. In outline, then, there had to be three major steps: excision of a stretch of single-stranded DNA carrying a dimer; filling in of the gap by nucleotides; tying together of the old and new portions of DNA. Fig. 14.2 shows a schematic picture of this 'cut-and-patch' mechanism. A slightly modified model – the 'patch-and-cut' mechanism – assumes that the first cut leads to a peeling back of the cut strand and that new nucleotides are inserted into each space as it is formed.

Experiments by Hanawalt and his collaborators brought proof for the patching step of the repair process. Essentially, their object was to

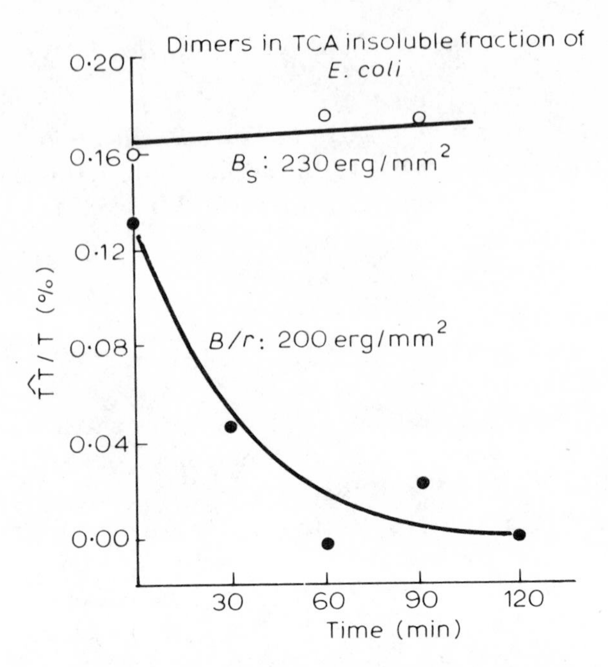

Fig. 14.1. The excision of dimers from the DNA of resistant cells (*B/r*) and their conservation in sensitive cells (B_S). (Fig. 6, Setlow (7). Courtesy Wistar Inst. of Anatomy and Biology.

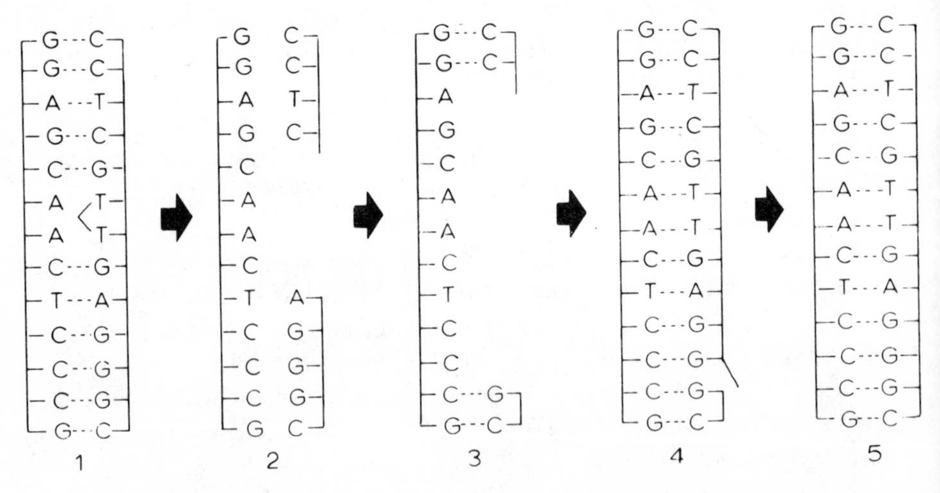

Fig. 14.2. Excision repair of pyrimidine dimers produced by UV in the DNA of *E.coli* ('cut-and-patch' model). 1) Thymine dimer in DNA; 2) section of DNA strand bearing dimer excised. 3) gap enlarged by exonuclease degradation. 4) gap filled by repair synthesis. 5) single-strand break sealed by DNA ligase. (Fig. 2, Witkin (8). Courtesy The Science Council of Japan.

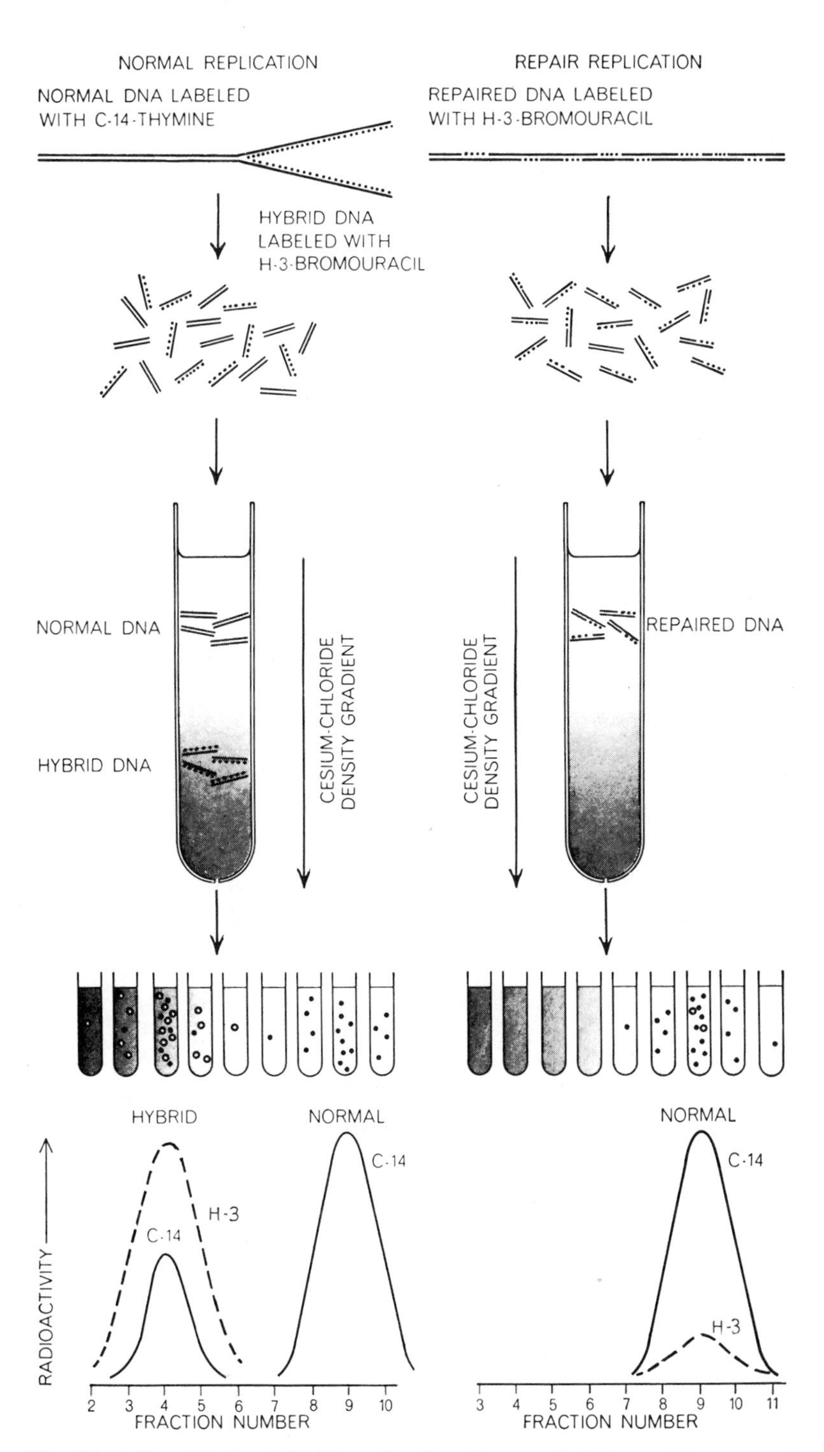

Fig. 14.3. Two kinds of DNA replication (see text). (Fig. p. 42, Hanawalt (10). Courtesy Scientific American.

show that 'repair synthesis' of UV-damaged DNA differs from normal semi-conservative DNA-replication by occurring in small, randomly placed patches. Fig. 14.3 shows one of their experiments. A thymine-requiring strain of bacteria was grown in radioactively tagged thymine so as to label its DNA. After irradiation, thymine was replaced by its analogue, 5-bromouracil (BU), which carried a different radioactive label. 5-BU is chemically so similar to thymine that it is readily taken up into bacterial DNA instead of thymine. It differs from thymine in having a higher molecular weight. As a result, when unirradiated cells are grown for a short time in the presence of BU and their DNA is then broken into fragments that can be banded in a density gradient, two peaks are found. The lighter one contains the still unreplicated DNA, while the heavier one consists of fragments with newly replicated, 'hybrid' DNA (left part of Fig. 14.3). When the same procedure was applied to fragments of newly repaired DNA, the result was quite different (right part of Fig. 14.3). The fragments now banded all in the same region whether or not they contained BU. This shows (*a*) that BU had been inserted randomly into DNA and (*b*) that it had been inserted in amounts that were too small to change the normal density noticeably.

This method has since been used to test for the presence of excision repair in widely different cell systems. In synchronized cultures, it is often accompanied or replaced by a test for 'unscheduled DNA synthesis', i.e. for DNA synthesis outside the S-period or under conditions, such as the presence of hydroxyurea, when normal semi-conservative replication is inhibited. A particularly interesting result was the finding that, while human cells in general are capable of excision repair, this ability is lacking in cells from patients with Xeroderma pigmentosum, a disease in which the skin develops malignant tumours under the influence of sunlight (11).

UV-sensitive strains that, by various tokens, appear to lack excision repair have been reported for a number of fungal species e.g. (12,13). This implies the existence of an excision system in the wild-type strains of these species. Evidence of excision has been obtained in *Drosophila* (14) and yeast (15). Because of lack of thymidine kinase, *Neurospora* DNA cannot be unambiguously labelled with thymidine. A novel technique for the labelling of cytosine has shown that cytosine dimers are rapidly excised during the first hour after exposure to UV of wavelength 254 nm (16).

Obviously, several enzymes have to be postulated for successful completion of excision repair: an endonuclease for making the first cut near a dimer; an exonuclease for peeling off a stretch of DNA containing the dimer; a polymerase for inserting new nucleotides into the gap; a ligase for tying the old and the new portions together. Such enzymes have been isolated, and bacterial strains have been found that are UV-sensitive through lack of one or several of them (17-20). A detailed study of the functions of Kornberg's DNA-polymerase I showed that this enzyme acts *in vitro* both as an exonuclease and a polymerase (21). *In vivo*, however, the situation is more complex and other enzymes too are engaged in excision.

The excision-repair enzymes are not species-specific. Extracts from the radio-resistant bacterium *Micrococcus lysodeikticus* (*M. luteus*) produce dark repair in two-stranded bacteriophage and in isolated DNA (22). Bacterial strains able to carry out excision repair are also able to carry out host-cell-reactivation (HCR), i.e. to increase the survival of certain kinds of UV-irradiated phages with which they are infected (hcr^+ strains) (23). In contrast, bacterial strains lacking excision repair cannot reactivate UV-irradiated phages (hcr^- strains). As would be expected from the template function of the undamaged strand, there is no HCR of single-stranded phage. Presence or absence of HCR is a convenient way of detecting the presence or absence of excision repair in a bacterial strain. UV-sensitivity alone is no sufficient indication for, as we shall see presently, strains that can carry out excision repair may be sensitive for other reasons.

Certain chemicals, in particular caffeine and acriflavine, affect UV-induced mutation frequencies by interfering with repair. We have seen in Chapter 13 (p. 219) that acriflavine given *before* UV protects the DNA against dimer formation. Given *after* UV, it has the opposite effect (1). By binding to the damaged sites or, as has been suggested (24), to DNA replicating in contact with the bacterial plasma membrane, it reduces the efficiency of excision repair and thus increases the frequency of mutations. Recently, it has been found that 8-methoxypsoralene, which sensitizes DNA to short-wave visible light (p. 221), resembles acriflavine in protecting *E. coli* against the mutagenic action of far-UV (254nm) when given *before* irradiation, while enhancing this action when present *after* UV in the plating medium (25). The action of caffeine (26,27) is more complex. Like acriflavine, it increases mutation frequencies by interfering with excision repair, probably through enzyme inhibition;

but this is partially offset by its interference with another type of repair which, as we shall see presently (p.245), is *required* for mutation. The net effect on mutation frequency depends on dose of UV and other experimental conditions. (See article by Witkin in 'Mutation as a Cellular Process', Bibliography).

Liquid holding recovery. Photoprotection

The efficiency of excision repair can be increased by allowing more time for repair before the first post-treatment division. This is the basis of two related phenomena: liquid holding recovery (LHR) and photoprotection (PP). When UV-irradiated bacteria are kept for some time in buffer before plating, survival increases with time of liquid holding (28). Previous photoreactivation reduces the magnitude of this LHR; vice versa, photoreactivation is less effective after a period of LHR. Either of the two mechanisms, when carried to completion, prevents occurrence of the other (29). Thus LHR, like photoreactivation, appears to act on pyrimidine dimers, presumably removing them by excision. As expected on this assumption, LHR does not occur in excision-deficient strains of bacteria (30) nor in bacteriophages inside an *hcr*$^{-}$ host (31). In yeast, LHR has also been observed (32).

Other means of delaying growth after irradiation have the same restorative effect. One of these is exposure to near-UV, i.e. to the range of UV present in sunlight (300 to about 360 nm) (33). When cells are exposed to near UV *before* exposure to far UV (less than 300 nm), the lethal and mutagenic effects of the latter are reduced. Obviously, this 'photoprotection' cannot be due to the splitting of dimers, which are not yet present at the time of exposure to near UV. Moreover, PP is also effective in cells lacking photoreactivating enzyme. The facts that PP acts only in excision-competent cells (34) and that its action completely overlaps that of LHR (35) is evidence that both act via excision repair. PP can be obtained also by exposing cells to visible light immediately *after* exposure to far-UV. Without further analysis, it may then be mistaken for enzymatic photorepair, a source of error that has been pointed out before (p. 223). While photoprotection increases the efficiency of excision repair through extending the time available for it, enzymatic photorepair may in certain conditions achieve the same result by drastically reducing the number of photolesions with which excision repair has to deal. We shall return to this point later on in this chapter (p. 244).

Mutation frequency decline

MFD was the first phenomenon to suggest the existence of dark repair systems. It appears that it is due to excision repair of a rather special kind. Unlike other types of dark repair, it acts specifically on supersuppressors (36), i.e. suppressors of nonsense mutations acting by means of mutant t-RNA. Lethality is not affected by MFD, nor are other kinds of mutation, such as true reverse mutations to prototrophy or mutations to streptomycin resistance. The conclusion that MFD acts via excision of dimers is supported by a number of observations, most of all by the absence or low efficiency of MFD in strains without excision repair, and by the greatly reduced rate of dimer excision in a strain (MFD^-) that lacks MFD (37). The specificity of MFD for nonsense suppressors is open to various interpretations, none of which has so far been proven by experiment. The final explanation will have to take account of the specificity of MFD for mutation versus killing, for hcr^+ versus hcr^- strains, and for supersuppressors versus other mutations. Under different conditions, (38,39) and with different bacterial strains (40), less specific broth effects have been found. A tentative model for the effects of broth on UV mutagenesis has been proposed (41).

Post-replication repair

Excision does not usually remove all pyrimidine dimers from irradiated DNA, but most bacterial strains can tolerate quite a large number of unexcised dimers. Thus, in K12, an excision-deficient strain of *E. coli*, 37% of the cells can form colonies when their DNA contains an average of 50 dimers (Table 14.1, p. 243). On the other hand, the most highly UV-sensitive strains cannot form colonies from cells that carry only one dimer per strand of DNA. From these findings, two conclusions can be drawn. (1) In addition to the excision repair system, there exists a second repair system that can deal with unexcised dimers. (2) Without this second repair system, replication of DNA is prevented by even one dimer. Presumably, since a dimer cannot serve as template for replication, it leads to the formation of a gap in the new strand and this prevents further replication. The question arises by what means the second type of dark repair can fill in these 'replication-gaps' (in contrast to the 'excision gaps' that are filled in before replication).

Rupp and Howard-Flanders (43) studied the mechanism of

post-replication repair in the excision-deficient strain uvrA6, using a technique that allows the isolation of long pieces of single-stranded DNA and the determination of their molecular weight by sedimentation in a sucrose gradient. Originally, this technique had been devised for studying the production of single-strand breaks by X-rays and their repair by ligase (44). It was now used to show that, after UV-irradiation, single strands of newly formed DNA were broken into fragments that were much smaller than the fragments produced by the same extraction technique in control DNA, their length corresponding roughly to the estimated distance between UV-induced dimers. During further incubation, the fragments again became longer until they had reached control length. Thus, the gaps left by replication past dimers were filled in by a post-replication repair process. Evidently, this cannot have been done in the same way as in excision repair. In the latter, the gap-filling mechanism has a template available on the complementary strand; in post-replicative repair, the complementary strand carries a pyrimidine dimer opposite the gap. The fact that the first strain to show lack of post-replication repair was characterized by extreme reduction of recombination (45), suggested that recombination is involved in post-replication repair. The term recombination repair is used as an alternative to post-replication repair; we shall discuss presently to what extent this is justified.

Table 14.1 shows survival after UV-irradiation of four strains of *E.coli* K12: a wild-type strain, an excision-deficient mutant (uvrA), a recombination-deficient mutant (recA), and a strain carrying both mutant genes and, therefore, deficient both for excision and recombination repair. The survival level chosen for comparison is 37%, corresponding to a dose of UV that produces on the average one lethal hit per cell. It will be seen that of the more than 3000 dimers tolerated by the wild-type strain, the majority are removed either by excision repair or by recombination repair. It appears that dimers not removed by excision can be removed after replication: if both types of repair are defective, even one dimer is usually lethal. It should be noted that, as would be expected, only the excision-competent strain shows LHR; in Table 14.1 this is seen from the numbers with the suffix ‡.

Experiments in which parental and newly synthesized DNA could be distinguished by label have provided convincing evidence that some kind of recombination is involved in post-replicative repair (46a).

Table 14.1
UV doses for 37% survival of mutant strains of *E.coli* K12*. (Table 1, Howard-Flanders (42). Courtesy Annual Reviews Inc.)

Strain number	*Relevant genotype*		*Characteristics*	*UV doses for 37% survival*	
	uvrA	*recA*		*ergs/mm*^{2}	*dimers per 10*^{7} *bases*†
AB1157	+	+	wild type	500	3200
AB1886	–	+	excision defective	8	50
AB2843	+	–	recombination deficient	3	20
				50†	300‡
AB2480	–	–	excision defective and recombination deficient	0.2	1.3
				0.2†	1.3‡

* Cells were grown overnight in yeast extract tryptone broth, exposed to UV light, plated on complete medium, and incubated at 37^0. The UV doses reducing the colony-forming ability to 37% or unirradiated controls are listed. For the significance of 37% survival see legend to Table 13.1.

† The *E.coli* genome contains slightly less than 10^7 bases.

‡ These cells were incubated in buffer for 6 h before plating on complete medium.

When the fragments of newly synthesized DNA have been repaired into coherent daughter strands, these contain insertions of parental DNA, and the number of insertions is more or less equal to the number of original gaps. Very recent experiments (46b) have shown that dimers which are included in the exchange regions are transferred to daughter strands and, in excision-deficient strains, may become gradually diluted out over successive generations.

The connection between post-replication repair and genetic recombination is not yet clear. A correlation between UV-sensitivity, due to lack of post-replicative repair, and ability for genetic recombination has been found in several bacterial and fungal strains; but there are also exceptions (47-50). At the moment, no more can be said about the relation between post-replicative repair and genetic recombination than that they appear to have one or more steps in common. Moreover, there are indications of a second type of post-replicative repair which does not proceed via recombination but fills gaps directly from the nucleotide pool in the cell, probably using the complementary strand of the sister double helix as template (51-53). In mammalian cells, caffeine has been shown to inhibit post-replicative repair (54-55), and it is also likely to do so in bacteria.

The mechanism of UV-mutagenesis

We now have to ask how the nature of the UV-induced lesions in DNA and the mechanisms by which they are repaired can account for the production of mutations. The first question to ask is whether pyrimidine dimers are the only source of mutation. There is no doubt that they are an important source (56,57). In cells that have been photosensitized to black light by acetophenone, practically all mutations arise from pyrimidine dimers, and in UV-irradiated hcr^- cells of *E.coli* the vast majority do so. But this does not necessarily apply to mutations produced by UV in hcr^+ cells. It is true that also in these cells excision repair, by removing dimers, photorepair, by resolving them, and acridine pretreatment by interfering with their formation reduce mutation frequency. Excision repair, however, is also known to remove other types of damage in DNA, and it might be able to do this more efficiently when it has fewer pyrimidine dimers to deal with. Thus, in hcr^+ strains, photorepair and acridine protection might act indirectly by making excision repair more effective (p. 240). The possibility that photoproducts other than pyrimidine dimers of the cyclobutane type result in mutations can, therefore, not be dismissed. Indeed, Witkin has come to the conclusion that, in hcr^+ strains of *E.coli,* suppressor mutations resulting in auxotrophy may be mainly due to unknown photoproducts, and that their apparent reparability by visible light is of the indirect type (see *Rad. Res.* Suppl., bibliogr.). For bacteriophage T4, Meistrich and Drake (58) have shown that, although thymine dimers are mutagenic, other photolesions provide a sizeable contribution to the total mutation yield.

Next let us consider the role of repair mechanisms in UV-mutagenesis. If pyrimidine dimers or other photoproducts were the direct cause of mutations, every kind of repair that eliminates these products should reduce the frequency of mutations. This, however, is not the case. The possibility that, under certain circumstances, repair may produce mutations rather than erase them was mooted by several scientists. In 1967 Witkin (59) provided convincing evidence that this may indeed happen. Her distinction between error-proof (mutation-proof) and error-prone (mutation-prone) repair has been generally accepted. Both types of repair remove lethal lesions and increase survival; but while the former restores the original nucleotide sequence and thus removes potential mutations, the latter tends to insert a wrong base sequence and thus produces mutations. Photorepair is error-proof; it repairs lethal lesions as well as mutational ones, although not necessarily to the same degree. The bulk of excision repair, too, is error-proof and reduces mutation frequency as well as lethality. Thus, generally, excision gaps are filled accurately in accordance with the template on the complementary strand. A small fraction of excision gaps however are filled incorrectly and give rise to mutations. This was concluded from the observation that, in *hcr*$^+$ but not in *hcr*$^-$ strains, mutations to streptomycin resistance lose their photoreversibility within 20 min after irradiation and before the onset of DNA synthesis (60,61). The simplest interpretation is that there are two kinds of excision repair, one – the prevalent one – being error-proof, the other – much rarer one – being error-prone.

Only a small fraction of UV-induced mutations arise from errors in prereplicative excision repair. The bulk are produced by an error-prone post-replicative repair process. Mutagenesis by either type of repair requires the normal functioning of two genes: *recA*$^+$ and *lex*$^+$ (or *exr*$^+$)*. When either of these genes has been inactivated by mutation, UV produces no mutations whatever. While *recA* also prevents recombination, *lex* does not. Both genes sensitize the cells to the killing action of UV, but *lex* does so only slightly. Moreover, Bridges has found that the repair ability of *lex*$^+$ and its function in UV-mutagenesis can be separated by their different responses to suppressors and to physiological conditions (62). This makes it likely that the error-prone step is connected with the *lex*$^+$ function, while the *recA*$^+$ function is important for some earlier step in the repair pathway. The normal recombining ability of *lex* strains speaks against recombination as a source of mutations in bacteria. A hypothesis that agrees best with the available evidence is that the same error-prone mechanism is involved in the faulty filling of excision gaps and replication gaps. Cooper and Hanawalt (63,64) have found that excision gaps may be filled by 'short' or 'long' patches and that the latter occur only in cells that carry the normal alleles of *recA*

* *lex*$^+$ and *exr*$^+$ are abbreviations for the same gene in two different strains of *E.coli.*

and *lex*. This suggests that the long patches may serve as sources of mutation. Recently Witkin and others have adduced evidence that the error-prone mechanism requires an inducible enzyme for its action (65-67). A normally functioning exr^+ (lex^+) gene is also required for the production of mutations by ionizing radiation (68a). This indicates that mutagenesis by UV and by X-rays have one or more steps in common.

Genes that resemble *recA* and *lex* in their action on induced mutations have been found in bacteriophage T4 (68b). Also in fungi, there is evidence for an error-prone type of repair. Strains that are sensitive to killing by UV but refractory to its mutagenic action have been found in various species (50, 69). In yeast, mutations at three different loci reduce UV-induced mutability while increasing sensitivity to killing (70).

The possibility has been considered that post-replicative repair may be interfered with by transcription. In an operator-constitutive mutation of the histidine operon in *Salmonella*, UV induced many more ochre reversions and frameshifts than in strains carrying the normal operator. The effect was confined to the histidine operon and did not depend on excision repair. It was thought to be due to the permanently derepressed state of the operon (70a).

The molecular nature of UV-induced mutations

It is obvious that mutagenesis by ultraviolet light is a highly complex process involving a variety of photoproducts and a variety of repair mechanisms that deal with them. A complete picture has not yet emerged; the present, provisional, one is likely to change as new data are obtained. A full explanation will also have to take account of the peculiar idiosyncrasies of UV mutagenesis at the molecular level. Although thymine dimers are the most frequent type of photolesion and the majority of excision gaps will be opposite to a dimer, UV-induced mutations are rarely transversions or transitions from AT to another base pair. Drake (71,72) applied reversion analysis (Chapter 17) to UV-induced mutations in the rII region of phage T4. About half of them were classified as GC to AT transitions, the remainder as frameshifts; a few transversions might have gone undetected by the technique used. The result was the same whether the phage was irradiated *in vitro* or inside its host; it applied also to phage that had been exposed to acetophenone-sensitized irradiation (73), when the vast majority of photoproducts are thymine-dimers. Psoralen-sensitized irradiation gave a different spectrum (74), as would be expected from the different primary lesion it induces (p. 221). In the single-stranded DNA of phage S13 (75), most – or perhaps all – of 16 UV-induced mutants were due to transitions from C to T. This confirms Drake's assumption that, in the GC to AT transitions of double-stranded DNA, the pyrimidine is the primarily affected base.

In bacteria, an analysis carried out by Person's method (p.196) confirmed the conclusions drawn for phage, at least to the extent that about half the UV-induced mutants were classified as transitions from GC to AT, and very few as transitions from AT to GC (76,77). In yeast, on the contrary, there seems to be a lack of preference for one or the other type of transition. This was concluded from a study of interconversions between amber and ochre mutants (p. 196), which occurred with similar frequencies in both directions (78), and from the polypeptide sequencing of UV-induced revertants of nonsense mutants in the iso-1-cytochrome *c* gene (79). Instead, this latter analysis revealed unexpected and interesting complexities. True reversions of UAA may produce six different amino acids (see Fig. 11.6), and all of these were indeed found among spontaneous revertants at two different sites of the gene. UV, however, induced only one of these at one of the sites, and two different ones at the other. In *Neurospora,* analysis of UV-induced *ad*-3B mutants by complementation pattern and reversion analysis led to the conclusion that about half of them were frameshifts (80).

In addition to gene mutations, UV produced deletions in *Salmonella* (81) and *E.coli* (82,84), and duplications in *E.coli* (83). The mechanism by which these chromosomal changes arise in prokaryotes is not yet fully understood, but it is known that the *recA* function is an indispensable part of it. It is unlikely to resemble the breakage-reunion mechanism by which chromosome or chromatid rearrangements are formed in eukaryotes. In *Neurospora,* on the contrary, this latter mechanism may account for UV-induced deletions (85). The fact that deletions produced in this way require two independent primary lesions might conceivably explain why they show a stronger response to photorepair than do intracistronic changes.

While we are still far from understanding the mechanism of UV-mutagenesis, some essential conclusions can already be drawn. (1) It involves cellular processes acting on primarily induced lesions in DNA. (2) There is no discernible correlation between the kind of primary lesion and the final mutational change. Thymine dimers produce both frameshifts and base changes. Moreover, the overall spectrum of mutations to rII in phage T4 was very much the same whether treatment was given by direct irradiation with far-UV or by acetophenone-sensitized irradiation with near-UV (p. 220), although the latter treatment produces pyrimidine dimers almost exclusively, while the former also yields other mutagenic photoproducts. This implies that photoproducts other than pyrimidine dimers give rise to the same types of mutation in approximately the same proportions as do dimers. (3) In at least one case it has been shown (p.247)that the molecular nature of UV-induced mutations depends not only on the affected base but also on its position in the nucleotide chain of the gene.

It seems that frequencies and types of UV-induced mutation, although ultimately dependent on the production of photoproducts in DNA, are to a large extent expressions of those cellular processes that intervene between the primary lesions and the appearance of mutant cells. This is a lesson to be kept in mind when we next deal with chemically induced mutations in which the connection between primary damage in DNA and final mutational change seems, at first sight, very close.

References

1. Witkin, E.M. (1961), 'Modification of mutagenesis initiated by ultraviolet light through post-treatment of bacteria with basic dyes', *J. Cell. Comp. Physiol.*, **58**, 135-144.
2. Lieb, M. (1961), 'Enhancement of ultraviolet induced mutations in bacteria by caffeine', *Zeitschr. Vererbungslehre* **92**, 416-429.
3. Hill, R.F. (1958), 'A radiation sensitive mutant of *E.coli*', *Biochim. Biophys. Acta.*, **30**, 636-637.
4. Setlow, J.K. and Duggan, D.E. (1964), 'The resistance of *Micrococcus radiodurans* to ultraviolet radiation. I. Ultraviolet induced lesions in the cell's DNA', *Biochim. Biophys. Acta.* **87**, 664-668.
5. Setlow, R.B. and Carrier, W.L. (1964), 'The disappearance of thymine dimers from DNA; an error-correcting mechanism', *Proc. Nat. Acad. Sci. U.S.A.* **51**, 226-231.

6. Boyce, R.P. and Howard-Flanders, P. (1964), 'Release of ultraviolet light-induced thymine dimers from DNA in *E.coli* K-12', *Proc. Nat. Acad. Sci. U.S.A.* **51**, 293-300.
7. Setlow, R.B. (1967), 'Physical changes and mutagenesis', *J. Cell. Comp. Physiol.* **64** (suppl. 1), 51-68.
8. Witkin, E.M. (1969), 'The role of DNA repair and recombination in mutagenesis', *Proc. 12th Int. Cong. Genetics* **3**, 225-245.
9. Pettijohn, D. and Hanawalt, P. (1964), 'Evidence for repair-replication of ultraviolet damaged DNA in bacteria', *J. Mol. Biol.* **9**, 395-410.
10. Hanawalt, P.C. and Haynes, R.H. (1967), 'The repair of DNA', *Scient. Am.*, **216**, (2), 36-43.
11. Cleaver, J.E. (1968), 'Defective repair replication of DNA in Xeroderma pigmentosum', *Nature* **218**, 652-656.
12. Kilbey, B.J. and Smith, S. (1969), 'Similarities between a UV-sensitive mutant of yeast and bacterial mutants lacking excision-repair ability', *Molec. Gen. Genetics* **104**, 253-257.
13. Nakai, S. and Matsumoto, S. (1967), 'Two types of radiation-sensitive mutant in yeast', *Mutation Res.* **4**, 129-136.
14. Trosko, J.E. and Wilder, K. (1973), 'Repair of UV-induced pyrimidine dimers in *Drosophila melanogaster* cells *in vitro*', *Genetics* **73**, 297-302.
15. Unrau, P., Wheatcroft, R. and Cox, B.S. (1971), 'The excision of pyrimidine dimers from DNA of ultraviolet irradiated yeast', *Molec. Gen. Genetics* **113**, 359-362.
16. Worthy, T.E. and Epler, J.L. (1973), 'Biochemical basis of radiation-sensitivity in mutants of *Neurospora crassa*', *Mutation Res.* **19**, 167-173.
17. Kaplan, J.C., Kushner, S.R. and Grossman, L. (1969), 'Enzymatic repair of DNA. 1. Purification of two enzymes involved in the excision of thymine dimers from ultraviolet-irradiated DNA', *Proc. Nat. Acad. Sci. U.S.A.* **63**, 144-151.
18. Friedberg, E.C. and King, J.J. (1971), 'Dark repair of ultraviolet-irradiated deoxyribonucleic acid of bacteriophage T4: purification and characterization of a dimer-specific phage-induced endonuclease', *J. Bacter.* **106**, 500-507.
19. Fareed, G.C. and Richardson, C.C. (1967), 'Enzyme breakage and joining of deoxyribonucleic acid. II. The structural gene for polynucleotide ligase in bacteriophage T4', *Proc. Nat. Acad. Sci. U.S.A.* **58**, 665-672.

20. De Lucia, P. and Carins, J. (1969), 'Isolation of an *E.coli* strain with a mutation affecting DNA polymerase', *Nature* **224**, 1164-1166.
21. Kelly, R.B., Atkinson, M.R., Huberman, J.A. and Kornberg, A. (1969), 'Excision of thymine dimers and other mismatched sequences by DNA polymerase of *Escherichia coli*', *Nature* **224**, 495-501.
22. Heijneker, H.L., Pannekoek, H., Oosterbaan, R.A. Pouwels, P.H., Bron, S., Arwert, F. and Venema, G. (1971), '*In vitro* excision-repair of ultraviolet-irradiated transforming DNA from *Bacillus subtilis*', *Proc. Nat. Acad. Sci. U.S.A.* **68**, 2967 - 2971.
23. Ellison, S.A., Feiner, R.R. and Hill, R.F. (1960), 'A host effect on bacteriophage survival after ultraviolet irradiation', *Virology* **11**, 294-296.
24. Forage, A.J. and Alper, T. (1973), 'Evidence for differing modes of interaction of acriflavine with ultraviolet-induced lesions in an *hcr*$^+$ bacterial strain' *Molec. Gen. Genetics* **122**, 89-100.
25. Bridges, B.A. (1971), 'Genetic damage induced by 254 nm ultraviolet light in *Escherichia coli*: 8-Methoxypsoralen as protective agent and repair inhibitor', *Photochem. Photobiol.* **14**, 659-662.
26. Lieb, M. (1961), 'Enhancement of ultraviolet-induced mutation in bacteria by caffeine', *Zeitschr. Vererblehre* **92**, 416-429.
27. Sideropoulos, A.S. and Shankel, D.M. (1968), 'Mechanism of caffeine enhancement of mutations induced by sublethal ultraviolet dosages', *J. Bacter.* **96**, 198-204.
28. Roberts, R.R. and Aldous, E. (1949), 'Recovery from ultraviolet irradiation in *Escherichia coli*', *J. Bacter.* **57**, 363-375.
29. Castellani, A., Jagger, J. and Setlow, R.B. (1964), 'Overlap of photoreactivation and liquid holding recovery in *Escherichia coli* B., *Science* **143**, 1170-1171.
30. Ganesan, A.K. and Smith, K.C. (1969), 'Dark recovery process in *Escherichia coli* irradiated with ultraviolet light. II. Effect of *uvr* genes on liquid holding recovery', *J. Bacter.* **97**, 1129-1133.
31. Martignoni, S. and Harm, W. (1972), 'Liquid-holding effects in UV-irradiated phage T3', *Photochem. Photobiol.* **14**, 659-662.

32. Parry, J.M. and Cox, B.S. (1968), 'The effects of dark holding and photoreactivation on ultraviolet light-induced mitotic recombination and survival in yeast', *Genetic Res.* **12**, 187-198.
33. Jagger, J. (1960), 'Photoprotection from ultraviolet killing in *Escherichia coli* B', *Radiation Res.* **13**, 521-539.
34. Lakchaura, B.D. (1972), 'Photoprotection from killing in ultraviolet sensitive *E.coli* K-12 mutants; involvement of excision-resynthesis repair', *Photochem. Photobiol.* **16**, 197-202.
35. Jagger, J., Wise, W.C. and Stafford, R.S. (1964), 'Delay in growth and division induced by near ultraviolet radiation in *Escherichia coli* B and its role in photoprotection and liquid holding recovery', *Photochem. Photobiol.* **3**, 11-24.
36. Bridges, B.A., Dennis, R.E. and Munson, R.J. (1967), 'Differential induction and repair of ultraviolet damage leading to true reversions and external suppressor mutations of an ochre codon in *Escherichia coli B/r* WP2', *Genetics* **57**, 897-908.
37. Witkin, E.M. (1966), 'Radiation-induced mutations and their repair', *Science* **152**, 1345-1353.
38. Clarke, C.H. and Hill, R.F. (1972), 'Mutation frequency decline for streptomycin-resistant mutations induced by ultraviolet light in *Escherichia coli B/r', Mutation Res.* **14** 247-249.
39. Witkin, E.M. and Wermundson, J.E. (1973), 'Do ultraviolet-induced mutations to streptomycin resistance exhibit susceptibility to mutation frequency decline?' *Mutation Res.,* **19**, 261-264.
40. Green, M.H.L., Rothwell, M.A. and Bridges, B.A. (1972), 'Mutation to prototrophy in *Escherichia coli* K12: effect of broth on UV-induced mutation in strain AB1157 and four excision-deficient mutants', *Mutation Res.* **16**, 225-234.
41. Bockrath, R. and Cheung, M.K. (1973), 'The role of nutrient supplementation in UV mutagenesis of *E.coli', Mutation Res.* **19**, 23-32.
42. Howard-Flanders, P. (1968), 'DNA repair', *Ann. Rev. Biochem.* **37**, 175-200.
43. Rupp, W.D. and Howard-Flanders, P. (1968), 'Discontinuities in the DNA synthesised in an excision-defective strain of *Escherichia coli* following ultraviolet irradiation', *J. Mol. Biol.* **31**, 291-304.
44. McGrath, R.A. and Williams, R.W. (1966), 'Reconstruction

in vivo of irradiated *Escherichia coli* deoxyribonucleic acid; the rejoining of broken pieces', *Nature* **212**, 534-535.

45. Clark, A.J. and Margulies, A.D. (1965), 'Isolation and characterization of recombination-deficient mutants of *Escherichia coli*', *Proc. Nat. Acad. Sci. U.S.A.* **53**, 451-459.

46a. Rupp, W.D., Wilde III, C.E., Reno, D.L. and Howard-Flanders, P. (1971), 'Exchanges between DNA strands in ultraviolet-irradiated *Escherichia coli*', *J. Mol. Biol.* **61**, 25-44.

46b. Ganesan, A. (1974), 'Persistence of pyrimidine dimers during post-replication repair in ultraviolet light irradiated *E.coli* K12', *J. Mol. Biol.* **87**, 103-120.

47. Jansen, G.J.O. (1970), 'Abnormal frequencies of spontaneous mitotic recombination in *uvs B* and *uvs C* mutants of *Aspergillus nidulans*', *Mutation Res.*, **10**, 33-41.

48. Holliday, R. (1967), 'Altered recombination frequencies in radiation sensitive strains of *Ustilago*', *Mutation Res.* **4**, 275-288.

49. Fabre, F. (1972), 'Relation between repair mechanisms and induced mitotic recombination after UV-irradiation in the yeast *Schizosaccharomyces pombe*', *Molec. Gen. Genetics* **117**, 153-166.

50. Rodarte-Ramon, U.S. and Mortimer, R.K. (1972), 'Radiation-induced recombination in *Saccharomyces*: isolation and genetic study of recombination-deficient mutants', *Radiation Res.* **49**, 133-147.

51. Leclerc, J.E. and Setlow, J.K. (1972), 'Post-replication repair of ultraviolet damage in *Haemophilus influenzae*', *J. Bacter.* **110**, 930-934.

52. Lehmann, A.R. (1972), 'Post-replication repair of DNA in ultraviolet-irradiated mammalian cells', *J. Mol. Biol.* **66**, 319–337.

53. Buhl, S.N., Setlow, R.B. and Regan, J.D. (1972), 'Steps in DNA chain elongation and joining after ultraviolet irradiation of human cells', *Int. J. Rad. Biol.* **22**, 417-424.

54. Cleaver, J.E. and Thomas, G.H. (1969), 'Single strand interruptions in DNA and the effects of caffeine in Chinese hamster cells irradiated with ultraviolet light', *Biochem. Biophys. Res. Comm.* **36**, 203-208.

55. Lehmann, A.R. and Kirk-Bell, S. (1974), 'Effects of caffeine and theophylline on DNA synthesis in unirradiated and UV-irradiated mammalian cells', *Mutation Res.* **26**, 73-82.

56. Mennigmann, H.D. (1972), 'Pyrimidine dimers as premutational lesions in *Escherichia coli* WP2 *hcr*$^-$', *Molec. Gen. Genetics* **117**, 167-186.
57. Meistrich, M.L. and Shulman, R.G. (1969), 'Mutagenic effect of sensitized irradiation of bacteriophage T4', *J. Mol. Biol.* **46**, 157-167.
58. Meistrich, M.L. and Drake, J.W. (1972), 'Mutagenic effects of thymine dimers in bacteriophage T4', *J. Mol. Biol.* **66**, 107-114.
59. Witkin, E.M. (1967), 'Mutation-proof and mutation-prone modes of survival in derivatives of *Escherichia coli* B differing in sensitivity to ultraviolet light', *Brookhaven Symp. Biol.* **20**, 17-55.
60. Nishioka, H. and Doudney, C.O. (1969), 'Different modes of loss of photoreversibility of mutation and lethal damage in ultraviolet light resistant and sensitive bacteria', *Mutation Res.* **8**, 215-228.
61. Bridges, B.A. and Mottershead, R. (1971), '*Rec A*$^+$ dependent mutagenesis occurring before DNA replication in UV- and γ-irradiated *Escherichia coli', Mutation Res.* **13**, 1-8.
62. Bridges, B.A., Gray, W.J.H., Green, M.H.L., Rothwell, M. and Sedgwick, S.G. (1973), 'Genetic and physiological separation of the repair and mutagenic functions of the *exrA* gene in *Escherichia coli', Genetics, Suppl.* **73**, 123-129.
63. Cooper, P.K. and Hanawalt, P.C. (1972), 'Heterogeneity of patch size in repair replicated DNA in *Escherichia coli', J. Mol. Biol.* **67**, 1-10.
64. Cooper, P.K. and Hanawalt, P.C. (1972), 'Role of DNA polymerase I and the *rec* system in excision-repair in *Escherichia coli', Proc. Nat. Acad. Sci. U.S.A.* **69**, 1156-1160.
65. Witkin, E.M. and George, D.L. (1973), 'Ultraviolet mutagenesis in *polA* and *uvrA polA* derivatives of *Escherichia coli B/r:* evidence for an inducible error-prone repair system', *Genetics Suppl.*, **73**, 91-108.
66. George, J., Devoret, R. and Radman, M. (1973), 'Indirect ultraviolet-reactivation of phage λ', *Proc. Nat. Acad. Sci. U.S.A.* **71**, 144-147.
67. Witkin, E.M. (1974), 'Thermal enhancement of ultraviolet mutability in a *tif-1 uvrA* derivative of *Escherichia coli B/r*: evidence that ultraviolet mutagenesis depends upon an inducible function. *Proc. Nat. Acad. Sci. U.S.A.* **71**, 1930-1934.

68a. Bridges, B.A., Law, J. and Munson, R.J. (1968), 'Mutagenesis in *Escherichia coli.* II. Evidence for a common pathway for mutagenesis by ultraviolet light, ionizing radiation and thymine deprivation', *Molec. Gen. Genetics,* **103**, 266-273.

68b. Green, R.R. and Drake, J.W. (in press), 'Misrepair mutagenesis in bacteriophage T4', *Genetics,* **78**, 81-89.

69. Nasim, A. (1968), 'Repair-mechanisms and radiation-induced mutations in fission yeast', *Genetics* **59**, 327-333.

70a. Savic, D.J. and Kanazir, D.T. (1972), 'The effect of a histidine operator-constitutive mutation on UV-induced mutability within the histidine operon of *Salmonella typhimurium', Molec Gen. Genetics,* **118**, 45-50.

70b. Lemontt, J.F. (1973), 'Genes controlling ultraviolet mutability in yeast', *Genetics Suppl.,* **73**, 153-159.

71. Drake, J.W. (1963), 'Properties of ultraviolet-induced rII mutants of bacteriophage T4', *J. Mol. Biol.,* **6**, 268-283.

72. Drake, J.W. (1966), 'Ultraviolet mutagenesis in bacteriophage T4. Irradiation of extracellular phage particles', *J. Bacter.* **91**, 1775-1780.

73. Meistrich, M.L. and Drake, J.W. (1972), 'Mutagenic effects of thymine dimers in bacteriophage T4', *J. Mol. Biol.* **66**, 107-114.

74. Drake, J.W. and McGuire, J. (1967), 'Properties of *r* mutants of bacteriophage T4 photodynamically induced in the presence of thiopyronin and psoralen', *J. Virology* **1**, 260-267.

75. Howard, B.D. and Tessman, I. (1964), 'Identification of the altered bases in mutated single-stranded DNA. III. Mutagenesis by ultraviolet light', *J. Mol. Biol.* **9**, 372-375.

76. Osborn, M., Person, S., Phillips, S. and Funk, F. (1967), 'A determination of mutagen specificity in bacteria using nonsense mutants of bacteriophage T4', *J. Mol. Biol.* **26**, 437-447.

77. Cheung, M.K. and Bockrath, R.C. (1970), 'On the specificity of UV mutagenesis in *E.coli.', Mutation Res.* **10**, 521-523.

78. Sora, S., Panzeri, L. and Magni, G.E. (1973), 'Molecular specificity of 2-aminopurine in *Saccharomyces cerevisiae, Mutation Res.* **20**, 207-213.

79. Sherman, F. and Stewart, J.W. (1974), 'Variation of mutagenic action on nonsense mutants at different sites in the iso-1-cytochrome *c* gene of yeast', *Genetics* **78**, 97-113.

80. Kilbey, B.J., de Serres, F.J. and Malling, H.V. (1971), 'Identification of the genetic alterations at the molecular level of ultraviolet light induced *ad*-3B mutants in *Neurospora crassa', Mutation Res.* **12**, 47-56.
81. Demerec, M. (1960), 'Frequency of deletions among spontaneous and induced mutations in *Salmonella', Proc. Nat. Acad. Sci. U.S.A.* **46** 1075-1079.
82. Goldschmidt, E.P., Cater, M.S., Matney, T.S., Butler, M. A. and Greene, A. (1970), 'Genetic analysis of the histidine operon in *Escherichia coli* K12', *Genetics,* **66**, 219-229.
83. Hill, C.W. and Combriato, G. (1973), 'Genetic duplications induced at very high frequency by ultraviolet irradiation in *Escherichia coli', Molec. Gen. Genetics* **127**, 197-214.
84. Ishii, Y. and Kondo, S. (1972), 'Spontaneous and radiation-induced deletion-mutations in *Escherichia coli* strains with different DNA repair capacities', *Mutation Res.* **16**, 13-25.
85. Kilbey, B.J. and de Serres, F.J. (1967), 'Quantitative and qualitative aspects of photoreactivation of premutational ultraviolet damage at the *ad*-3 loci of *Neurospora crassa', Mutation Res.* **4**, 21-29.

CHAPTER FIFTEEN

Chemical mutagens: alkylating agents. I: Genetical effects

Tests for mutagenic effects of chemicals are almost as old as modern genetics (see article by Auerbach in Vol. 1 of 'Chemical Mutagens'; Bibliography). In 1914, T.H. Morgan had already tried to produce mutations in *Drosophila* by treatment with alcohol and ether, but without success. In the twenties and thirties, Muller's techniques for the determination of mutation rates in *Drosophila* (Chapter 5) were applied to a variety of chemicals. Russian workers claimed success for ammonia, iodine, potassium permanganate and copper sulphate; for the last-named substance, confirmatory data were published in U.S.A. In most of these experiments, the differences between the frequencies of sex-linked lethals in controls and treated flies were statistically significant, but treatment effects were small. Lethal frequencies in the treated series rarely exceeded or even reached 1%, which is roughly the upper limit for untreated flies. A fact that was not known at the time but meanwhile has been clearly established, is the dependence of spontaneous mutation frequency on breeding procedure and on the brood tested (1). This raises the possibility that the chemicals used in these early experiments had yielded increased mutation frequencies not by inducing mutations but by affecting the rate of sperm utilization in such a way that the samples from treated males contained a higher proportion of mutation-bearing spermatozoa than did those from the control flies. It would be interesting to retest some of the more promising chemicals with a rigorous breeding and sampling technique.

Discovery of the mutagenic action of mustard gas

The first undoubted mutagens were discovered during the Second World War: mustard gas in Scotland, urethan in Germany. In this chapter we shall deal with the former, which is by far the more potent and universally effective mutagen. It belongs to a large group of substances, called alkylating agents, which act by transferring alkyl groups to macromolecules of biological importance. We shall consider their chemical action in the next chapter, after having discussed their genetical effects in the present one (formulae in Table 16.1). A full review up to 1966 is found in the book by Loveless (Bibliography).

Mustard gas was tested because of its pharmacological properties, which are very similar to those of X-rays. Mustard gas burns, like X-ray burns, heal only with difficulty and, after apparent healing, tend to break open again, sometimes after years. This 'radiomimetic' property of mustard gas had already been noted in the First World War. By the time of the Second World War it had been realized that X-ray burns owe their peculiar behaviour in large part to their chromosomal effects. It seemed at least possible that the same might be true for mustard gas, and pharmacologists in Edinburgh suggested that it should be tested for mutagenic action in *Drosophila.* Already the first experiments brought convincing proof for the hypothesis. Up to 24% sex-linked lethals were produced in tests carried out in 1940 and 1941; but the data were published only several years after the war because all research on war gases was 'classified' until 1947 (2). Further tests showed that mustard gas can induce the whole array of mutagenic effects known to occur after X-radiation: dominant lethal and visible mutations, recessive sex-linked and autosomal lethals and visibles, large and small deletions, inversions, and translocations. Later work has shown that mustard gas, like other alkylating agents, is superior to X-rays in the ability to produce duplications. In addition, it resembles X-rays in producing crossing-over in somatic cells and in male germ cells.

Comparison with X-rays

The similarity between the effects of mustard gas and X-rays was very striking indeed. It led some geneticists to discredit the target theory on the ground that X-rays, too, probably act via chemical intermediates and not through direct 'hits' on the chromosomes. This reasoning overlooked a number of important points. First, the biophysical analysis that culminated in the target theory was undertaken only when it had been shown that X-rays do *not* act by means of stable chemical intermediates, because no mutations were induced when unirradiated spermatozoa were introduced into irradiated eggs of *Drosophila* or *Habrobracon* (Chapters 5 and 7). When this experiment was repeated with mustard gas on *Drosophila,* it gave the same negative result. Thus, the effects of X-rays and mustard gas resemble each other not because X-radiation acts via a chemical intermediate, but because mustard gas, like X-rays, acts directly on the target. Second, the target theory is based on a vast number of radiation data and, if the term target is used in its wider sense (Chapter 6), it yields a consistent picture of radiation mutagenesis. It cannot be disproved by observations from an entirely different field. Third, similarity of results does not necessarily imply similarity of causes, and mutation and chromosome breakage may be the end result of many different primary mechanisms. Fourth, the genetic material has only a limited array of responses to genetic injury, and any agent that can change the genes and break the chromosomes is likely to produce most or all of the known types of mutation and rearrangement.

Moreover, differences between the effects of mustard gas and X-rays had already been discovered in the earliest experiments; others turned up in later work with mustard gas itself or with other alkylating agents. The most important ones are (*a*) a relative shortage of large rearrangements coupled with a relative excess of small ones, in particular duplications, (*b*) a tendency for delayed effects and instabilities. We shall deal with them in turn although, as we shall see, it is probably the last peculiarity which is the key to the others.

The shortage of large rearrangements. Storage effect on rearrangement frequencies

Table 15.1 presents data from seven experiments in which translocations and/or large sex-linked deletions were scored among the progeny of mustard gas treated males (for techniques of scoring see Chapters 2 and 5). Since these rearrangements are exceedingly rare in untreated flies, the data leave no doubt that mustard gas can produce both translocations and deletions. Inversions, for which no good genetical screening tests are available, were detected cytologically or were found incidentally as crossing-over inhibitors in mapping tests.

Table 15.1

Comparison between the frequencies of large rearrangements in the progeny of males exposed to mustard gas or X-rays. (After Auerbach (3) and Nasrat (4))

*Dose** *(in % of sex-linked lethals)*		*Deletions*		*Ratio (approx)*		*Translocations †*		*Ratio (approx)*
	n‡	*Expected §*	*Observed*		*n‡*	*Expected §*	*Observed*	
5	8 331	21	2	10:1	2 655	54	5	11:1
7	4 523	20	4	5:1	3 858	123	12	10:1
7	5 052	18	9	2:1				
9	6 635	41	16	3:1				
9					816	56	7	8:1
10	4 926	31	14	2:1	3 060	172	26	7:1
15					981	98	21	5:1
Mean	29 467	131	45	3:1	11 370	503	71	7:1

* Determined in tests on aliquots of the treated males.

† Involving chromosomes I, II, III; in the last experiment only II and III.

‡ Number of X-chromosomes in the deletion test; number of chromosome sets in the translocation test.

§ From a dose of X-rays yielding the same frequency of sex-linked lethals.

Quantitative comparisons between large rearrangement frequencies in mustard gas-treated and X-rayed flies met with a difficulty that has already been discussed in regard to UV (Chapter 10): the difficulty of comparing effects of mutagens that act in different ways. We have seen there that the best, although by no means perfect, method is to standardize doses by reference to some undoubted genetic effect, which in *Drosophila* is usually the frequency of sex-linked lethals. A dose of mustard gas was considered 'mutagenically equivalent' to a dose of X-rays when both produced the same frequencies of sex-linked lethals. On this basis, it was possible to compare the observed frequency of rearrangements with that expected from a dose of X-rays yielding the same lethal frequency. The expectations were derived from published X-ray data; extrapolations from one dose to another were made by means of the 3/2 power dose-effect rule for large rearrangements (p. 94). Although these calculations give only very rough estimates, the relative shortage of rearrangements after mustard gas is obvious throughout the table. It decreases with increasing dose of mustard gas; the reason for this will become clear when we consider the kinetics of mustard gas-induced rearrangements in the next chapter. The shortage is more pronounced for translocations than for deletions. This agrees with experiments on the cytological effects of alkylating agents in *Drosophila* (5) and plants (6), where the ratio of interchromosomal to intrachromosomal rearrangements was always lower than after irradiation.

Two interpretations were considered (7). (1) Mustard gas, being a weaker mutagen than X-rays, produces fewer chromosome breaks. (2) Mustard gas is not inferior to X-rays in break production, but the breaks are less likely to rejoin into rearrangements. If most of the breaks restituted in the original way, the result would be indistinguishable from a shortage of primary breaks. Alternatively, mustard gas-induced breaks might have a tendency to remain open, in which case they would be eliminated as dominant lethals in *Drosophila.* Fig. 15.1 is in support of the last assumption. It shows that the frequency of mustard gas-induced dominant lethals equals that of X-rays at low doses, when most dominant lethals are due to single breaks, but exceeds it at high ones when an increasing proportion of dominant lethals is formed by inviable rearrangements. This would be expected if, at high doses of mustard gas, the frequency of dominant lethals were increased by the 'loss' of potentially viable rearrangements as unjoined breaks. Moreover, the ease with

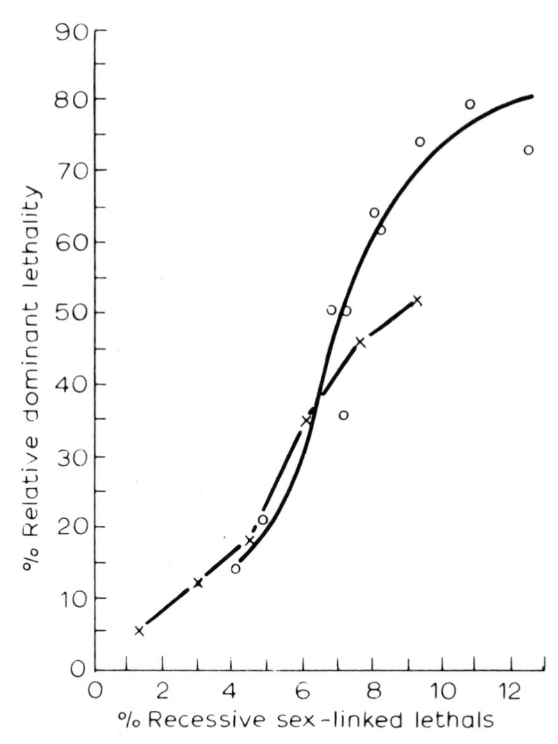

Fig. 15.1. Dominant lethality plotted against the frequency of sex-linked lethals in experiments with mustard gas (o) and X-rays (x). Relative dominant lethality = 1-hatchability of eggs in the treated series in percent of control hatchability. (Fig. 3, Nasrat (4). Courtesy Springer-Verlag.)

which mustard gas and the closely related nitrogen mustard were shown to break the chromosomes of plants testified to their efficiency in this regard (8,9).

A special version of the second hypothesis – low rejoining ability of mustard gas induced breaks – was put forward tentatively quite early (3) and has since gained strong support from experiments with various alkylating agents including mustard gas. It assumes that mustard gas breaks, once formed, have the same rejoining ability as do X-ray induced breaks, but that the primary chemical lesions mature only slowly into open breaks, so that no more than a small fraction of breaks is available for rejoining before the next mitosis; in *Drosophila,* this means before the first cleavage division. If, in any particular zygote, only one out of several latent breaks is open at that time, a potential rearrangement is lost as a dominant lethal. If several breaks open simultaneously *after* the first cleavage division and form a rearrangement, this will appear as a mosaic and will be missed by most genetical test methods. Cytologically, mosaic rearrangements can be scored and have indeed been found to form a high proportion among chemically induced changes, but a very low proportion among X-ray induced ones (5). We shall return to this presently.

Proof for the first part of the hypothesis – i.e. normal rejoining ability of mustard gas induced breaks – was provided by experiments in which *Drosophila* males were exposed in succession to X-rays and mustard gas, or vice versa (10). As usual, mustard gas by itself produced far fewer translocations than did radiation by itself. The frequency of translocations among the mixed population of breaks, however, was not additive as it would have been if mustard gas breaks had joined only with mustard gas breaks, and X-ray breaks with X-ray breaks. Instead, it was much higher and approximately that expected if all available breaks had rejoined equally readily with each other (see argument on p. 126).

The second part of the hypothesis – delayed opening of latent breaks – was proved in experiments in which treated *Drosophila* spermatozoa were stored for days or weeks in the seminal receptacles of untreated females (5,11,12). Since, in *Drosophila,* reunion of breaks occurs only after fertilization, this allows varying amounts of time for potential breaks to mature.

When irradiated males are mated to unirradiated females, the frequency of rearrangements is the same whether the females are allowed to lay eggs at once or whether they are prevented from doing so for several days or weeks. When the same kind of experiment is carried out on males that have been treated with mustard gas or some other alkylating agent, there is a dramatic increase in the frequency of rearrangements, while that of recessive lethals changes only very slightly. With increasing storage time, this leads to a gradual approximation of the chemically induced rearrangement frequencies to those induced by equivalent X-ray doses.

Table 15.2 shows the storage effect of the powerful alkylating agent triethylenemelamine (TEM). The striking increase in the frequency of rearrangements is obvious. The very much smaller increase in the frequency of lethals is probably entirely due to the fact that, especially at high doses, a proportion of lethals is connected with rearrangements and that these are subject to a storage effect. We find here another example for the caution that has to be exercised

Table 15.2

The effect of storage in untreated females on the frequencies of sex-linked lethals and large rearrangements in TEM-treated *Drosophila* sperm.

Method of scoring	*Types of rearrangement*	*Days of storage at 25°*	*Lethals (%)*		*Rearrangements (%)*	
			pre-stored	*post-stored*	*pre-stored*	*post-stored*
Genetical*	Translocations	6	6.7	9.3	0.8	10.3
	Y, II, III	6	6.7	8.7	0.9	7.5
Genetical†	Y, II, III	9	10.0	23.6	3.0	36.5
Cytological‡	Translocations and inversions in salivary gland chromosomes of F_1 larvae	9	10.5	16.9	*per 100 spermatozoa* 5.4 (+ 2.2.)§	47.4

* Watson (12), 1964;

† Snyder (11), 1963;

‡ Slizynska (5), 1969;

§ In brackets – mosaic changes

when standardizing genetically effective doses by one selected effect (compare discussion to Table 10.1). The last line of Table 15.2 shows that mosaics formed about one third of all rearrangements but were absent from the much larger stored sample. It is of interest that the different ways in which point mutations and chromosome breaks respond to storage after chemical treatment was the first clear proof of an essential difference between them.

It thus appears that the shortage of rearrangements after chemical treatment is mainly – or wholly – attributable to the slow tempo at which chemically induced breaks 'mature' into a state of readiness for rejoining. Under the usual experimental conditions there is not enough time for all potential breaks to reach that stage and many are lost as dominant lethals because other breaks in the same cell have not yet reached maturity. It has, indeed, been possible to boost the frequency of X-ray induced translocations in *Drosophila* by pretreatment with an alkylating agent which by itself had produced no translocations but, presumably, had provided some open breaks for reunion with singly occurring X-ray breaks (13).

The finding (Table 15.1) that more translocations than large deletions are lost as dominant lethals indicates that two potential breaks in a cell have a better chance of opening simultaneously when they have occurred in the same chromosome than when they have occurred in different ones. This has been taken to point to a connection between chromosome replication in the zygote and the final step in the maturation of breaks (14). It may be relevant that, in the experiments of Table 15.1, all deletions but none of the translocations involved the X-chromosome; asynchrony of replication between X and autosomes has been reported (15,16). With storage, the difference between intrachromosomal and interchromosomal frequencies tends to become smaller, presumably because fully matured breaks no longer require the stimulus of replication.

Other observations strengthen the assumption that chromosome replication speeds up the acquisition of rejoining ability by potential chromosome breaks. One observation concerns the distribution of chemically-induced breaks over euchromatin and heterochromatin. In plants, most chemicals produce relatively more heterochromatic breaks than do X-rays (Kihlman; Bibliography); in *Drosophila,* the opposite is true. Chromosome rearrangements in the F_1 of irradiated larvae contain about 20% breaks in the heterochromatin; chemically induced ones contain from 5-10%. Like the shortage of large

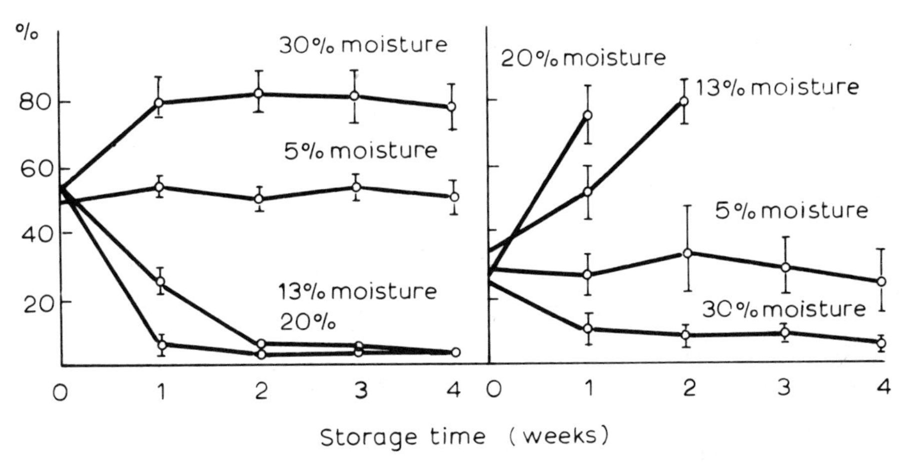

Fig. 15.2. Seedling height (% of control) and frequency of aberrant anaphases (fragments and/or bridges) after storage at various moisture contents of ethyl methane sulphonate-treated barley seeds. (Fig. 1, Gichner (17). Courtesy Verl Georg Thieme.)

rearrangements, the relative shortage of heterochromatic breaks is reduced by storage: among the rearrangements scored in the last experiment of Table 15.2 the proportion of heterochromatic breaks was 8.2 before and 14.5 after storage. It seems likely that this difference between plants and *Drosophila* is apparent rather than real. In plants, all heterochromatic breaks can be scored; in *Drosophila*, the clumping together of all heterochromatic regions in the chromocentre prevents the detection of rearrangements with both breaks in heterochromatin. Thus, even if heterochromatic breaks should form the same high proportion of chemically-induced breaks in *Drosophila* as they do in plants, many of them will remain undetected if the asynchronous replication of euchromatin and heterochromatin favours reunion between breaks of the same type. With storage, both types of break mature, rearrangements with mixed eu- and heterochromatic breaks become more frequent, and more heterochromatic breaks are detected.

In plants, marked storage effects have been obtained when chemically treated seeds were kept dormant for days or weeks before allowing germination (17). Moisture content during storage was found to play a critical role. This can be seen from Fig. 15.2, which illustrates results obtained on barley seeds treated with the alkylating agent ethylmethane sulphonate (EMS). At 5% moisture, when metabolic processes are more or less suspended, storage had no effect. At 13% and 20% moisture, the frequency of chromosome aberrations was more than doubled during the first two weeks of storage. At a still higher water content, storage had the opposite effect: aberration frequency decreased: this is attributed to repair (Chapter 16).

These findings show that chemically-induced chromosome breakage is a slow process. The usual experimental conditions may not allow sufficient time for more than very few potential breaks in dividing cells to become actual before the first post-treatment mitosis. Conceivably this may account for the fact that most chemicals produce only chromatid aberrations even when applied in G1, where only chromosome aberrations would be expected. This would be an alternative to the more usual, but so far unproved, explanation that chromosome breakage by alkylating agents arises as an error of DNA replication at one of the first mitoses after treatment (6).

Various experimental conditions influence the production of chromosome breaks by alkylating agents. In plants, divalent metal cations have been found highly effective co-factors for chromosome breakage by EMS. Fig. 15.3 shows that EMS, which in pure buffer solution produced hardly any breaks in *Vicia* chromosomes, produced many in the presence of low concentrations (about 10^{-3} mM) of copper or zinc ions. Since distilled water often contains traces of metal, even treatment in water yielded different break frequencies depending on the source of the water. Moreover, the effect of the metal ions was strongly pH-dependent.

Obviously, what is true for X-rays is even more true for chemicals: although the primary effects are chromosomal lesions, how many of these will result in breaks and how many and what kinds of rearrangement will be formed from these breaks depends on a variety of cellular and environmental conditions. Inferences from the final results on the primary events is at best hazardous and may be misleading. Research into the molecular nature of the primary lesions is going on in many laboratories; but it will not provide the key to the whole process of rearrangement formation by chemical agents. For this, further analysis of the secondary steps in living

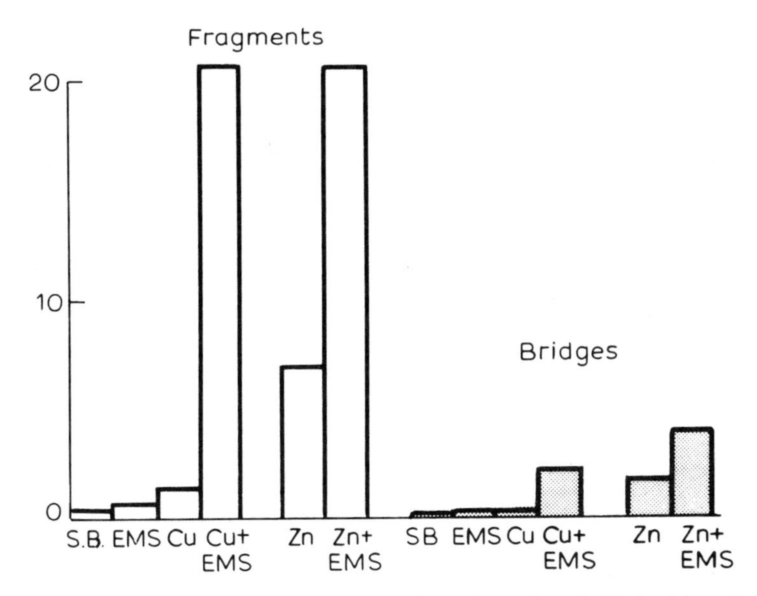

Fig. 15.3. Percent of anaphase aberrations in cells of *Vicia faba* after treatment of seeds with EMS in the presence of Zn(*a*) or Cu(*b*). (EMS concentration in mg/100ml) SB = Sørenson buffer (Fig. 1, Moutschen-Dahmen (18). Courtesy Birkhäuser Verlag.)

cells remains indispensable.

The excess of small deficiencies and duplications

While alkylating agents produce fewer large rearrangements than do mutagenically equivalent doses of X-rays, they produce more small deficiencies. This suggests that at least some of the chemically-induced deficiencies arise by a different mechanism from the breakage-fusion process that leads to large rearrangements. We shall see that this is only partially correct. Both large and small rearrangements appear to require breakage and reunion; the difference lies in the timing relation between chromosome replication and the opening of potential breaks.

The clue to the special mechanism by which chemicals may produce deficiencies was provided by the finding that chemically treated chromosomes quite frequently carry duplications in the form of repeats. This was first observed in cells of *Vicia faba* that had been treated with nitrogen mustard (9). The interpretation favoured at the time was that chemical treatment could induce 'unequal' or 'oblique' crossing-over between neighbouring but not homologous loci, thus transferring a small section of one chromosome into its homologue. In view of the extreme precision of matching that characterizes crossing-over, this interpretation did not seem convincing. A plausible interpretation has been based on a very thorough analysis of structural changes in the F_1 salivary glands of treated *Drosophila* males (5,14,19,20). The treatment agents included not only alkylating chemicals but also formaldehyde, which will be dealt with more fully in Chapter 18. Since, in their essentials, the effects of all these chemicals were the same, I shall draw on all of the data in what follows.

Duplications comprising from several up to more than 100 bands were found to form a sizeable proportion of chemically-induced changes, in contrast to their extreme rarity after irradiation. Like other peculiarities of chemical mutagens, the frequency of duplications diminishes with storage. In the last experiment of Table 15.2, their number per 100 spermatozoa was 6 before and 0 after storage. Most of the duplications occurred as mosaics together with the complementary deficiency, i.e. the repeat was carried by about half the cells of the salivary gland, while the homologous chromosome in the remaining cells lacked the duplicated segment altogether. Superficially, this is just what would be expected from unequal crossing-over; yet analysis of the banding pattern invalidated this interpretation. Theoretically, four different types of repeat are possible. Representing individual bands as letters, they are as follows. Tandem repeat (ab*cdcd*ef), reverse tandem repeat (ab*dcdc*ef), reverse repeat (ab*cddc*ef or ab*dccd*ef). All four were found among chemically-produced duplications. Crossing-over, which is always polarized in regard to the centromere, cannot give rise to inverted segments and therefore cannot have produced the three last types. Fig. 15.4 shows how all four types can arise from breakage and fusion between homologous chromatids. Numerical comparisons between their expected and observed proportions after chemical treatment, showed that the majority of the first type must have had the same origin as the other three. This obviates the necessity for invoking unequal crossing-over as a cause of repeats. Once a repeat is present, it provides opportunity for oblique crossing-over and, through this, for further amplification of the duplicated segment (21,22).

It will be realized that very special conditions are required for the formation of a repeat. (1) The chromosome that donates the

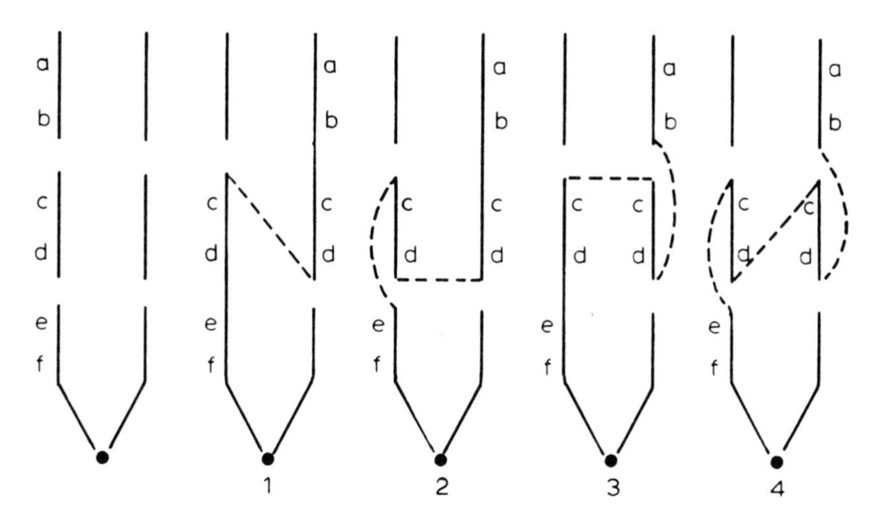

Fig. 15.4. The mechanism underlying the formation of repeats. All types of repeat have the same complementary deficiency ab-ef. Dotted lines mark new rejoinings. (Fig. 1, Slizynska (20). Courtesy Cambridge Univ. Press.)

segment must be within rejoining distance of the chromosome that receives it and (2) the breaks in the two chromosomes must occur at identical sites. Slizynska has pointed out that opening of potential breaks at replication creates just these conditions. The high frequency of duplications after chemical treatment and its decline with storage furnish the best support for the assumption that replication is the final stimulus for near-mature breaks to open and take part in rejoinings.

It is important to point out that, in this hypothesis, replication plays a very different role from that assigned to it in the hypothesis that chromatid breaks arise through replication errors. On the latter assumption, only one chromatid will be broken and the formation of the isochromatid breaks needed for repeats requires additional assumptions. On the former assumption, the initial lesion affects the whole chromosome; the type of break formed – chromosome, chromatid, isochromatid – depends on the relation between the time of chromosome replication and that of the effective opening of breaks.

Economy of hypotheses suggests that the mechanism of chromosome breakage and rearrangement formation by alkylating agents should be essentially similar in all eukaryotic cells. Indeed, the storage effect in plant seeds (17), the production of duplications in *Vicia* by nitrogen mustard (9), the prevalence of chromatid (mosaic) over chromosome (complete) aberrations even when plant cells are treated in G1 (6), and the important role of the S-phase in the production of rearrangements point to just such essential similarities between plants and *Drosophila.* Regrettably, there is very little contact and comprehension between cytologists working on *Drosophila* and on plants, and there have as yet been no serious attempts to formulate a comprehensive theory of chemical chromosome breakage in eukaryotes.

Also in *Escherichia coli,* alkylating agents were shown to produce high frequencies of duplications, probably mainly of the tandem type (22a). Several mechanisms for their origin, applicable also to UV-induced similar duplications (Chapter 14), have been tentatively suggested.

Delayed mutation. Replicating instabilities

Storage effects in Drosophila sperm are restricted to chromosome breaks and rearrangements; they do not affect point mutations. Yet evidence that chemically-induced point mutations may also appear after a delay was obtained long before the discovery of the storage effect. The first pointer in this direction was the high frequency of mosaics for visible mutations in the F_1 of mustard gas treated *Drosophila* males (23). Also in maize, pollen treatment with mustard gas yielded about 50% mosaic kernels (24). From the point of view of modern genetics, mosaics can no longer be considered evidence for delayed mutation. Most chemicals produce lesions in only one strand of DNA (Chapter 16), and these will automatically yield mosaics after segregation of the two strands. On the contrary, it is the complete mutations that nowadays require explanation. We shall deal with this problem in Chapter 19.

A stronger argument for mutational delay came from experiments in which lethals were scored in the F_3 of treated *Drosophila* males. These experiments followed the scheme laid out by the early X-ray workers (Fig. 5.4). It will be recalled that lethals that arise with a delay give rise to mosaic gonads in F_1 and are detected as full lethals in F_3. In contrast to what had been found for X-rays, such delayed lethals were found after treatment of spermatozoa with mustard gas. Table 15.3 shows the relevant data from two experiments.

Table 15.3

Frequencies of sex-linked lethals in the F_2 and F_3 of mustard gas treated *Drosophila* males. (For scheme of matings see Fig. 5.4). (After Auerbach (25).

Expt.	*% lethals in F_2*	*Number of tested F_1 ♀♀*	*Lethals in F_3* *n*	*%*
I.	13	828	21	2.5
II.	10	1049	45	4.5
Controls		2222	3	0.1

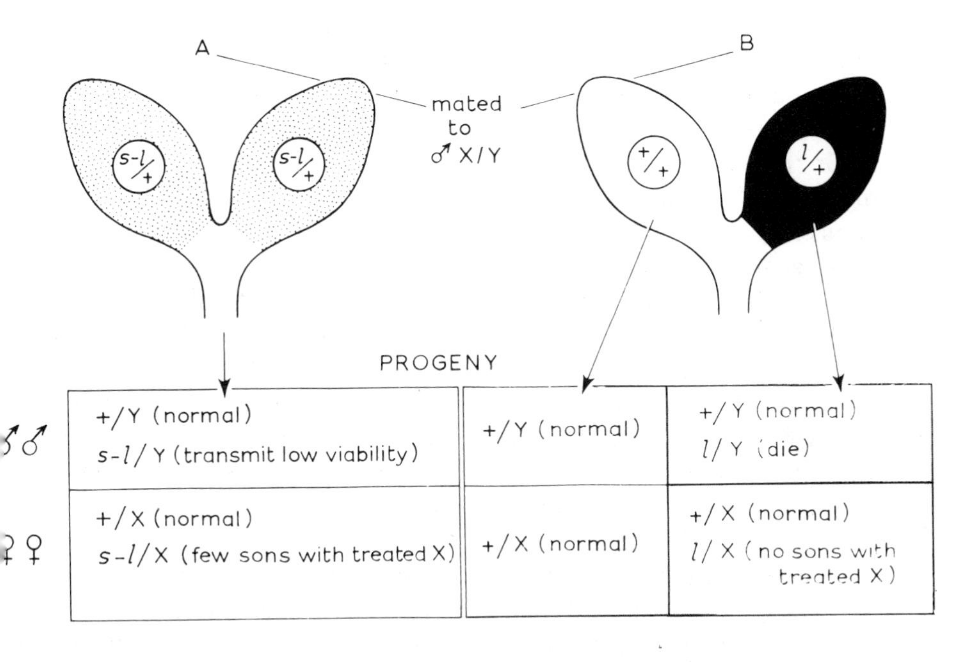

Fig. 15.5. Distinction, by breeding, between a female that is heterozygous for a semi-lethal (A), and one that is a gonadic mosaic for a full lethal (B).

In the second experiment, pedigrees of individual F_1 females were kept so that those with mosaic gonads could be identified. It appeared that all 45 lethals in F_3 traced back to three gonadic mosaics among the 77 F_1 females that had provided the 1049 tested F_2 females.

More gonadic mosaics were detected by a technique that has subsequently been applied frequently – though not always cautiously – to tests for delayed action of mutagens. It is based on the consideration that a gonadic mosaic for a lethal will produce live sons with the treated X only from the non-mutant part of her ovaries. While the F_2 culture produced by such a female will contain some males with the treated X and will, therefore, not be scored as a lethal, the number of these males will be smaller than expected. Such cultures are usually classified as 'semi-lethals' and are attributed to viability mutations that kill a certain proportion of the males carrying them. It is indeed impossible to distinguish a semi-lethal mutation from a gonadic mosaic by inspection of the F_2 vials. They can, however, be readily distinguished by progeny tests on those F_2 females that carry the treated X (Fig. 15.5). If the F_1 female was heterozygous for a semi-lethal on the treated X (A), all these daughters will have inherited it and will again produce too few sons with this X-chromosome. If the F_1 female was a gonadic mosaic for a full lethal (B), the daughters will be of two types. Those tracing back to the non-mutant part of the ovary will have the normal number of sons with the treated X; those tracing back to the mutant part will have none at all. Table 15.4 shows data from an experiment in which seemingly semi-lethal cultures in the F_2 of mustard gas treated males were analysed by this technique. The particular F_1 female to which these data refer was clearly a gonadic mosaic for a full lethal. It should be noted that all her daughters could be classified unambiguously as either carrying a lethal or its normal allele. This is a crucial condition for identification of a lethal mosaic. Where it is not met, i.e. where the proportion of males with the treated X varies more or less continuously between different cultures, one is in all probability dealing not with a mosaic for a lethal but with a semi-lethal of varying expression (26).

In the F_2 of X-rayed males, shortage of males with the treated X is usually due to a semi-lethal. In the F_2 of males that have been treated with alkylating agents, it is very often due to gonadic mosaicism for a full lethal. It is true that even gonadic mosaicism is no absolute proof for delayed mutation because it, too, might conceivably arise at the first cleavage division through segregation of a mutant from a non-mutant strand of DNA. However, the average number of independent cleavage nuclei that participate in formation of the ovaries is so low (about 1.5. for the whole germ line) (27) that most mosaics are likely to have arisen later during the development of the gonads. This conclusion has also been drawn from an analysis of mosaic patterns in the soma and the gonads of flies whose

Table 15.4

Detection of gonadic mosaicism for a lethal. (Table II, Auerbach (25). Courtesy Roy. Soc. Edinburgh.)

♀ *No.*	*Number of daughters*	*Number of sons carrying*		*Classification of treated X*
		untreated X	*treated X*	
1	33	10	0	lethal
2	31	10	11	normal
3	39	17	0	lethal
4	27	9	17	normal
5	26	11	0	lethal
6	50	15	0	lethal
7	24	6	0	lethal
8	22	6	7	normal
9	26	8	12	normal
10	36	19	16	normal
11	38	16	0	lethal
12	37	7	0	lethal
13	35	6	11	normal
14	46	25	27	normal
15	28	10	7	normal
16	41	9	0	lethal
17	42	11	0	lethal
18	39	13	0	lethal
				10 lethal 8 normal

P_1 ♂ treated with mustard gas.
F_1 ♀ had only 9 sons with the treated X as against 85 daughters.
18 of the daughters were progeny-tested.

fathers had been treated with the alkylating agent EMS (28).

Final proof for a delayed mutagenic action of mustard gas was obtained when gonadic mosaics for a sex-linked lethal or visible mutation were found among the F_2 females of mustard gas treated males (25). Subsequently, this finding has been confirmed for other alkylating agents (29). Since gonadic mosaicism in F_2 females can, at the earliest, have arisen during their first cleavage division, the primary lesion in the spermatozoon and the actual mutation must have been separated by the 30 or more cell cycles that separate the first cleavage division in F_2 from that in F_1 The nature of these long-persisting potential mutations is as little understood as that

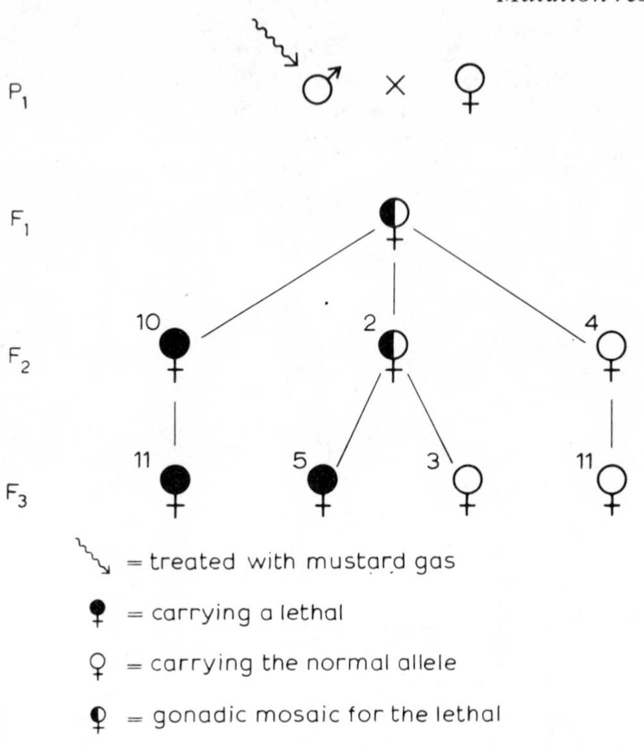

Fig. 15.6. Pedigree of a replicating instability in *Drosophila.* The numbers refer to the flies in successive generations.

of the potential chromosome breaks. Some light is thrown on it by experiments in which pedigrees of F_1 mosaics were analysed. One of these is shown in Fig. 15.6. Several points should be noted. (1) Location tests made it very likely that the lethals in F_1 and F_2 had occurred at the same locus. This has been found to be the case whenever recurrent mutations in the same pedigree were compared by mapping or phenotype. Thus, chemical injury to a gene may create a localized instability which breaks down repeatedly to yield a mutation. (2) Each gonadic mosaic gave off stable lines for both the lethal and its normal allele. Thus the instability can become stabilized in either direction. (3) In the present pedigree, one gonadic mosaic in F_1 gave rise to two in F_2. The production of two or more F_2 mosaics by one F_1 mosaic has been observed repeatedly in such pedigrees. It shows that the instability, in addition to throwing off stable mutant and non-mutant cell lines, can also replicate *as such.* In order to emphasize this, we speak of 'replicating instabilities'.

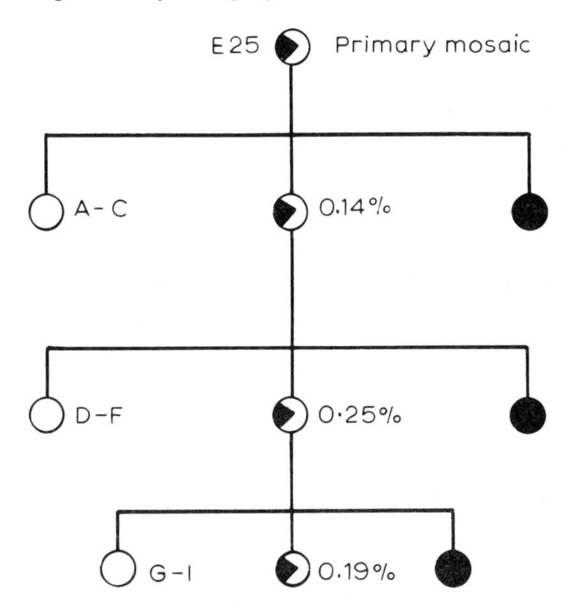

Fig. 15.7. A replicating instability from red to white in fission yeast (*Schizosaccharomyces pombe*) ● = parental red colony; ○ = complete white mutant; ◉ = mosaic mutant colony. The numbers refer to the percentage of mosaics in each 'plating generation' (Fig. 1, in Nasim, 1973 (31). Courtesy Elsevier)

Pedigree tests on *Drosophila* mosaics are laborious and their numerical scale is necessarily low. With the introduction of colour mutations in micro-organisms, tests for delayed mutagenesis and replicating instabilities could be carried out on a large scale and over many cell generations. This has been done especially in yeast, where both the red-to-white and the white-to-red system (p.184) have been used. In both cases, the technique consists of plating treated cells at high dilutions on adenine-supplemented medium, letting them grow into colonies and scoring complete and mosaic mutant colonies by their colour. For a study of replicating instabilities, mosaic colonies – or parts of them – are dispersed into their component cells and replated, and this procedure can be continued as long as mosaics are found. Complications arising from meiosis are avoided in this method; all divisions are mitotic. Fig. 15.7 shows a pedigree obtained for the red-into-white system after treatment of fission yeast with EMS. Like one of the gonadic mosaics in the F_1 of treated *Drosophila* males (Fig. 15.6), the primary mosaic gave rise to three types of progeny: stable non-mutants (red), stable mutants (white), and mosaics for red and white, and this pattern was repeated in each 'plating generation'. The proportion of mosaics remained more or less the same over the generations. Although mutations from red to white may occur at five different loci, every pedigree yielded mutations at only one of these (30). Moreover, complementation tests for one of the loci showed that the recurrent mutations had always occurred at the same site within the gene (31). This agreed with earlier observations on instabilities created by nitrous acid in the white-to-red system (32). Thus, the experiments on yeast confirmed the essential features of chemically induced replicating instabilities as derived from experiments on *Drosophila*. They added the finding that the instabilities are site-specific within the unstable gene.

References

1. Lamy, R. (1947-48), 'Observed spontaneous mutation rates in relation to experimental technique', *J. Genetics* **48**, 223-236.
2. Auerbach, C. (1948), 'Chemical induction of mutations', *Proc. 8th Int. Cong. Genetics Hereditas suppl.* 128-147.
3. Auerbach, C. and Robson, J.M. (1947), 'The production of mutations by chemical substances', *Proc. Roy. Soc. Edinburgh B.* **62**, 271-283.
4. Nasrat, G.E., Kaplan, W.D. and Auerbach, C. (1954-55), 'A quantitative study of mustard gas induced chromosome breaks and rearrangements in *Drosophila melanogaster', Zeitschr indukt. Abst. Vererb. Lehre* **86** 249-262.
5. Slizynska, H. (1969), 'The progressive approximation with storage of the spectrum of TEM-induced chromosome changes in *Drosophila* sperm to that found after irradiation', *Mutation Res.* **8**, 165-175.
6. Evans, H.J. and Scott, D. (1969), 'The induction of chromosome aberrations by nitrogen mustard and its dependence on DNA synthesis', *Proc. Roy. Soc. B* **173**, 491-512.
7. Auerbach, C. (1947), 'Nuclear effects of chemical substances', *Proc. 6th Int. Cong. Cytology. Exp. Cell Res. Suppl* **1**, 93-96.
8. Darlington, C.D. and Koller, P.C. (1947), 'The chemical breakage of chromosomes', *Heredity* **1**, 187-221.
9. Ford, C.E. (1948), 'Chromosome breakage in nitrogen mustard

treated *Vicia faba* root tip cells', *Proc. 8th Int. Cong. Genetics Hereditas, Suppl.* 570-571.

10. Oster, I.I. (1958), 'Interactions between ionizing radiation and chemical mutagens', *Zeitschr. indukt. Abst. Vererb. Lehre* **89**, 1-6.
11. Snyder, L.A. (1963), 'Evidence of an essential difference between point mutations and chromosome breaks induced by triethylene melamine in *Drosophila* spermatozoa', *Zeitschr. Vererb. Lehre.* **94**, 182-189.
12. Watson, W.A.F. (1964), 'Evidence of an essential difference between the genetical effects of mono- and bi-functional alkylating agents', *Zeitschr. Vererb. Lehre.* **95**, 374-378.
13. Sharma, R.P. and Grover, R.P. (1970), 'Interaction of mutational lesions induced by ethyl methanesulphonate and γ-rays in males of *Drosophila', Mutation Res.* **10**, 221-226.
14. Slizynska, H. (1973), 'Cytological analysis of storage effects on various types of complete and mosaic change induced in *Drosophila* chromosomes by some chemical mutagens', *Mutation Res.* **19**, 199-213.
15. Lima-de-Faria, A. (1959), 'Differential uptake of tritiated thymidine into hetero- and euchromatin in *Melanoplus* and *Secale', J. Biophys. Biochem. Cytology* **6**, 457-466.
16. Taylor, J.H. (1960), 'Asynchronous duplication of chromosomes in cultured cells of Chinese hamster', *J. Biophys. Biochem. Cytology* 7, 455-463.
17. Gichner, T., Veleminsky, J. and Zadražil, S. (1972), 'Storage effects in barley after the action of ethyl methanesulphonate and their relation to single strand breaks in DNA', *Biol. Zbl.* **91**, 81-89.
18. Moutschen-Dahmen, J. and M. (1963), 'Influence of Cu^{2+} and Zn^{2+} ions on the effects of ethyl methanesulphonate (EMS) on chromosomes', *Experientia* **19**, 144-145.
19. Slizynska, H. (1956-57), 'Cytological analysis of formaldehyde induced chromosomal changes in *Drosophila melanogaster', Proc. Roy. Soc. Edinburgh B.* **66**, 288-304.
20. Slizynska, H. (1963), 'Origin of repeats in *Drosophila* chromosomes', *Genetic Res. C.* **4**, 154-157.
21. Slizynska, H. (1968), 'Triplications and the problem of nonhomologous crossing over', *Genetic Res. C.* **11**, 201-208.
22. Parma, D.H., Ingraham, L.J. and Snyder, M. (1972), 'Tandem duplications of the rII region of bacteriophage T4D', *Genetics* **71**, 319-335.

22a. Hill, C.W. and Combriato, G. (1973), 'Genetic duplications induced at very high frequency by ultraviolet irradiation in *E.coli*', *Molec. Gen. Genetics* **127**, 197-214.

23. Auerbach, C. (1946), 'Chemically induced mosaicism in *Drosophila melanogaster*', *Proc. Roy. Soc. Edinburgh B.* **62**, 211-222.

24. Gibson, P.B., Brink, R.A. and Stahmann, M.A. (1950), 'The mutagenic action of mustard gas on *Zea mays*', *J. Heredity* **41**, 232-238.

25. Auerbach, C. (1947), 'The induction by mustard gas of chromosomal instabilities in *Drosophila melanogaster*', *Proc. Roy. Soc. Edinburgh B.* **62**, 307-320.

26. Ondrej, M. (1971), 'The proportion of complete mutations, mosaics and instabilities induced by ethylnitrosourea in *Drosophila melanogaster*', *Mutation Res.* **12**, 159-169.

27. Lee, W.R., Kirby, C.J. and Debney, C.W. (1967), 'The relation of germ line mosaicism to somatic mosaicism in *Drosophila*', *Genetics* **55**, 619-634.

28. Lee, W.R., Sega, G.A. and Bishop, J.B. (1970), 'Chemically induced mutations observed as mosaics in *Drosophila melanogaster*', *Mutation Res.* **9**, 323-336.

29. Alexander, M.L. (1967), 'Mosaic mutations induced in *Drosophila* by ethylenimine', *Genetics* **56**, 273-281.

30. Nasim, A. (1967), 'The induction of replicating instabilities by mutagens in *Schizosaccharomyces pombe*', *Mutation Res.* **4**, 753-763.

31. Nasim, A. and Grant, C. (1973), 'Genetic analysis of replicating instabilities in yeast', *Mutation Res.* **17**, 185-190.

32. Loprieno, N., Abbondandolo, A., Bonatti, S. and Guglielminetti, R. (1968), 'Analysis of the genetic instability induced by nitrous acid in *Schizosaccharomyces pombe*', *Genetic Res. C.* 12, 45-54.

CHAPTER SIXTEEN

Chemical mutagens: alkylating agents. II: Chemistry. Molecular analysis of mutants. Influence of cellular processes. Kinetics

Chemical structure of alkylating agents

Biological alkylating agents are chemicals that transfer alkyl groups to biologically important macromolecules under physiological conditions. Ross (1) has reviewed their general chemistry; Lawley (2) their reactions with nucleic acids. Table 16.1 lists the main classes of alkylating agent; each class is represented by one or several well-known mutagens. Cutting across this classification by structural group are differences in two properties whose role in mutagenicity has been much discussed. One is the type of alkyl group transferred, whether it is, e.g. a methyl or ethyl group or a more complex one like $-CH_2COCH_3$. The other is the number of alkyl groups that a single molecule can donate. This property is called the 'functionality' of the compound. Thus, among the nitrogen mustards, $H_2N(CH_2CH_2Cl)$ is monofunctional, $HN(CH_2CH_2Cl)_2$ is bifunctional, and $N(CH_2CH_2Cl)_3$ is trifunctional. The term polyfunctional may be used for all compounds whose functionality is greater than one. It should be noted that the degree of functionality cannot be inferred simply from the number of alkyl groups carried by a compound. The alkyl alkanesulphonates, for example, are mono-functional, EMS donating only its ethyl group and MMS only one of its two methyl groups. Even when two alkyl groups are bound in the same way, only one of them may be available for alkylation; this applies to the dialkyl sulphates, which act as monofunctional agents. Nitroso compounds require chemical activation, which for some compounds can take place *in vitro* or *in vivo*. Alkyl nitrosamines become mutagenic only after enzymatic conversion

Table 16.1

Main classes of alkylating agents, each represented by one or more well-known mutagens.

Class	Formula	Example
I Sulphur mustards	$S(CH_2CH_2Cl)_2$	mustard gas
II Nitrogen mustards	$HN(CH_2CH_2Cl)_2$	nitrogen mustard (HN_2)
III Epoxides	CH_2—CH_2 bridged by O (three-membered ring)	ethylene oxide (EO)
	CH_2–CH–CH–CH_2, each end ring-closed by O	diepoxybutane (DEB)
IV Ethylene imines	CH_2—CH_2 bridged by NH (three-membered ring)	ethyleneimine (EI)
	triazine ring (C, N alternating) with three aziridinyl groups, N(CH_2–CH_2), on the C atoms	triethylenemelamine (TEM)
V Alkyl alkanesulphonates	$C_2H_5OSO_2CH_3$	ethyl methane-sulphonate (EMS)
	$CH_3OSO_2CH_3$	methyl methane-sulphonate (MMS)
VI Dialkyl sulphates	$SO_2(OC_2H_5)_2$	diethylsulphate (DES)
VII β-lactones	H_2C—CH_2 ring with O—C=O	β-propiolactone
VIII Diazo compounds	$CH_3N{=}N$	diazomethane
IX Nitroso compounds	(ON)(H_3C)N–$COOC_2H_5$	*N*-nitroso-*N*-methyl urethane (NMU)
	(CH_3–CH_2)$_2$N.N=O	diethylnitrosamine (DEN)
	CH_3.N(NO).C(NH).NH.NO_2	*N*-methyl-*N′*-nitro-*N*-nitrosoguanidine (NG or NTG or MNNG

in vivo (3,4). Another compound that requires chemical or enzymatic conversion before becoming an alkylating mutagen is the antibiotic mitomycin-C (5) (not listed in Table 16.1).

Reactions with DNA

Many sites in DNA are open to attack by alkylating agents. Lawley (6) has discussed which of these reactions are probable sources of mutation. The most frequent point of attack is the N-7 of guanine, and originally this was considered the main source of mutations. Alkylation in this position favours ionization, and ionized guanine should have a tendency to mispair with thymine instead of cytosine, thus yielding transitions from GC to AT. We shall see presently that such transitions are, indeed, the main type of mutational change after treatment with alkylating agents; but the hypothesis that they result from ionization has not been supported by experimental evidence and is now more or less abandoned. On present evidence, it seems most likely that alkylations of O-6 of guanine and O-4 of thymine are the main sources of mutation. Other alkylated sites that might play a role are N-3 of guanine, N-1, N-3 and N-7 of adenine, N-3 of cytosine, and N-3 and O-4 of thymine. It should be kept in mind that mutation is a very rare event and that only a small fraction of all alkylations in DNA gives rise to mutations. The relative frequencies of the chemical reactions between DNA and a given mutagen are not good pointers to the relative importance of these reactions in mutagenesis. This applies to other mutagens as well as to alkylating agents.

A secondary effect of the N-7 alkylation of guanine is a loosening of the bond between the base and the sugar-phosphate chain, leading to a gradual leaching out of guanine from alkylated DNA. The 'apurinic gaps' thus created might, if filled in wrongly, produce transitions or transversions, but there is no evidence that this happens. What does appear to happen, although rarely, is the production of a deletion mutation of the frameshift type at the site of the gap. More important seems the fact that apurínic gaps result in scission of the chain in which they occur and, eventually, in double-strand scission (7). These reactions proceed more rapidly for methylated than for ethylated guanines, and this may explain why methylating compounds tend to be more toxic than the corresponding ethylating ones (e.g. MMS compared with EMS). Early experiments on bacteriophage had led to the conclusion that methylation is not mutagenic at all (8). While this has turned out to be incorrect, it seems true that the ratio of lethal to mutagenic effects is often higher for methylating agents than for the corresponding ethylating ones e.g. (9,10). This is, however, not always the case. In bacteria, MMS was mutagenically as effective as EMS (11), and its mutagenic action on transforming DNA was five times that of EMS (12).

Depurination leading to strand breakage is a conceivable candidate for the potential lesions that, in the chromosomes of *Drosophila* and plants, mature gradually into chromosome breaks (Chapter 15).It is suggestive in this respect that, in *Drosophila,* the storage effect on chromosome breakage requires weeks for monofunctional agents, but only days for polyfunctional ones (13-15). This is paralleled by the fact that the sequence from apurinic gap to double-strand scission occurs more rapidly after treatment with the bifunctional mustard gas than after treatment with a monofunctional derivative of it, probably because only the former has cross-linking ability (16). Cross-links between neighbouring guanines on the same or on opposite strands of DNA have been found in prokaryotes and in mammalian cells (17). In bacteriophage, about 20% of the alkylations produced by mustard gas were in the form of di-alkylations, and about ¼ of these linked opposite strands (18).

The effect of functionality

When alkylating agents were introduced into cancer therapy, it was found that only bi- or polyfunctional ones were effective carcinostats. Since it is very probable that their carcinostatic action is due to chromosome breakage, this difference between mono- and polyfunctional agents was taken to mean that only the latter, because of their cross-linking ability, are effective chromosome breakers. This turned out not to be true. Some monofunctional agents are excellent chromosome breakers (19,20); moreover, their efficiency of breakage may be greatly increased by various means, e.g. by the presence during treatment of certain metal ions (p.266), by pre- or post-treatment with the metabolic inhibitor dinitrophenol, or by prolonged storage before DNA replication (p. 262). It is true that often – although not always – higher amounts of a monofunctional than of a homologous polyfunctional agent are required to produce the same frequency of chromosome breakage; but this applies also

Table 16.2

The frequencies of translocations in *Drosophila* males that had been injected with one of two monofunctional agents (EO, EI) or one of two closely related polyfunctional ones (DEB, TEM). The genetically effective doses were measured in terms of sex-linked lethals

Expt.	*Agent*	*% s-1-1**	*Translocations (Y, II, III)* No. genomes tested	No. transloc.	% transloc.	*Ratio lethals/ transloc.*
I†	EO	5.3	1074	3	0.3	17
	DEB	6.5	1151	5	0.4	13
II ‡	EO	7.1	433	3	0.7	10
	DEB	9.9	434	4	0.9	11
III §	EI	3.4	926	6	0.6	6
	TEM	6.7	718	6	0.8	11
IV §	EI	7.3	589	15	2.5	3
	TEM	6.7	1666	15	0.9	7

* Percent sex-linked lethals in samples ranging from 500-1000 chromosomes.

† After Table 1, Nakao (23).

‡ After Table 1, Watson, 1966 (14).

§ After Table 1, Watson, 1964 (13).

to other measures of toxicity, although even here there are exceptions. Thus for *Paramecium* EI is much more toxic than TEM (21), and *Neurospora* conidia are more easily killed by EO than by DEB (22). Comparisons between the effects of chemical mutagens on chromosomes should not be based on molar concentrations in the treatment solution but on estimates of the amount that effectively reached the genetic material. Table 16.2 presents data from experiments on *Drosophila* in which the mutagenically effective dose was measured in terms of frequencies of sex-linked lethals.

The chromosome breaking ability within two pairs of closely related chemical compounds, one monofunctional (EO, EI) the other polyfunctional (DEB, TEM), was compared by determining the ratios of translocations to lethals. In Table 16.2 the reciprocal of this ratio is given, so that higher ratios mean lower breakage abilities. In the comparisons between EO and DEB, the ratios were very similar, indicating similar efficiency of breakage. In the comparisons between EI and TEM, the ratios for the monofunctional compound were about half those for the polyfunctional one, suggesting that EI is superior to TEM in chromosome breaking ability. This agrees with the cytological analysis of rearrangements in the salivary glands of the F_1 (24): non-mosaic aberrations (the only ones that would be detected genetically) were about four times as frequent among the progeny of males treated with EI than among that of males treated with a mutagenically equivalent dose of TEM. Thus, when dose is measured in units of mutagenicity, monofunctional agents are not generally inferior to polyfunctional ones in their ability to produce chromosome breakage and may even be superior to them. The picture changes when treated spermatozoa are stored in the seminal receptacles of untreated females. Under these circumstances, the frequency of rearrangements but not that of lethals increases rapidly and drastically when a polyfunctional agent has been used for treatment (Table 15.2). Equally strong storage effects can be obtained after treatment with monofunctional agents, but they require a much longer time and are hardly apparent during the first week of storage (15). As a result, when scoring is carried out after 5-6 days of storage, polyfunctional agents yield much higher ratios of rearrangements to lethals than do monofunctional ones. When scoring is delayed for two or three weeks, this difference disappears again. Possibly, the rapid storage effect for polyfunctional agents contributes to their superiority in cancer therapy; we shall return to this point in Chapter 23.

Since it is no longer believed that cross-linkage between the two strands of DNA is the main source of chromosome breakage, there is in fact no reason to postulate an effect of functionality on chromosome breakage. Even less is there reason to assume an influence of functionality on the production of gene mutations and, in fact, none was found. In *Neurospora,* where EO is much more toxic than DEB, it is also much more mutagenic (22), so that, in the net result, both substances produce similar frequencies of adenine-reversions per survivors. This was not so for barley, where the concentration of EO had to be about 50 times that of DEB for producing comparable frequencies of chlorophyll mutations (25). Since, however, DEB is several hundred times as toxic as EO for barley seeds, the monofunctional compound allowed higher mutation frequencies to be obtained at tolerable levels of lethality and sterility. The Swedish plant breeders have called this property of a mutagen its 'efficiency' as opposed to its 'effectiveness', which is a measure of mutation frequency per dose. Because of their low toxicity, some monofunctional compounds have very high efficiencies and are therefore especially useful for mutation breeding (Chapter 23). Rapoport (26) has introduced the term 'supermutagens' for these compounds, which include EMS and NG.

Molecular analysis of mutants

Reversion analysis (Chapter 17) as well as direct or indirect sequencing of gene products (Chapter 11) have been used for deriving the molecular types of alkylation-induced mutations. On the whole, the results were in reasonable agreement with expectations, the more so, the simpler the system and the more direct the application of the mutagen to DNA. But discrepancies were also obtained, even in the simplest systems and increasingly in more complex ones. These have led to modification of the chemical hypotheses and to the consideration of the effects of structural and cellular factors on the primary reactions with DNA and on the secondary processes that produce mutant clones from premutational lesions.

In bacteriophage T4, reversion analysis was applied to rII mutants that had been obtained by *in vitro* treatment with EMS or EES (ethylethane-sulphonate) (27). Most of them were inferred to be transitions from GC to AT, while transitions from AT to GC, transversions and frameshifts (probably all of the deletion type) were rare. Thus, most of the mutants that were produced by either agent should have AT at the mutant site and should be refractory to reversion by the agent that produced them. This expectation, too, was fulfilled in T4: of twelve EMS-induced rII mutants, only one—possibly two – could be made to revert with EMS (28).

Reversion analysis in single-stranded DNA phages can be used to determine which base, in contrast to which base-pair, reacts with a given mutagen. Results obtained on the phages S13 and ϕX174 differed from those obtained on T4 (29,30). EMS, while again acting on purines, also induced the pyrimidines to mutate, each of them with approximately the same frequency as its complementary purine. NG, which in T4 is even more specific than EMS for GC to AT transitions, produced equal proportions of GC to AT and AT to GC transitions in single-stranded phage. While these results may be attributed to the better accessibility of the bases in single-stranded DNA, they raise the question whether reactions with pyrimidines are not in part also responsible for the base-pair changes in double-stranded DNA. Indeed, in denatured DNA nitrogen mustard alkylated cytosine and guanine to the same extent (30a); it may also do this in locally denatured regions of two-stranded chromosomal DNA. In any case, the structural configuration of DNA obviously plays a role in its reactions with mutagens. The same conclusion was drawn by Singer and Fraenkel-Conrat (31), who studied the chemical reactions of several mutagens with the RNA inside the tobacco mosaic virus (TMV) with mononuclotides, with single- and double-stranded synthetic polynucleotides, and with single and double-stranded nucleic acids. They concluded that 'no extrapolation from one system to another ... is justified, since in many instances the difference in the behaviour of nucleic acids in different milieus is not only quantitative but qualitative'. Particularly striking was the contrasting effect of formamide on the chemical and mutagenic effects of NG (32). In the absence of formamide, NG methylated guanine and, to a lesser extent, adenine in synthetic polynucleotides, but had only a weak mutagenic effect on TMV. In the presence of formamide, it failed to methylate either purine, but turned into a strong mutagen for TMV.

In bacteria, sequencing of gene products yielded results that agreed with those obtained on T4. Indirect sequencing of EMS-induced nonsense suppressors (p. 196) led to the conclusion that over 90% were due to transitions from GC to AT (33). Direct sequencing of the mutant *t*-RNA in nine EMS-induced nonsense suppressors showed that all had either T or A at the mutant site, as would be expected from GC to AT transitions (34). The results of reversion analysis were contradictory. In *E.coli* none of 23 EMS-induced *lac*-auxotrophs could be reverted with EMS (35). On the contrary, in *Salmonella typhimurium* the vast majority of auxotrophs that had been induced by DES or NG could be reverted by either of these chemicals (36). No doubt, these inconsistencies are at least in part due to the fact that the specificity of the standard mutagens, which forms the basis of reversion analysis, may be modified or masked in cellular organisms by processes that take place before or after the reaction with DNA.

Such complexities are bound to play an even greater role in eukaryotes than in prokaryotes. In *Neurospora*, Malling and de Serres (38) applied complementation analysis (Chapter 11) to *ad*-3B mutants that had been produced by EMS and NG. They concluded that 60% or more of the EMS-induced mutants and at least 80% of the NG-induced ones were due to base-pair substitutions. From reversion analysis of the EMS-induced mutants, they concluded further that the majority of these were transitions from AT to GC, in marked contrast to what had been found for prokaryotes. While it is quite possible that conditions in eukaryotic cells may modify the reaction of EMS with DNA, one should not forget that the same may be true for the standard mutagens used as testers, and that this must diminish the reliability of reversion analysis. For *Saccharomyces cerevesiae,* Sherman and his collaborators (39) showed that even sequence analysis may reveal unexpected complexities in the pattern of reactions between mutagens and DNA. They studied reversions in 11 mutants of the *cy1* gene, whose product, *iso*cytochrome-*1c,* can be sequenced. Two of the mutants were nonsense codons within the cistron, the other nine were known base changes in the initiation codon AUG. Theoretically, the only mutant that could revert by transition from GC into AT – and in no other way – was one of the initiation mutants, *cy1-131,* which carried a GUG triplet in place of the normal AUG. The other mutants could revert by a variety of base changes, transitions as well as transversions, but not by a change of GC into AT. A mutagen that specifically produces GC into AT transitions would therefore be expected to act preferentially on *cy1-131;* less specific mutagens would revert all 11 mutants with similar frequencies. Among the alkylating agents tested in this system, EMS, DES and NG acted specifically on *cy1-131*; three others – MMS, DMS (dimethylsulphate) and HN2 – acted unspecifically. In view of the finding that, in the same system, UV shows different specificities depending on the mutant site within the cistron (p.247), it is not clear whether the specificity of some of these alkylating agents refers to a general preference for the production of GC into AT transitions, or for a preference of this transition at this particular site, i.e. the initiation codon. Results obtained with nitrous acid in the same system suggest that the latter may be true (Chapter 17).

Influence of cellular processes

Repair

In 1948, Bryson (40) selected strains of *E.coli* for resistance to the nitrogen mustard HN2. All resistant strains had also become resistant to UV. Since at that time the action of neither mutagen was understood, this observation was not followed up. With hindsight we see that it suggests a repair mechanism that can handle damage to DNA whether it has been produced by UV or HN2. In 1965, Reiter and Strauss (41) observed that phage that had been treated with HN2 survived better in wild-type hosts than in *hcr*$^-$ hosts that lack excision repair. Soon after this, analysis of mustard gas treated DNA from *hcr*$^+$ and *hcr*$^-$ strains proved that in the former but not in the latter diguaninyl derivatives are excised (42,43). Excision has also been found in mammalian cells that had been treated with mustard gas (44). On the other hand, MMS, EMS and NG are equally toxic for *hcr*$^+$ and *hcr*$^-$ strains, although some of the damage they cause is repaired in bacterial and mammalian cells (45). MMS, for example, shows cross-resistance with X-rays in bacteria. The probable nature of repair from monoalkylation has been discussed by Strauss (46). In barley seeds, recovery from damage by monofunctional agents was accompanied by repair of single-strand breaks but not by excision of alkylated bases (47). Although our knowledge in this area is still very incomplete, it is already clear that – as for UV – repair may both erase and create mutations. The potentiating effect of caffeine on alkylation-induced chromosome breakage and mutations in eukaryotes (48-50) indicates that alkylation damage in DNA can be removed in an error-proof way. The same follows from Kimball's (21) observation that, in *Paramecium*, TEM-induced mutations vary in frequency with the stage in the cell cycle in a similar – although not identical – way as do X-ray induced mutations (p.135). This is taken to mean the existence of a premutational stage which is open to repair before replication. In bacteria, Kondo and his co-workers (51) have studied the effects of an array of chemical mutagens on strains lacking known repair systems. Among alkylating agents, EMS does not depend on the *rec*-function for its mutagenic action, but MMS does. Mitomycin-C is so far the only mutagen that requires not only *rec*-repair but also excision repair for mutagenesis, possibly because excision of the cross-links (5) is not error-proof. The whole subject is still very much in the state of experimentation. Kondo has put forward a scheme for the classification of mutagens on the basis of their relation to repair processes and has used it for speculations on their various modes of action. Evans (52) has suggested that chromosome breakage, whether induced by radiation or chemicals, always proceeds by misrepair at the next replication, but there is so far no evidence for this hypothesis. It certainly cannot apply to *Drosophila* e.g. (13,14) and mice (53), in which both X-rays and alkylating agents, acting on mature sperm, produce chromosome breaks that comprise both strands of DNA. It is, however, evident that repair mechanisms of various kinds may be intercalated between alkylation damage to DNA and the appearance of mutations and chromosome breaks and that their idiosyncrasies may modify or mask the primary pattern of damage.

Replication and transcription

The role of these processes in mutagenesis by any agent is obvious. With one exception (Chapter 18), no known chemical mutagen can directly change a normal base into a different normal one. Since only normal bases can be replicated, abnormal ones can produce mutations only by a replication process that results in the incorporation of a *wrong normal* base. In a similar way, replication is required for the perpetuation in DNA of errors due to frameshifts or larger deletions. It is equally trivial to state that mutational lesions in DNA have to be transcribed before yielding mutated cells.

There are, however, certain special means by which replication and transcription can enter the mutagenic pathway. One is the monitoring of mismatched strands by DNA-polymerase at the time of replication (54). The efficiency of this process depends on the type of lesion, so that there is an interaction between primary damage and the amount of erroneous replication to which it gives rise. As this interaction does not seem strong for alkylating agents, we shall defer its discussion to the next chapter. A different kind of connection with replication has been found for the action of NG on *E.coli.* Cerdá-Olmedo and his group (55,56), have brought convincing evidence that in this organism NG acts specifically at the replication fork (Fig. 16.1). In synchronous cultures, different genes show peaks of response to NG at times that correspond to their position on the chromosome, so that a replication map based on the times when peaks of mutability appear for different genes corresponds well with the standard map of *E.coli.* In non-synchronized

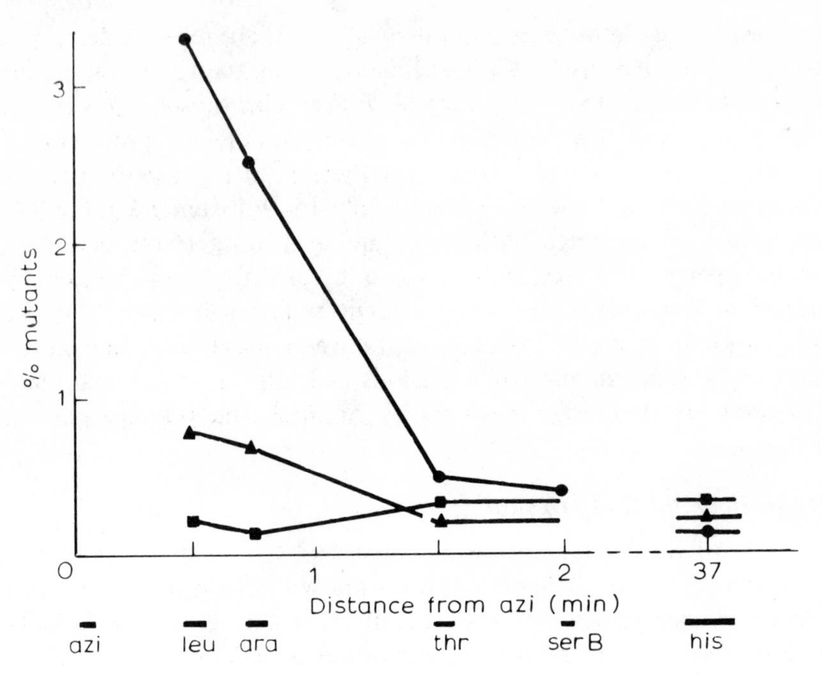

Fig. 16.1. Frequencies of NG-induced auxotrophic mutations among azide-resistant mutants in *E.coli.*

● cultures grown exponentially prior to treatment; clones tested for auxotrophy were previously selected for resistance to sodium azide.

▲ cultures starved of amino acids prior to mutagenesis; selection for azide-resistance.

■ cultures grown exponentially prior to treatment; no selection for azide-resistance.

(Fig. 3, from Guerola (56). Courtesy Macmillan Journals Ltd.)

cultures, different genes respond with similar frequencies to NG (■ in Fig. 16.1). When, however, resistance to sodium azide was used as a selective marker (●), the frequencies of unselected auxotrophs among resistant clones showed a close correlation with distance from the locus for azide-resistance, indicating that most of them occurred in a replicating region. Deletions could be excluded as origin of the double mutants, because most of the NG-induced mutations in bacteria are revertable. When, in the same type of experiment, the number of replication points was reduced by previous amino acid starvation (▲), the frequency of auxotrophs declined, but still showed a weak correlation with distance from the azide-resistance locus.

The basis for the preferential action of NG at the replication point is not clear. Since treatment occurred in buffer, with no synthesis of DNA, erroneous replication in the presence of NG does not appear to be involved. It rather seems that NG acts directly on DNA, but that this action is in some way facilitated by conditions near the replication point. A direct mutagenic action of NG on DNA has been shown to occur after treatment of transforming DNA (57) and of tobacco mosaic virus (32). In *Drosophila* (58), NG produces point mutations and rearrangements in spermatozoa, in which no replication takes place. As regards the special conditions that may facilitate mutation at the replication fork, one obvious possibility is that strand-separation makes the nucleotide bases more accessible to the action of NG and, perhaps, of other alkylating agents. If this were true, then regions in which the strands have opened up for transcription should show a similar strong response to NG. Preliminary tests by Cerdá-Olmedo spoke against this explanation, but subsequently Brock found pronounced effects of derepression on the response of the β-galactosidase locus in *E.coli* to NG, EMS and DES (59).

Kimball (60) has suggested that NG-lesions may be very rapidly repaired, so that only those near the replication point have a good probability of becoming fixed. If this were true, one might find an inverse correlation between the repair capacity of an organism and the proportion of NG-induced mutations that occur *outside* the replication point. In the *E.coli* strains studied by Cerdá-Olmedo, such mutations formed about 20% of the total. In some other organisms, they appear to be so prevalent that no correlation between replication and NG mutagenesis was apparent (61,62). On the contrary, in *Chlamydomonas* and yeast, this correlation is so strong for both chromosomal and non-chromosomal DNA that it can serve as the basis for selecting mutations in one or the other system (63,64), The rule that, in biology, extrapolation from one system to another is very hazardous, clearly applies also to mutagenesis by NG.

Metabolism

Again, it is trivial to state that metabolism must play a role in those processes that decide whether alkylation lesions in DNA will become repaired or fixed and, once fixed, will lead to expressed mutations. Supersuppressors in *B. subtilis* that had been induced by EMS showed a dependence on post-treatment media that paralleled mutation frequency decline and mutation fixation after UV-irradiation (Chapter 12), although the mechanisms may well have been different (65). Of special interest is the question whether alkylating agents may modify steps in the mutation process in ways that affect the fate of those lesions that they themselves have produced; in other words, whether alkylating agents can act at the same time as primary mutagens and as selective sieves that regulate the quantities and types of realized mutations. Certainly, alkylating treatment may affect cellular processes even more than does UV. Nitrogen mustard, for example, was found to interfere to different extents with the coding ability of poly-A, poly-U, and poly-C; to stimulate polypeptide synthesis on ribosomes at low concentrations and to inhibit it at high ones; and to inhibit enzymes engaged in protein synthesis (66). Ethylene imine inactivates amino acid acceptance by *E.coli t*-RNA's to very different degrees, lysine acceptance being strongly inhibited, isoleucine acceptance not at all (67). The mutagen nitrosomethylguanidine inhibits the acceptor activity of various *t*-RNA's to different degrees (67a). Such cellular effects – and doubtless they will be produced to varying extents by all alkylating agents – are bound to influence the mutation process.

This area of mutation research has so far received very little attention. There has been some interest in the dependence of *repair* on strength and conditions of treatment; treatment effects on *expression* are rarely considered beyond the necessity of providing a good expression medium. The reason for this attitude is not far to seek. While we are still trying to unravel the intricacies of the first, general steps in mutagenesis, the more remote and specific ones that enter into expression are of secondary interest. Yet, neglecting even the possibility of their importance may mean – and repeatedly has meant – that the burden of accounting for vagaries of expression has been put onto the framework that should carry only the analysis of the common first steps in mutagenesis, thus overloading and distorting it. The few data that deal with alkylation effects on repair and expression will be accommodated best in the next section.

Dose-effect curves. Dose rate. Storage effect on gene mutations

In transforming DNA (68) and in phage treated *in vitro* (28), alkylating agents produce mutations with linear kinetics; this indicates that a 'hit' on a single base is sufficient to produce a mutation. When DNA is treated inside cells or cellular organisms, the dose response curve reflects not only the kinetics of the reaction between mutagen and DNA but includes many components, such as membrane permeability, reactions with cellular components, effects on enzymes involved in transcription, in repair and expression, and others. It is indeed surprising that sometimes linear dose-effect curves have been found even for eukaryotes; one suspects that these may be due to the mutual cancelling out of deviations from linearity.

More definite conclusions can be drawn from dose-effect curves in which the dose is measured by the frequency of an effect that is likely to arise from single hits on DNA. This was done for the production of translocations by mustard gas in *Drosophila* (69). Plotted against sex-linked lethals, which could be shown cytologically to include few gross rearrangements, the frequency of translocations increased very nearly as the square of the dose. The conclusion was drawn that mustard gas-induced translocations, like those induced by X-rays, arise from the rejoining of two independently produced breaks, and that the events producing these breaks are similar in nature to those that, singly, produce recessive lethals. We now can define these events more precisely as alkylations, although the types of alkylation that produce breaks need not be identical with those that produce mutations. The close approximation of the dose coefficient to 2 makes it similar to the dose coefficient for very low X-ray doses which yield comparably low frequencies of translocations. The dose-effect curve for mustard gas-induced translocations thus comes closer to the X-ray induced one with increasing dose, and this explains why in Table 15.1 the shortage of mustard gas induced translocations diminishes with dose.

Because of its very real significance for the theory of chromosome rearrangements, the quadratic curve has been endowed with almost mystical properties, and when mutation frequencies show a steeper than linear dose-response, attempts are usually made to fit them to a square-law curve. For a small number of points, each with a fairly large standard error, these attempts are often successful in the sense that the deviation from a square-law curve is not significant. However, if there is no *a priori* reason for expecting a two-hit event (as there is for the formation of chromosome rearrangements), there is also no special reason for fitting the data to just the particular dose exponent of 2. A different one, or a curve of an altogether different shape, may give an equally good or even a better fit. Two examples of alkylation mutagenesis in *E.coli* illustrate this point. Turtoczky and Ehrenberg (70) obtained a very steep dose-effect curve for the production of streptomycin-resistant mutations by EMS. On closer analysis, the curve turned out to be biphasic, with a linear portion at the beginning, followed by a much steeper portion at higher doses. A less powerful mutagen *iso*propylmethane-sulphonate (iMPS) yielded a uniformly linear dose-effect curve over the same survival range. Among the interpretations suggested by the authors is the possibility that higher doses of EMS increasingly inhibit a repair mechanism so that more and more premutational lesions are turned into mutations. This interpretation agrees with the shapes of the survival curves. The curve for iMPS was exponential throughout, as expected for killing by single 'hits'. The curve for EMS became exponential only after a flat 'shoulder' at low doses when most of the lethal damage is probably repaired. The second set of experiments was carried out by Tarmy *et al.* (71), who produced tryptophan-reversions by an alkylating carcinogen. In their experiments, a curve that bent upwards and might well have been fitted by a 'multihit' equation, was resolved into a sub-threshold region with hardly any increase in mutation frequency, followed by a portion in which mutation frequency increased linearly with dose as measured either by the concentration of mutagen in the treatment medium or by the number of alkylations in DNA. The authors assumed that the repair system was able to cope with the majority of mutagenic lesions up to the threshold, when it became saturated. Taken together, these two sets of experiments show that mutation frequencies followed linear kinetics both when repair was complete (initial part of the EMS-curve) and when it was saturated (final part of the carcinogen-curve). In between, there must be a dose range when repair gradually loses efficiency; in the experiments with the carcinogen, this region formed only a small upward bend between the two parts of the curve; in those with EMS, it formed the greater part of the curve beyond its first linear portion.

Neurospora workers (22,72), have used the same interpretation for the very steep dose-effect curve of DEB-induced adenine-reversions.

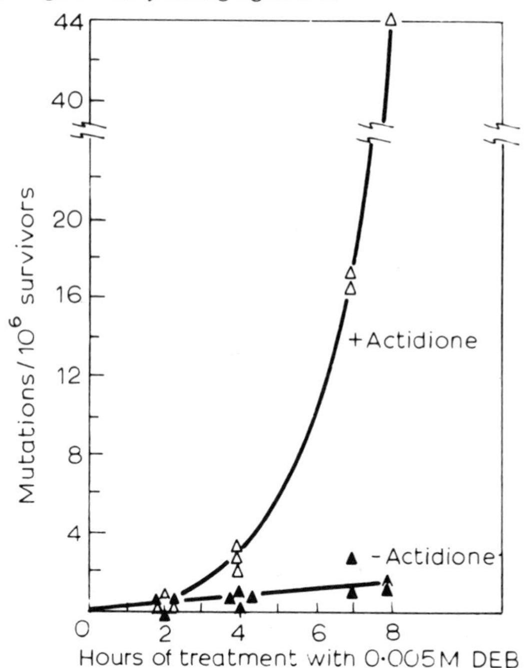

Fig. 16.2. The effect of actidione on the induction of mutations in *Neurospora* by DEB given at a low dose rate. Survival was very high after all treatments. The shape of the upper curve is that typically obtained by DEB given at a high dose rate. (Fig. 1, Kilbey (73). Courtesy Springer-Verlag, Berlin. Heidelberg, New York.)

Compared with the two bacterial experiments, their dose range was intermediate, comprising mainly the portion in which an intrinsically linear curve is bent upward by decreasing repair efficiency. Support for this interpretation came from a number of independent observations. (1) Given at a low dose-rate, DEB produced a linear dose-effect curve; this would be expected if the slow treatment allowed recovery of the repair mechanism during exposure. (2) In the presence of actidione, an inhibitor of protein synthesis, the dose-effect curve bent upward even for treatment given at a very low dose rate (Fig. 16.2) (73); this would be expected if recovery of the repair mechanism required protein synthesis. (3) The very small amounts of DEB that cannot be removed by centrifuge-washing continued to produce mutations over several hours ('storage effect') (Fig. 16.3A) (74); this would be expected if the first treatment had damaged the defences of the cell against the very weak after-treatment. It should be pointed out that this storage effect is

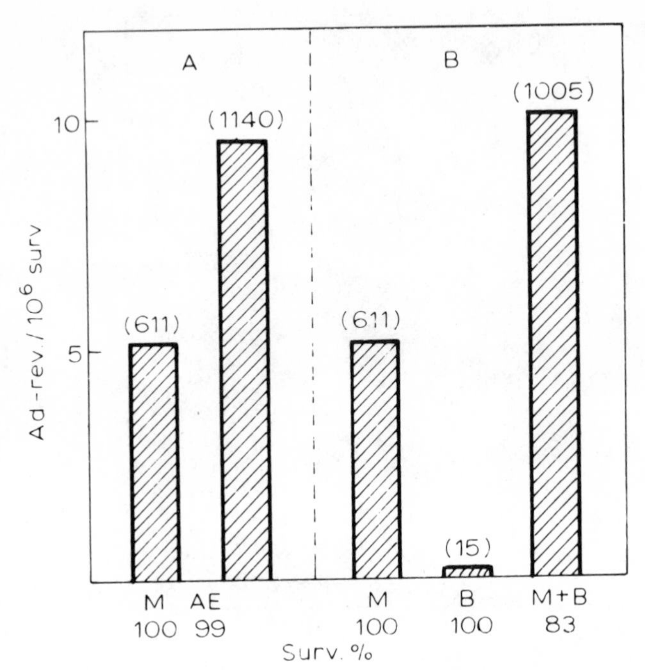

Fig. 16.3. Sensitization by DEB to further very small doses of DEB. A. Storage effect: Main treatment (M) followed by 4 h standing at 25° (AE). B. Main treatment (M) followed by a second weak treatment (B). The numbers on top of the columns give the colony counts. (Fig. 3, Auerbach (75). Courtesy Springer-Verlag, Berlin. Heidelberg. New York.

probably different in kind from the storage effect of alkylating agents on chromosome breakage in plants and *Drosophila* (Chapter 15). (4) Spores that had been treated with normally mutagenic doses of DEB became exceptionally sensitive to its action (75). They yielded high mutation frequencies after exposure to a second dose, which hardly affected spores that had not been pre-treated (Fig. 16.3B). We may say that they had become 'sensitized' to extraneously administered small amounts of DEB, just as the cells in the storage experiments had become sensitized to internally remaining traces of DEB. (5) Sensitization extended also to treatment with UV (76). Spores that had been pre-treated with barely mutagenic doses of DEB, yielded up to twice as many reversions after UV-treatment than did non-pretreated spores (Fig. 16.4 .)This would be expected if DEB inhibits a repair mechanism that can also remove UV-induced lesions. The fact that the excess mutations after double treatment were UV-induced was established by their photoreactivability (77).

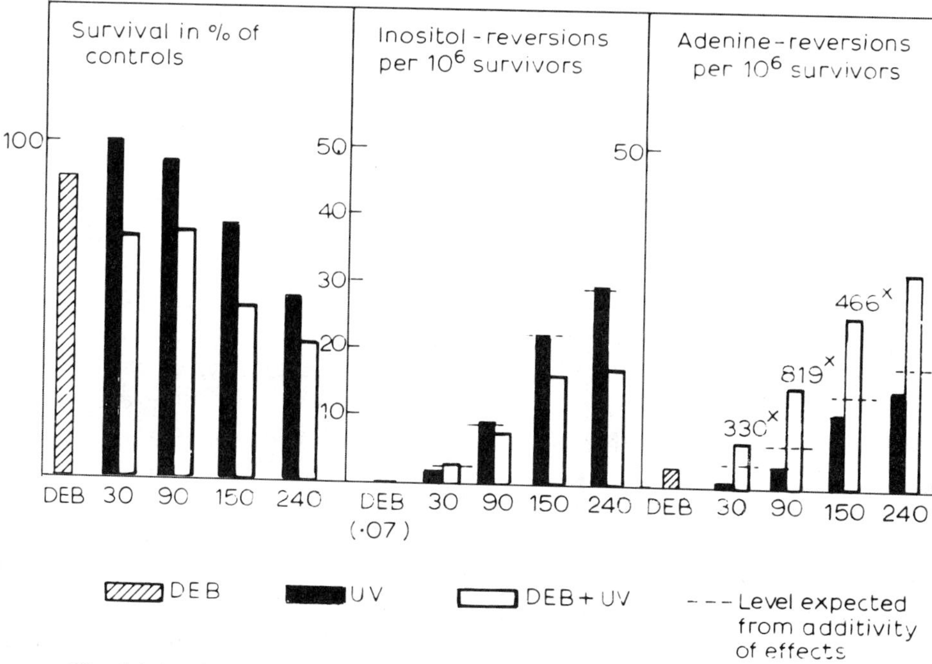

Fig. 16.4. The effects of pre-treatment with DEB on the frequencies of UV-induced adenine- and inositol-reversions in *Neurospora*. The numbers on top of the columns are the numbers of excess colonies after double-treatment; after the longest exposure to UV, survival was too low for an increase in colony count. (Fig. 1, Auerbach (76). Courtesy Springer-Verlag, Berlin. Heidelberg. New York.)

(6) The normally occurring gradual decay of sensitization to either UV or very small doses of DEB could be prevented by actidione (73); this would be expected if sensitization is due to inhibition of a repair mechanism that requires protein synthesis for recovery. It is known that wild-type *Neurospora* contains a repair enzyme that carries out excision of UV-induced lesions (78). While it has not yet been established whether this or indeed any other repair enzyme is inhibited by DEB, the bulk of the evidence suggests that this is the case.

Inositol-reversions, studied simultaneously in the same DEB-treated cells, arose with very different kinetics and differed from adenine-reversions in all the points that have just been discussed. The dose-effect curve, instead of bending steeply upward, rose very slowly and tended to bend downward at higher doses (72). The dose-exponent was usually less than 1 even in the first ascending portion of the curve, suggesting that DEB-treatment in some way interferes with the production of inositol-reversions from premutational lesions in DNA. This was borne out by the observation that the frequency of UV-induced inositol-reversions was decreased to about one half by pretreatment of the spores with DEB (Fig. 16.4) (76). Trivial explanations such as killing of DEB-treated parental cells on inositol-free medium ('inositol-less death') or preferential killing of inositol-revertants by DEB could be excluded. An explanation in terms of repair would have to make the unlikely assumption that the same treatment which inhibits repair of adenine-reversions stimulates that of inositol-reversions. A more plausible assumption is that DEB in some way interferes with the expression of inositol-revertants, perhaps because of their unique effect on the chemistry of membranes. It seems possible that slow growth after treatment counteracts this inhibition of expression, for there was no shortage of inositol-reversions among DEB-treated spores of the very slow-growing respiratory-deficient 'poky' strain (79). It is also suggestive that all of three tested inositol-alleles were recalcitrant to DEB, whereas three different *ad*-3B alleles gave different degrees of response, although always with characteristically steep kinetics (80).

While these results require biochemical corroboration and explanation, they encourage the hope that the analysis of dose-effect curves in eukaryotes may aid in the analysis of steps in the mutation process: the common steps like repair as well as the more specific ones involved in the expression of individual mutants. The latter analysis, in turn, should be useful for an understanding of gene action and its response to cellular disturbances.

A point that may have occurred to the reader is that the difference in the slopes of the adenine- and inositol-reversion curves must result in a strongly dose-dependent 'specificity' of DEB for the production of adenine- as opposed to inositol-reversions. This is indeed true; we shall discuss it more fully in Chapter 20.

References

1. Ross, W.C.J. (1962), 'Biological Alkylating Agents', Butterworths, London.
2. Lawley, P.D. (1966), 'Effects of some chemical mutagens and carcinogens on nucleic acids', *Progr. Nucleic Acid Res. Mol. Biol.* 5, 89-131.

3. Magee, P.N. and Farber, E. (1962), 'Toxic liver injury and carcinogenesis. Methylation of rat liver nucleic acids by dimethylnitrosamine *in vivo'*, *Biochem. J.* **83**, 114-124.
4. Malling, H.V. (1971), 'Dimethylnitrosamine: formation of compounds by interaction with mouse liver microsomes', *Mutation Res.* **13**, 425-429.
5. Iyer, V.N. and Szybalski, W. (1964), 'Mitomycins and porfiromycin: Chemical mechanism of activation and cross-linking of DNA', *Science* **145**, 55-58.
6. Lawley, P.D. and Brookes, P. (1974), *Alkylation of nucleic acids and mutagenesis,* in *Molecular and Environmental Aspects of Mutagenesis',* L. Prakash *et al.* (ed.), Thomas, Springfield, Ill., pp 17-31.
7. Lawley, P.D. and Brookes, P. (1963), 'Further studies on the alkylation of nucleic acids and their constituent nucleotides', *Biochem. J.* **89**, 127-138.
8. Loveless, A. (1959), 'The influence of radiomimetic substances on deoxyribonucleic acid synthesis and function in *Escherichia coli*/ phage systems. III. Mutation of T_2 bacteriophage as a consequence of alkylation *in vitro*: the uniqueness of ethylation', *Proc. Roy. Soc. Lond.* **150**, 497-508.
9. Kølmark, G. (1956), 'Mutagenic properties of certain esters of inorganic acids investigated by the *Neurospora* back-mutation test', *C.R. Labor. Carlsberg. Ser. Physiol.* **26**, 206-220.
10. Loprieno, N. (1966), 'Differential response of *Schizosaccharomyces pombe* to ethyl methanesulfonate and methyl methanesulfonate', *Mutation Res.* **3**, 486-493.
11. Strauss, B.S. (1961), 'Specificity of the mutagenic action of the alkylating agents', *Nature* **191**, 730-731.
12. Rhaese, H.Y. and Boetker, N.K. (1973), 'The molecular basis of mutagenesis by methyl and ethyl methanesulfonates', *Eur. J. Biochem.* **32**, 166-172.
13. Watson, W.A.F. (1964), 'Evidence of an essential difference between the genetical effects of mono- and bifunctional agents', *Zeitschr. Vererb. Lehre* **95**, 374-378.
14. Watson, W.A.F. (1966), 'Further evidence of an essential difference between the genetical effects of mono- and bifunctional alkylating agents', *Mutation Res.,* **3**, 455-457.
15. Šram, R.J. (1970), 'The effect of storage on the frequency of translocations in *Drosophila melanogaster', Mutation Res.* **9**, 243-244.

16. Laurence, D.Y.R. (1963), 'Chain breakage of deoxyribonucleic acid following treatment with low doses of sulphur mustard', *Proc. Roy. Soc. Lond. A* **271**, 520-530.
17. Flamm, W.G., Bernheim, N.Y. and Fishbein, L. (1970), 'On the existence of intrastrand crosslinks in DNA alkylated with sulphur mustard', *Biochim. Biophys. Acta* **224**, 657-659.
18. Lawley, P.D., Lethbridge, J.H., Edwards, P.A. and Shooter, K.V. (1969), 'Inactivation of bacteriophage T_7 by mono- and difunctional sulphur mustards in relation to crosslinking and depurination of bacteriophage DNA', *J. Mol. Biol.* **39**, 181-198.
19. Smith, H.H. and Lotfy, T.A. (1955), 'Effects of β-propiolactone and Ceepryn on chromosomes of *Vicia* and *Allium*', *Amer. J. Bot.* **42**, 750-758.
20. Swanson, C.P. and Merz, T. (1959), 'Factors influencing the effect of β-propiolactone on chromosomes of *Vicia faba*', *Science* **129**, 1364-1365.
21. Kimball, R.F. (1965), 'The induction of reparable premutational damage in *Paramecium aurelia* by the alkylating agent triethylene melamine', *Mutation Res.* **2**, 413-425.
22. Kølmark, H.G. and Kilbey, B.J. (1968), 'Kinetic studies of mutation induction by epoxides in *Neurospora crassa*', *Molec. Gen. Genetics* **101**, 89-98.
23. Nakao, Y. and Auerbach, C. (1961), 'Test of a possible correlation between cross-linking and chromosome breaking abilities of chemical mutagens', *Zeitschr. Vererb. Lehre.* **92**, 457-461.
24. Slizynska, H. (1973), 'Cytological analysis of storage effects on various types of complete and mosaic change induced in *Drosophila* chromosomes by some chemical mutagens', *Mutation Res.* **19**, 199-213.
25. Ehrenberg, L. and Gustafsson, A. (1957), 'On the mutagenic action of ethylene oxide and diepoxybutane in barley', *Hereditas* **43**, 595-602.
26. Rapoport, J.A., Zoz, N.N., Makarova, S.I. and Salnikova, T.V. (Eds.) (1966). 'Super-mutagens', Nauka, Moscow (in Russian).
27. Bautz, E. and Freese, E. (1960), 'On the mutagenic effect of alkylating agents', *Proc. Nat. Acad. Sci. U.S.A.* **46**, 1585-1594.
28. Krieg, D.R. (1963), 'Ethyl methanesulfonate-induced reversion of bacteriophage T4 rII mutants', *Genetics* **48**, 561-580.

29. Tessman, I., Poddar, R.K. and Kumar, S. (1964), 'Identification of the altered bases in mutated single-stranded DNA. I. *In vitro* mutagenesis by hydroxylamine, ethyl methanesulfonate and nitrous acid', *J. Mol. Biol.* **9**, 352-363.
30. Baker, R. and Tessman, I. (1968), 'Different mutagenic specificities in phages S13 and T4: *in vivo* treatment with N-methyl-N-nitro-N-nitroso-guanidine', *J. Mol. Biol.* **35**, 439-448.
30a. Salganik, R.I. (1972), 'Some possibilities of mutation control concerned with local increase of DNA sensitivity to chemical mutagens', *Biol. Zbl.* **91**, 49-59.
31. Singer, R. and Fraenkel-Conrat, H. (1969), 'The role of conformation in chemical mutagenesis', *Progr. Nucleic Acid Res. Mol. Biol.,* **9**, 1-29.
32. Singer, B., Fraenkel-Conrat, H., Greenberg, J. and Michelson, A.M. (1968), 'Reaction of nitrosoguanidine (N-methyl-N′-nitro-N-nitrosoguanidine) with tobacco mosaic virus and its RNA', *Science* **160,** 1235-1236.
33. Osborn, M., Person, S., Phillips, S. and Funk, F. (1967), 'A determination of mutagen specificity in bacteria using nonsense mutants of bacteriophage T4', *J. Mol. Biol.* **26**, 437-448.
34. Smith, J.D., Barnett, L., Brenner, S. and Russell, R.L. (1970). 'More mutant tyrosine transfer ribonucleic acids', *J. Mol. Biol.* **54**, 1-14.
35. Schwartz, N.M. (1963), 'Nature of ethyl methanesulfonate induced reversions of lac$^-$ mutants of *Escherichia coli'*, *Genetics* **48**, 1357-1375.
36. Eisenstark, A., Eisenstark, R. and Van Sickle, R. (1965), 'Mutation of *Salmonella typhimurium* by nitrosoguanidine', *Mutation Res.* **2**, 1-10.
37. Malling, H.V. and de Serres, F.J. (1968), 'Identification of genetic alterations induced by ethyl methanesulfonate in *Neurospora crassa', Mutation Res.* **6**, 181-193.
38. Malling, H.V. and de Serres, F.J. (1970), 'Genetic effects of N-methyl-N-nitro-N-nitrosoguanidine in *Neurospora crassa', Molec. Gen. Genetics* **106,** 195-207.
39. Prakash, L. and Sherman, F. (1973), 'Mutagenic specificity: Reversion of iso-1-cytochrome *c* mutants of yeast', *J. Mol. Biol.* **79**, 65-82.
40. Bryson, V. (1948), 'Reciprocal cross resistance of adapted *Escherichia coli* to nitrogen mustard and ultraviolet light', *Genetics* **33**, 99.

41. Reiter, H. and Strauss, B. (1965), 'Repair of damage induced by a monofunctional alkylating agent in a transformable ultraviolet-sensitive strain of *Bacillus subtilis*', *J. Mol. Biol.* **14**, 179-194.
42. Papirmeister, B. and Davison, C.L. (1964), 'Elimination of sulphur mustard induced products from DNA of *Escherichia coli*', *Biochem. Biophys. Res. Comm.* **17**, 608-617.
43. Lawley, P.D. and Brookes, P. (1965), 'Molecular mechanism of the cytotoxic action of difunctional alkylating agents and of resistance to this action', *Nature* **206**, 480-483.
44. Crathorn, A.R. and Roberts, J.J. (1966), 'Mechanism of the cytotoxic action of alkylating agents in mammalian cells and evidence for the removal of alkylated groups from deoxyribonucleic acid', *Nature* **211**, 150-153.
45. Cerdá-Olmedo, E. and Hanawalt, P.C. (1967), 'Repair of DNA damaged by N-methyl-N′-nitro-N-nitrosoguanidine in *Escherichia coli*', *Mutation Res.* **4**, 369-671.
46. Strauss, G., Coyle, M. and Robbins, M. (1968), 'Alkylation damage and its repair', *Cold Spring Harbor Symp. Quant. Biol.* **33**, 277-287.
47. Veleminsky, J., Zadraxil, S., Pokorny, V., Gichner, T. and Svachulova, J. (1973), 'Repair of single-strand breaks and fate of N-7-methylguanine in DNA during the recovery from genetical damage induced by N-methyl-N-nitrosourea in barley seeds', *Mutation Res.* **17**, 49-58.
48. Roberts, J.J. and Sturrock, J.E. (1973), 'Enhancement by caffeine of N-methyl-N-nitrosourea-induced mutations and chromosome aberrations in Chinese hamster cells', *Mutation Res.* **20**, 243-255.
49. Roberts, J.J., Sturrock, J.E. and Ward, K.N. (1974), 'The enhancement by caffeine of alkylation-induced cell death, mutations and chromosomal aberrations in Chinese hamster cells, as a result of inhibition of post-replication DNA repair', *Mutation Res.* **26**, 129-143.
50. Swietlinska, Z. and Zuk, J. (1974), 'Effect of caffeine on chromosome damage induced by chemical mutagens and ionizing radiation in *Vicia faba* and *Secale cereale*', *Mutation Res.* **26**, 89-97.
51. Kondo, S., Ichikawa, H., Iwo, K. and Kato, T. (1970), 'Base change mutagenesis and prophage induction in strains of *E.coli* with different repair capacities', *Genetics* **66**, 187-217.

52. Evans, H.J. (1967), 'Repair and recovery at chromosome and cellular levels: similarities and differences', *Brookhaven Symp. Biol.* **20**, 111-133.
53. Cattanach, B.M. (1957), 'Induction of translocations in mice by triethylene melanine', *Nature* **180**, 1364-1365.
54. Drake, J.W. and Greening, E.O. (1970), 'Suppression of chemical mutagenesis in bateriophage T4 by genetically modified DNA polymerases', *Proc. Nat. Acad. Sci. U.S.A.* **66**, 823-829.
55. Cerdá-Olmedo, E., Hanawalt, P.C. and Guerola, N. (1968), 'Mutagenesis of the replication point by nitrosoguanidine: Map and pattern of replication of the *Escherichia coli* chromosome', *J. Mol. Biol.* **33**, 705-719.
56. Guerola, N., Ingraham, J.L. and Cerdá-Olmedo, E. (1971), 'Induction of closely linked multiple mutations by nitrosoguanidine', *Nature New Biology*, **230**, 122-125.
57. Bresler, S.E., Kalinin, V.L. and Sukhodolova, A.T. (1972), 'Action of supermutagens on the transforming DNA of *B. subtilis*', *Mutation Res.* **15**, 101-112.
58. Browning, L.S. (1969), 'The mutational spectrum produced in *Drosophila* by N-methyl-N′-nitro-N-nitrosoguanidine', *Mutation Res.* **8**, 157-164.
59. Brock, R.D. (1971), 'Differential mutation of the β-galactosidase gene of *Escherichia coli*', *Mutation Res.* **11**, 181-186.
60. Kimball, R.F. (1970), 'Studies on the mutagenic action of N-methyl-N′-nitro-N-nitrosoguanidine in *Paramecium aurelia* with emphasis on repair processes', *Mutation Res.* **9**, 261-271.
61. Asato, Y. and Folsome, C.E. (1970), 'Temporal genetic mapping of the blue-green alga *Anacystis nidulans*', *Genetics*, **65**, 407-419.
62. Altenbern, R.A. (1973), 'Gene order in species of *Staphylococcus*', *Canad. J. Microbiol.* **19**, 105-108.
63. Gillham, N.W. (1965), 'Induction of chromosomal and non-chromosomal mutations in *Chlamydomonas reinhardi* with N-methyl-N′-nitro-N-nitrosoguanidine', *Genetics* **52**, 529-537.
64. Dawes, I.W. and Carter, B.L.A. (1974), 'Nitrosoguanidine mutagenesis during nuclear and mitochondrial gene replication', *Nature* **250**, 709-712.
65. Corran, J. (1968), 'The induction of supersuppressor-mutants of *B. subtilis* by ethylmethanesulphonate and the post-treatment modification of mutation yield', *Molec. Gen. Genetics* **103**, 42-57.

66. Johnson, J.M. and Ruddon, R.W. (1967), 'Interaction of nitrogen mustard with polyribonucleotides, ribosomes and enzymes involved in protein synthesis in a cell-free system', *Molec. Pharmacol.* **3**, 195-203.
67. Reid, B.R. (1968), 'Selective inactivation of *E.coli* *t*-RNA by ethyleneimine', *Biochem. Biophys. Res. Comm.* **33**, 627-635.
67a. Chandra, P., Wacker, A., Sussmuth, R. and Lingens, F. (1967), 'Wirkung von 1-Nitroso-3-nitro-1-methyl-guanidin auf die Matrizenaktivität der Polynucleotide bei der zellfreien Proteinsynthese'. *Zeitschr. Naturforsch.* **22b**, 512-517.
68. Bresler, S.E., Kalinin, V.L. and Perumov, D.A. (1968), 'Inactivation and mutagenesis in isolated DNA. II Kinetics of mutagenesis and efficiency of different mutagens', *Mutation Res.* **5**, 1-14.
69. Nasrat, G.E., Kaplan, W.D. and Auerbach, C. (1954), 'A quantitative study of mustard-gas induced chromosome breaks and rearrangements in *Drosophila melanogaster', Zeitschr. indukt. Abst. Vererb. Lehre* **86**, 249-262.
70. Turtoczky, J. and Ehrenberg, L. (1969), 'Reaction rates and biological action of alkylating agents', *Mutation Res.* **8**, 229-238.
71. Tarmy, E.M., Venitt, S. and Brookes, P. (1973), 'Mutagenicity of the carcinogen 7-bromomethylbenz-(a) anthracene', *Mutation Res.* **19**, 153-160.
72. Auerbach, C. and Ramsay, D. (1968), 'Analysis of a case of mutagen specificity in *Neurospora crassa.* I. Dose-response curves', *Molec. Gen. Genetics* **103**, 72-104.
73. Kilbey, B. (1973), 'The manipulation of mutation induction kinetics in *Neurospora crassa', Molec. Gen. Genetics* **123**, 73-76.
74. Auerbach, C. and Ramsay, D. (1973), 'Analysis of the storage effect of diepoxybutane (DEB)', *Mutation Res.* **18**, 129-141.
75. Auerbach, C. and Ramsay, D. (1970), 'Analysis of a case of mutagen specificity in *Neurospora crassa.* III. Fractionated treatment with diepoxybutane (DEB)', *Molec. Gen. Genetics* **109**, 285-291.
76. Auerbach, C. and Ramsay, D. (1970), 'Analysis of a case of mutagen specificity in *Neurospora crassa.* II. Interaction between treatments with diepoxybutane (DEB) and ultraviolet light', *Molec. Gen. Genetics* **109**, 1-17.

77. Rannug, J.U. (1971), 'Photoreactivation of mutational damage produced by the interaction of DEB and UV in *Neurospora*', *Molec. Gen. Genetics* **111**, 194-196.
78. Worthy, T.E. and Epler, J.L. (1973), 'Biochemical basis of radiation-sensitivity in mutants of *Neurospora crassa*', *Mutation Res.* **19**, 167-173.
79. Paterson, H.F. (1974), 'Investigations into a reversal of diepoxybutane specificity in *Neurospora crassa*', *Mutation Res.* **25**, 411-413.
80. Allison, M. (1969), 'Mutagen specificity at the *ad*-3A and inositol loci in *Neurospora crassa*', *Mutation Res.* **7**, 141-154.

CHAPTER SEVENTEEN

Chemical mutagens: purines; base analogues; acridines; hydroxylamine; hydrazine; bisulphite. Reversion analysis

Caffeine and related purines

The possibility that purines or pyrimidines might be mutagenic had already been considered at a time when most geneticists thought that the specificity of the gene resided in its protein moiety. A considerable number of compounds were tested. While the results with pyrimidines were negative, various purines were found to produce chromosome breaks in plants and mutations in fungi and bacteria. Special interest was aroused by the mutagenic effects of caffeine because of the large amount of it that civilized man consumes in tea or coffee. At present, this interest has been revived through the prevailing preoccupation with genetic hazards from the environment. A recent issue of *Mutation Research* is wholly devoted to the genetic effects of caffeine (1). The introductory article by Kihlman surveys what is known about its mutagenic and chromosome breaking activity; it should be consulted for references to earlier literature. The remainder of the issue deals with the interaction effects between caffeine and other mutagens; those with UV have been mentioned in Chapter 14, those with alkylating agents in Chapter 16. The mechanism by which caffeine and related compounds act as mutagens in their own right is not yet understood. It is likely that, as in their synergistic action with other mutagens, they act via inhibition of enzymes concerned with repair and, perhaps, with other processes such as fidelity of DNA replication and the rejoining of broken chromosomes. An observation that might give a clue to their action but has never been followed up was made twenty years ago on bacteria growing in the chemostat in the presence of caffeine or theophylline. The mutagenic action of these compounds was completely abolished by the addition of guanosine or adenosine; curiously enough, the deoxyderivatives of these substances were very much less effective as 'antimutagens'. Another clue lies in the fact that the formation of rearrangements by caffeine or 8-ethoxy-caffeine in plant cells, and in Chinese hamster cells at temperatures of 30° or less, requires oxidative phosphorylation and is closely correlated with the ATP level in the cells. At 37° another type of damage appears. The chromosomes in Chinese hamster cells are fragmented but do not rejoin; this effect is not dependent on ATP and is probably different in nature from that occurring at lower temperatures. We shall come back to caffeine in Chapter 23.

Base analogues

With the recognition of DNA as the essential component of the gene, the search for new mutagens became directed towards substances that may be presumed to react with DNA. Foremost among these were analogues of the normally occurring purine and pyrimidine bases in DNA. Several of these, in particular the thymine analogue 5-bromouracil (5-BU) and the purine analogue 2-aminopurine (2-AP), proved to be very effective mutagens in bacteriophage (2-4) and bacteria (5,6). Their uptake into the cells can be promoted by starvation for the normal analogue, either by using thymine- or purine auxotrophs, or by stopping the formation of thymidine from precursor molecules through folic acid inhibitors such as sulphanilamide or aminopterin. The thymidine analogue 5-bromodeoxyuridine (5-BUdR) is a much more powerful mutagen than 5-BU and does not require folic acid inhibition for its action.

Models for the mutagenic action of base analogues were at first based firmly on Watson and Crick's scheme of base-pair selection at DNA replication and on their suggestion that spontaneous mutations arise when the pairing properties of the bases are changed through rare tautomeric shifts from the keto- to the eno- form in the pyrimidines, and from the amino- to the imino- form in the purines. Chemical considerations suggest that these tautomeric shifts occur more frequently in the base analogues than in the normal bases. Fig. 17.1 shows the correct pairing of 5-BU in the keto-form with adenine, and its incorrect pairing with guanine in the eno-form. Similarly, 2-amino-purine in the amino-form pairs with thymine, in the imino-form, with cytosine.

On this basis, two possible origins of base-analogue mutagenesis

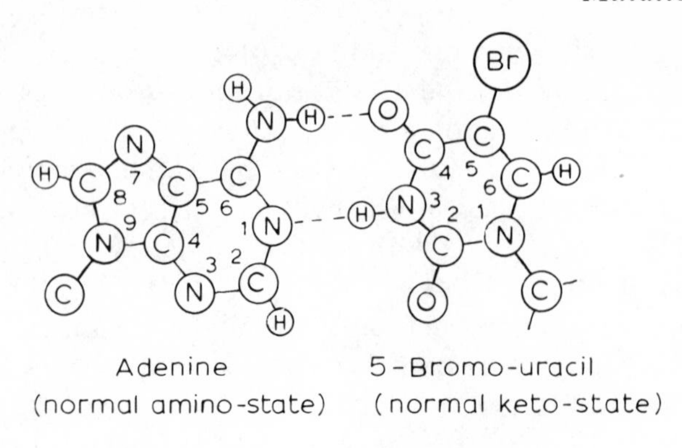

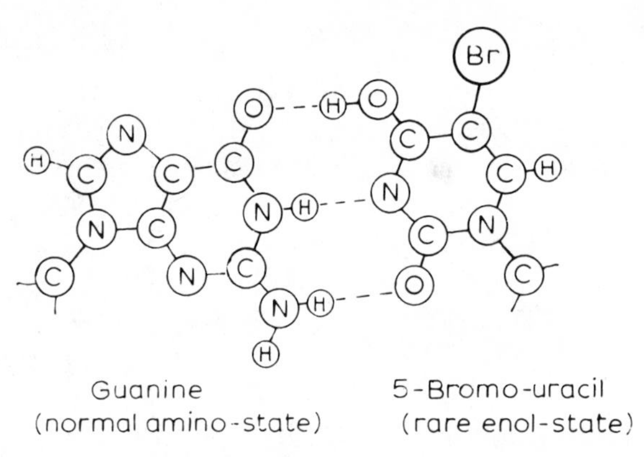

Fig. 17.1. Normal and rare base-pairing of bromouracil. (Fig. 8, in Freese, vol. 1. of 'Chemical Mutagens' (Bibliography). Courtesy Plenum Press.)

were considered. They are shown diagrammatically in Fig. 17.2 for 5-BU as mutagen. (1) Mistakes in incorporation. These occur when the base analogue is incorporated 'wrongly' opposite a normally non-complementary base, in this case, opposite guanine instead of adenine. Through steps shown in the upper diagram, this will result in a transition from GC to AT. (2) Mistakes in replication. These occur when the base analogue is incorporated 'correctly' opposite its complementary base, in this case opposite adenine. Through the tendency of the analogue to undergo tautomeric shifts, it may select wrongly at one of the following replications. In the case of 5-BU this will result in transitions from AT to GC. Several points should be noted about this scheme. (*a*) The formation of a mutant base pair requires two divisions in either case. The time

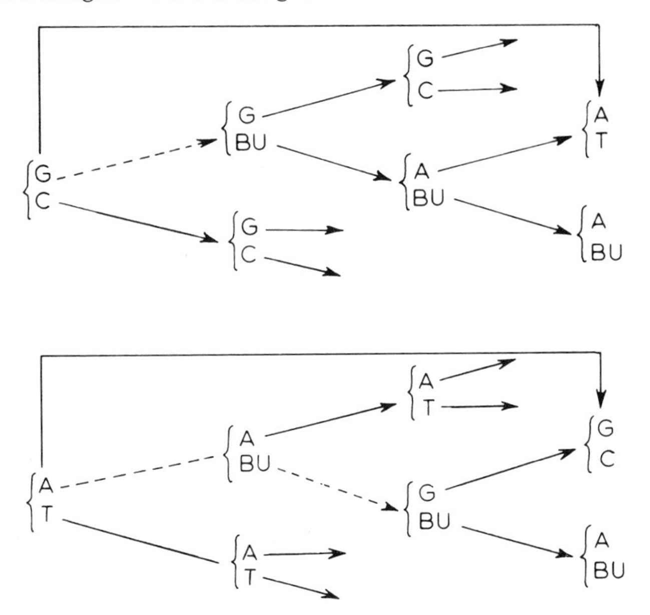

Fig. 17.2. The origin of transitions by base-pairing mistakes of bromouracil. Upper scheme: mistakes in incorporation. Lower scheme: mistakes in replication. (Fig. 9, in Freese, vol. 1 of 'Chemical Mutagens' (Bibliography). Courtesy Plenum Press.)

of phenotypic expression will depend on whether the analogue is incorporated in the transcribed or the non-transcribed strand and whether it will itself be read as normal or mutant. (*b*) Mutations affect only one strand and its derivatives; the resultant populations should therefore consist of normal and mutant progeny. This has often been found to be true, but there have been exceptions. These will be discussed in Chapter 19. (*c*) An error of *incorporation* is assumed to result in a replacement of G by A at the first replication following incorporation because, as shown by selective incorporation into DNA (see below), the pairing affinity of BU for A is almost as strong as that of T. No further mutations are expected from the incorporated BU; for even if mis-pairing did occur at subsequent replications it would only restore the original GC base pair. (*d*) In contrast, the same strong preference of BU for A makes it likely that mistakes in *replication* do not always occur at the first replication after incorporation but may continue to take place at later replications as long as the BU-molecule is present. Thus, while transitions from GC to AT are expected to require the presence of the analogue, those from AT to GC are expected to occur even after its removal. This expected difference between 'clean-growth' mutagenesis through incorporation errors and 'dirty-growth' mutagenesis through replication errors has been used repeatedly in attempts to assess the relative frequencies of the two types of transition. Although both clean- and dirty-growth mutagenesis were found to occur, interpretation by means of the model met with difficulties (7).

The evidence for incorporation of base analogues as a prerequisite of mutagenesis is indirect but cumulatively convincing. In bacteria and DNA phages, only analogues that can be incorporated into DNA are mutagenic. Among the halogenated uracils, this applies to chloro-, bromo- and iodo-substituted ones (8,9). Fluoro-uracil, which is not incorporated into DNA, is ineffective, but produces mutations in RNA-viruses (10,11) and in the transducing phage PBS2 in which thymine is normally replaced by uracil (12). In a cytosine-requiring strain of *E.coli*, the cytidine analogue N^4-hydroxycytidine is mutagenic (13). In phage T2, BUdR mutagenesis starts at the time of DNA-synthesis; thymidine inhibits mutagenesis when given simultaneously with the analogue but becomes progressively less effective with time after the initiation of DNA synthesis (14).

Yet, incorporation alone cannot account for mutagenesis, and subsequent research has supported the hypothesis put forward by Litman and Pardee in 1960 (8) 'that BrU mutagenesis is mediated through its incorporation into DNA, but that the incorporation *per se* is insufficient to cause mutagenesis'. Particularly striking was the difference between 5-BU which can replace thymine almost quantitatively in DNA, (15,16) and 2-AP which is incorporated in vanishing small amounts (17): yet both are efficient mutagens. It is true that, for stereochemical reasons, 2-AP would be expected to have a higher tendency than 5-BU to undergo tautomeric shifts; but in order to explain the differences in correlation, one would have to assume that this tendency is several hundred times as strong in 2-AP as in 5-BU. In any case, a lack of correlation between incorporation and mutational yield was also observed in experiments in which only BUdR was applied to phage T2 (8). Even when nearly 100% of the thymine in DNA had been replaced by 5-BU, and infectivity had been reduced to less than 10%, only about 10% of the surviving phage particles yielded plaque type mutants. This discrepancy was resolved subsequently when experiments with hydroxylamine, to be discussed presently, showed that incorporation of 5-BU instead of T produces very few mutations in T4; the majority of mutations are transitions from GC to AT and must be due to the small and chemically undetected amount of 5-BU that is incorporated instead of C.

Yet even this is not the general rule. Specificities of interaction at the level of individual bases are overlaid by regional and site specificities in DNA, and by differences between organisms. In the single-stranded phage S13, most BU-induced mutations are transitions from AT to GC (18). This striking difference between the response of two DNA phages, even when exposed in the same host, points to the importance of structural features of DNA, or of its mode of replication in single-stranded as compared with double-stranded phages. Results obtained with 2-AP also differed between organisms. The two phages T4 and S13 responded by transitions in both directions (18), and reversion analysis applied to *Salmonella* agreed with this conclusion (19). On the contrary, when nonsense reversions in *E.coli* were classified by Person's method (p.196)(20), at least 95% were inferred to be transitions in one direction only, from AT to GC. The same conclusion was drawn from 2-AP induced interconversions of nonsense suppressors in yeast (21). Whether these differences in specificity reside in the organisms, in the types of mutation scored, or in the mutational sites on the chromosomes, cannot be decided from the data. They do, however, warn against easy generalizations.

As regards mutational changes other than transitions, the induction of some transversions by base analogues cannot be excluded, but, for reasons discussed by Drake 1970 (Bibliography), their contribution can be only very small. Frameshifts are almost certainly never produced by base analogues; we shall come back to this presently. There is, however, good evidence that 5-BU produces breaks in euchromatic chromosomes (22) and mutations in *Arabidopsis* (22a); possibly it does so indirectly by sensitizing the chromosomes to visible light (p. 220). This can hardly have been the origin of the sex-linked lethals which were induced in *Drosophila* spermatozoa and spermatids by injection of 5-BUdR into adult males (22b); however, since the mutants occurred in non-replicating chromosomes, some secondary effect of the treatment appears to have been responsible.

The fact that response of a given base to mutagenesis by base analogues depends on its site within the nucleotide chain became strikingly evident when base analogue-induced rII mutants were placed on the genetic map. It was found that mutations induced by 2-AP and 5-BU are not distributed at random but show a strong tendency for clustering in certain 'hot spots' (23). The hot spots for 2-AP and 5-BU overlap to some extent with each other but not with those for spontaneous mutations. This distinguishes hot spots from the sites of generally high mutability in the β-galactosidase gene of *E.coli* that have been discussed in Chapter 3 (p. 40). While these probably represent sites of special importance for the functioning of the gene product, hot spots are indications of mutagen specificity. Under this aspect they will be discussed in Chapter 20. A step towards an explanation of hot spots was made when Koch (24) found that the frequency with which 2-AP produces transitions

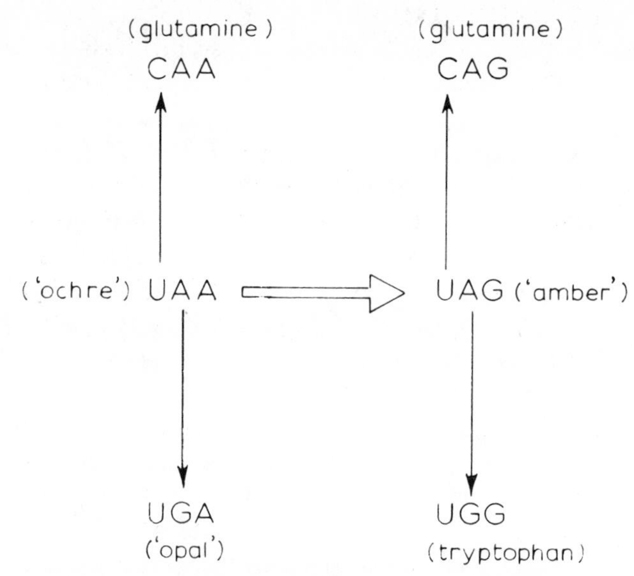

Fig. 17.3. The m-RNA codons that correspond to transitions from AT to GC in ochre and amber codons. Mutations to opal are recognized by their response to an opal-specific suppressor. Transitions to glutamine at this site are true revertants and wild-type in phenotype, while transitions to tryptophan are partial revertants due to intracodon suppression. (Fig. 1, in Koch (24). Courtesy Nat. Acad. Sci., U.S.A.)

from AT to GC at a given site depends on the neighbouring bases. The system he used is shown in Fig. 17.3. Transitions are scored at the first and second sites of an ochre or amber codon in which different transitions can be distinguished phenotypically. The second site in the ochre codon has as its right-hand neighbour an AT pair; the same site in the amber codon has a GC pair in its stead. 2-AP-induced transitions from amber to tryptophan were 33 times as frequent as transitions from ochre to opal; thus, the immediate neighbourhood of the GC pair strongly promoted the mutagenic action of AP. To a lesser extent this happened even when one base pair intervened between GC and the mutating pair; for transitions from amber to glutamine were about three times as frequent as transitions from ochre to glutamine. How far this kind of neighbourhood effect can account for hot spots is not known. In any case, both phenomena show that it is not the individual base but its position in the nucleotide sequence that determines the mutagenic effect of a base analogue. One possible explanation is that a certain sequence of nucleotides is required for recognition by an enzyme that is concerned with DNA-replication or repair. Drake's (25) finding that the frequency of base analogue-induced mutations in T4 is reduced in strains with a mutant DNA-polymerase points in this direction. So does the observation (26) that efficient induction of mutations in phage λ by 5-BU requires the normal functioning of the *recA* and *lex* genes in the host bacterium. On the contrary, more mutations were produced when the host strain lacked excision repair. It seems that bacteriophage repair mechanisms play a similar role in mutagenesis by BU as they do in UV-mutagenesis. This does not, however, apply to bacteria (26a). Treatment effects on expression are suggested by experiments on T4 which showed that three different types of mutation respond differently to the presence of chloramphenicol or of other phages during growth in BUdR (27). In mammalian cells, BUdR has been shown to affect the transcription process (28).

Clearly, mutagenesis by base analogues is no exception to the general rule that forms the central theme of this book: the rule that mutation, initiated by whatever physical or chemical means, is a *biological* process and, as such, deeply imbedded in the network of structural components and biochemical reactions that constitute life.

Dyes: pyronin, acridines

In 1947, Witkin (29) reported that pyronin and acriflavin (Fig. 17.4) produced mutations to phage resistance in *E.coli*. For lack of a plausible chemical interpretation, these results were not followed up at the time. Subsequently, both pyronin and acridines have been shown to produce mutations in other organisms but, while acridine

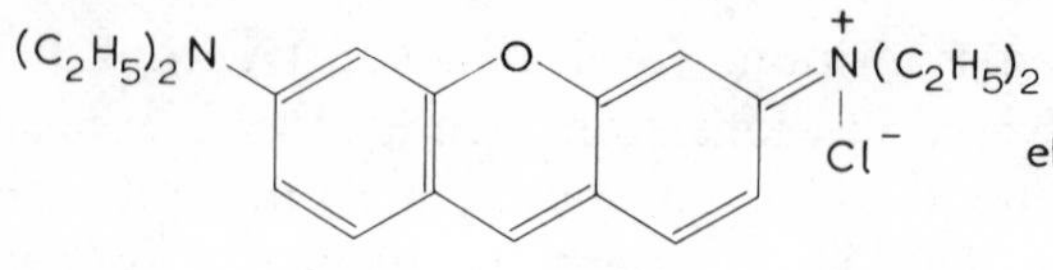

ethylpyronine

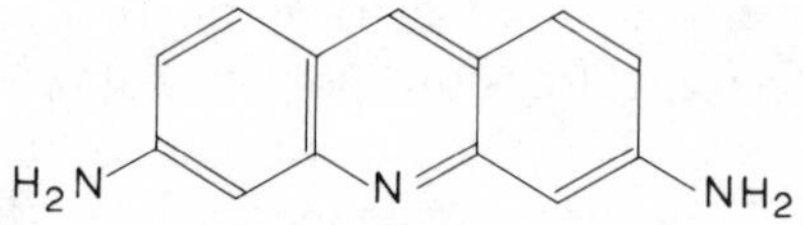

proflavine

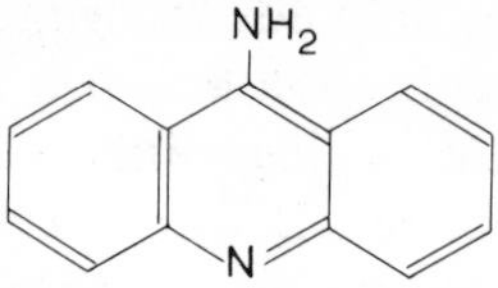

5-aminoacridine

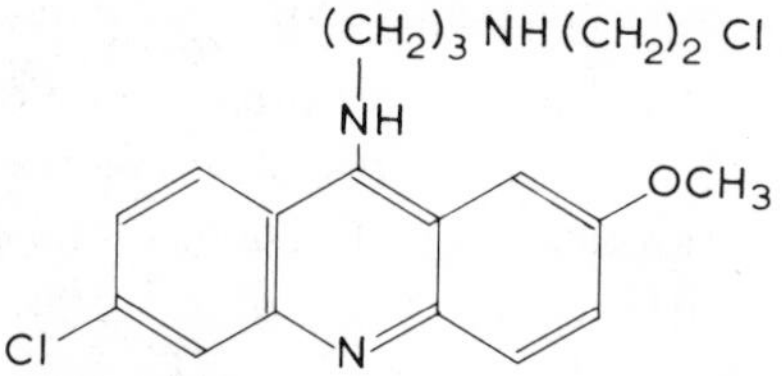

ICR 191

Fig. 17.4. Chemical formulae of pyronin and two mutagenic acridines.

mutagenesis has become one of the cornerstones of mutation research, the results with pyronin have received scant attention, probably because the relevant experiments were carried out on *Drosophila* in the pre-Watson-Crick era. In view of a possible connection with acridine mutagenesis, they will be briefly reviewed (30-32). Pyronin fed to larvae or injected into adults produced sex-linked and autosomal recessive lethals and recessive visibles. It acted only on spermatogonia, not on mature sperm; this suggests that replication is required for pyronin mutations to occur. Both methyl- and ethyl-pyronin were effective, but the latter much more so. The overall effect was weak, and this may have been the main reason why further work was abandoned. Yet, analysis of mutation frequencies in individual males showed that this weak overall effect masked an extremely strong mutagenic action on those males that responded to the dye. Thus, in a test for autosomal recessives – in which mutant clusters derived from the same treated spermatogonium can be distinguished from independent mutations in different spermatogonia – only 16 independent mutations were found in about 900 chromosomes, i.e. less than 2%; but 10 of these occurred in only three out of 48 males, each tested by 19-20 offspring. The cause of this heterogeneity could not be established; differential exposure to light, leading to photodynamic effects, could be excluded.

Acridine mutagenesis due to photodynamic action has been discussed in Chapter 13. In what follows, we shall deal with acridine mutagenesis in the dark. These studies received great impetus from the discovery in 1961 (33), that acridine, proflavin and acridine orange, all of which were known to be strong mutagens for phage T4 (34), intercalate between adjacent nucleotide pairs of the DNA. In the same volume of the journal in which this finding was published, there appeared a noted called 'The theory of mutagenesis', in which Brenner, Barnett, Crick and Orgel (35) put forward the idea of frameshift mutagenesis as an additional mechanism to the transitions and transversions of Freese's theory. The importance of frameshift mutations for inferring the nature of the genetic code, and the final molecular proof that these mutations and their intracistronic suppressors are indeed due to additions or deletions of bases have been discussed in Chapter 3. In the present chapter we are concerned with the role of acridines in producing these changes. Here we are on insecure ground. Although there is good evidence for the intercalation of acridines into the stacked bases of DNA (36,37), the subsequent steps by which the intercalation leads to mutation are still under debate. Lerman (37) has shown that intercalation by itself is not sufficient for mutagenesis. In tests on a variety of intercalating agents, he found no correlation between intercalating ability and mutagenic efficiency; some strongly intercalating agents did not yield mutations at all. As intermediaries in the production of mutations from intercalated acridines, replication (35), recombination (38) and repair (39) have been considered. There is evidence for and against every one of these. The position up to 1970 has been fully discussed by Drake (Bibliography). Meanwhile, there has been more evidence that, in phage T4 (40) and in bacteria (41), recombination is not required for acridine mutagenesis, although it may be connected with it. The finding that proflavin produces many more mutations in a ligase-deficient (42) than in a normal strain of T4 indicates that single breaks in DNA are the connecting link. One report links acridine mutagenesis in *E.coli* with DNA-replication (41), mutations in synchronized cultures occurring preferentially at or near replication points. The fact that this correlation disappeared in an excision-deficient strain suggests that is is due to a rapid repair system which removes most damage before it can be fixed at replication. It may be recalled that a similar suggestion has been made for the preferential action of nitrosoguanidine on replicating genes (Chapter 16).

Whatever the details of the mechanism by which acridines in the dark produce mutations it seems that these are always frameshifts. Since, barring external suppressors, frameshifts are reversed only by frameshifts, and base substitutions only by base substitutions (Chapter 4), one would expect that mutations produced by acridines would not be revertible by base analogues and vice versa. This was indeed found to be true. In a pioneer experiment by Orgel and Brenner (34), rII mutants of phage T4 that had been induced by either 5-BU or an acridine were tested for revertibility by these same substances. There was strict mutual exclusion between the two classes: while most BU-induced mutations reverted with BU, and most acridine-mutations reverted with acridines, no BU-induced mutation reverted with acridine and vice versa. This distinction between the two classes of mutagen has been confirmed in all subsequent experiments. It establishes at one stroke the specificity of base analogues for the production of base substitutions, and the specificity of acridines for the production of frameshifts. It forms one of the cornerstones of the reversion analysis that we shall discuss presently.

There is no reason to doubt that in cellular systems also acridines produce exclusively frameshifts. Unfortunately, they are poor mutagens for cells, although 5-amino acridine has been used successfully in bacteria (42) and yeast (43). The compounds generally used for producing frameshifts in cells are the so-called ICR compounds (produced by the Institute for Cancer Research in Philadelphia), which combine an acridine with an alkylating side chain (44). It is thought that the side chain, by reacting with DNA, promotes the intercalation of the acridine moiety between the stacked nucleotides. Some of these compounds are potent mutagens. Their efficacy differs between organisms; while ICR-170 is the most widely used compound for eukaryotes, ICR-191 (Fig. 17.4) is more effective in bacteria (45).

The diagnostic value of the ICR-compounds for frameshift mutations is diminished by the fact that they may also act as alkylating agents. Although it seems that at low doses frameshift mutagenesis predominates (46), this duality of action creates an ambiguity that has to be taken into account when mutations in cellular organisms are classified by means of their response to ICR-compounds. Estimates of the proportion of frameshifts among mutants produced by similar doses of the same compound in closely related organisms may vary widely depending on the type of mutation scored and the criteria for classifying a mutation as frameshift. Low doses of ICR-170 were estimated to produce almost 100% frameshifts among suppressor mutations in *Saccharomyces cerevisiae* (47), but only 20% and 30% among forward mutations at two adenine-loci in *Schizosaccharomyces pombe* (48). The type of mutation scored and the stringency of scoring procedure play a role in creating such differences. When leaky mutants are discriminated against, the proportion of frameshifts will be increased. Additional criteria that can be used for the identification of frameshifts are the lack among them of temperature-sensitive and osmotic-remedial mutants (but see p. 44), the lack of response to nonsense suppressors and, in complementing cistrons, the absence of the non-polarized type of complementation (p. 47). In *Saccharomyces,* the finding that ICR-compounds produce suppressor mutations much more readily in meiotic than in mitotic cells has been taken as further support for their frameshift nature (47). The inference is indirect and based on a similar 'meiotic effect' for certain classes of spontaneous mutations, for which an origin by errors of recombination could be established. We shall deal with this in Chapter 21. Until similar evidence has been obtained for the ICR-induced suppressors, it remains possible to speculate that their fixation or expression finds more favourable conditions in meiotic than in mitotic cells of yeast. Here, as in the case of other mutagens, one should not overlook the possibility that specificities of mutagenic action may have their roots in effects of the mutagen on cellular steps in the mutation process. Proflavin has been shown to inhibit the attachment of amino acids to *t*-RNA (49); conceivably, this might play a different role in the mutagenic pathways of meiotic and mitotic cells. In the latter, 5-aminoacridine was found to *reduce* the spontaneous frequency of the same types of mutation that it induces in meiotic cells (43).

Reversion analysis, first part

Reversion analysis has been mentioned repeatedly in previous chapters as one of the means for determining the molecular nature of individual mutations. We shall now see to what extent and in which systems it can do this reliably. The analysis proceeds in two steps, which we shall discuss separately. The first step is the distinction between frameshifts and base changes. It requires only mutagens with which we have dealt in this chapter and the preceding ones. The second part deals specifically with base changes and attempts to classify them into the various kinds of transition and transversion. It requires the use of diagnostic mutagens; a discussion of these will precede that of the second part of reversion analysis. We shall see that both parts need support from data outside reversion analysis, the second part much more than the first.

Table 17.1 summarizes data from a reversion analysis of 80 ICR-induced mutations to auxotrophy in the histidine-operon of *Salmonella.* 48 of them had been induced with ICR-190, 32 with a non-alkylating acridine-derivative. Although the latter produced fewer mutations, they did not differ in kind from those induced by ICR-190. They did, however, differ from them in location on the chromosome.

On the basis of their response to the four mutagens used, the mutants were allotted to five different classes. Class V consisted only of two mutants that, in contrast to all others, reverted with 2-AP but not with ICR. These were judged to be base substitutions; since they were found in a sample that had been treated with the non-alkylating compound, they may have been of spontaneous origin. All other mutants were assumed to be frameshifts. None of them responded to 2-AP and, with the exception of those in class I, all reverted spontaneously. In Chapter 21, we shall deal with the tendency of frameshifts to spontaneous reversion. The mutants in classes III and IV showed the expected behaviour of frameshifts: spontaneous and ICR-induced reversion, failure to revert with a base analogue. Those in class IV had the additional property of being revertible by alkylating agents. Since the apurinic gaps produced by alkylation may result in the loss of single bases (Chapter 16), class IV was assumed to represent the addition type of frameshift, which can be reverted by a nearby deletion (Fig. 3.4). The mutants in class III, which did not respond to alkylating agents, were classified as the deletion type of frameshift. This distinction

Table 17.1

Reversion analysis of 80 ICR-induced histidine-requiring mutants in *Salmonella*. (After Oeschger (50), Table 9)

Class	*Spontaneous reversions*	*Revert with*				*n*	*Inferred molecular change*
		ICR	*NG*	*DES*	*2AP*		
I	–	–	–	–	–	11	special type of FS* or Del.
II	+	–	–	–	–	11	special type of FS*
III	+	+	–	–	–	20	deletion of 1 or 2 bases
IV	+	+	+	+	–	36	addition of 1 or 2 bases.
V	+	–	+	+	+	2	BS

ICR = either one of two different ICR compounds (see text).
NG = nitrosoguanidine
DES = diethylsulphate
2AP = 2-aminopurine
FS = frameshift, BS = base substitution, Del = large deletion.
* see text

between classes III and IV remains speculative but seems plausible; exceptions would be expected to arise from the fact that, for example, a deletion of two bases can be corrected by deletion of a third. Classes I and II created difficulties of interpretation. The mutants in class I might have a frameshift in a very sensitive region of the operon where even a nearby complementary frameshift is unable to rescue gene function; alternatively, they might be large deletions. The refractoriness of the mutants in class II to ICR might have something to do with the special sites at which they had occurred; ICR compounds have a tendency to act on 'hot spots'. We shall return to this in Chapter 21 when dealing with spontaneous frameshifts.

The conclusion that, with two doubtful exceptions, the two ICR compounds had produced only frameshifts in the tested system was supported by additional evidence from tests on random samples of the mutants. None were leaky. There was no genotypic suppression by nonsense suppressors or phenotypic suppression by streptomycin (Chapter 4). When revertants of a mutation in one of the cistrons were located by transduction, all were found to have occurred in the same cistron as the original mutation, showing that reversion had occurred via intracistronic suppression. Mutants belonging to a complementing cistron showed the complementation pattern expected for frameshifts. All mutants showed polarity effects on distal but not on proximal cistrons of the operon as would be expected from the fact that frameshifts produce polarity by creating nonsense codons (Chapter 3).

The discussion of Table 17.1 shows both the strength and the weakness of reversion analysis. Its main weakness is its circularity. One of the chief supports for the conclusion that all ICR-induced mutations were frameshifts was the assumption that, in general, ICR compounds produce frameshifts and, therefore, revert frameshift mutations. Special assumptions were necessary to explain why about a quarter of the supposed frameshifts did *not* revert with ICR-compounds. The other piece of evidence, i.e lack of response to 2-AP, was independent but wholly negative. However, there was ample additional evidence for the frameshift nature of the tested mutations. Where this is lacking and, in particular, where there is reason to assume that an ICR-compound can act by alkylation as well as by intercalation, the results of reversion analysis lose much in cogency.

Often, attempts are made to subdivide base substitutions further into transversions and the two types of transition. In Chapter 11, we have discussed some special techniques for drawing such inferences from reversions and interconversions of nonsense codons. Reversion analysis by itself requires a fixed point for this purpose, i.e. at least one substance than can be relied upon to act always on the same base with the same result. Before completing the discussion of reversion analysis we shall deal with substances that have been considered to provide such fixed points.

Hydroxylamine; Hydrazine, maleic hydrazide; sodium bisulphite

1. *Hydroxylamine* (NH_2OH; abbrev. HA) is moderately effective as a mutagen for transforming DNA (51,52) and for extracellularly

treated bacteriophage (53,54), in which it produces mutations with linear kinetics. Much has been written about the reaction, or reactions, between hydroxylamine and nucleic acids and their relation to the lethal and mutagenic effects of the treatment (for reviews see 55, and Drake, 1970, Bibliography). Among the nucleotide bases, cytosine and uracil were found to react specifically with HA. This suggested that it is reactions with these bases that lead to mutations in the DNA of phages and the RNA of tobacco mosaic virus. This was supported by the parallelism between the pH dependence of the reactions with these nucleotides and the frequencies of mutation (56,57). Moreover, *in vitro* experiments (58) with polynucleotides showed that treatment of poly-C with HA affects its template properties for RNA, so that a certain proportion of A is incorporated instead of G. More recent analyses (59,60) have led to the conclusion that in virus RNA also it is the reaction with cytosine rather than that with uracil that leads to mutation.

In both double-stranded and single-stranded bacteriophage, the mutational data obtained by *in vitro* treatment were in fairly good agreement with the assumption that HA produces exclusively transitions from GC to AT (54,61); we shall discuss the evidence more fully when we return to reversion analysis. This specificity was confirmed also for reversions of nonsense codons in phage T4 (62): treatment with HA produced all three codons but reverted none of them. Reference to the genetic code (Fig. 3.1) shows that reversion to a sense codon requires either a transition from AT to GC or a transversion.

Yet, even for phage treated *in vitro* the specificity of HA is not absolute. In a transducing phage, HA produced reversions from UAA to a sense codon and conversions from UAA to UAG (63). Conceivably, this might be explained by the finding that HA *in vitro* also reacts with adenine, although at a lower rate than with cytosine (64). Alternatively, and more probably, the mutations in these cases were caused by the radicals that HA produces in the presence of oxygen (65). In agreement with this assumption, nitrogen or catalase greatly reduced mutation frequency, and no mutations were produced by methoxyamine, a derivative of HA which does not produce radicals. Radicals and hydrogen peroxide play the major role in the inactivation of bacteriophage and transforming DNA but are only weakly mutagenic (65). The efficiency of mutagenesis by HA is increased by all conditions that inhibit radical formation, such as a high concentration of HA, absence of oxygen, presence of NaCl.

It is thus not surprising that HA loses its specificity for cytosine when applied to bacteriophage inside its host (66). Both in single- and double-stranded phages, *in vivo* exposure produced mutations of all four nucleotides. Very recently, Budovsky (66a) has made the interesting suggestion that the unexpected types of transition may not be the primary HA-induced ones, but secondary spontaneous changes, which were selected because they restored the rigidity of a double-stranded region of DNA after it had been weakened by the primary change. This model, if supported by experimental evidence, might also apply to higher organisms in which the necessity to preserve the structures of nucleoprotein complexes or of proteins may provide the selective stimulus. A further complication for the interpretation of *in vivo* data is the recently discovered fact of phenotypic suppression (Chapter 4) by HA (67). Among 105 rII mutants of T4, eight were able to grow in the non-permissive strain in the presence of HA; their progeny showed that they had retained the rII genotype. Probably, suppression was due to the action of HA on m-RNA. When rII mutants are produced by HA treatment *in vivo*, the suppressible ones may escape detection. This would superimpose an additional specificity on that due to the reaction between HA and DNA.

In cellular organisms, there is little reason for assuming a specific action of HA on C. In *E.coli*, five out of 11 HA-induced auxotrophs could be reverted by HA (68). When HA-induced nonsense-revertants were typed by Person's method (p.196), transitions from AT to GC were even more frequent than transitions from GC to AT (69). In eukaryotes the mutagenic effect of HA is very weak. In *Drosophila* (70), HA appears to produce visible mutations in preference to lethals; we shall come back to this in Chapter 20. In fungi (71-73), the toxic action of HA is much more pronounced than its mutagenic one. Toxicity has many causes, for HA disrupts many metabolic processes. It inhibits the amino acid accepting ability of *t*-RNA (74) and the synthesis of DNA, RNA and protein (75). In various ways, it interferes with the action of a variety of enzymes (73). Under these circumstances, it is impossible to assess the role played in mutagenesis by the specific reaction between HA and cytosine. Putrament *et al.* (73) conclude that it is at best very small.

Hydroxylamine and its derivatives produce chromosome breaks in animal and plant cells (76,77). It is not clear whether they do this by direct action on DNA, via radical formation, or in some more indirect way.

2. *Hydrazine* (N_2H_4; abbrev. HZ) is a weak mutagen; it has produced mutations in transforming DNA (51), bacteria (78) and *Drosophila* (70). Like HA, it produces radicals and hydrogen peroxide in the presence of oxygen; at high concentrations, it reacts directly with pyrimidine bases (79). Some of its derivatives, in particular maleic hydrazide, are effective chromosome breakers (80).

3. *Sodium bisulphite* ($NaHSO_3$) is a newcomer to the field of mutation research. *In vitro* it reacts with pyrimidines in one-stranded polynucleotides and DNA but is inactive on two-stranded DNA; it specifically changes cytosine into uracil (81,82). It has produced mutations in phage (83,84), bacteria (85) and yeast (86). In bacteria and phage, it showed the same site specificity as did HA; this agrees with the assumption that it specifically produces transitions from GC to AT. Its action, like that of other mutagens, is not confined to DNA; by inhibiting specifically the amino acid acceptor activity of certain t-RNA's (86a), it may exercise an indirect influence on the mutation process.

Reversion analysis, second part

We have seen that the first part of reversion analysis serves to classify mutations into base substitutions and frameshifts, and mutagens into those that produce one or the other or both of these types of mutation. The second part of reversion analysis is specifically concerned with base substitutions. In regard to mutations, it attempts to identify the changed base at the mutant site and its origin through transition or transversion. In regard to mutagens, it tries to infer which substitutions they induce.

It has been pointed out above that reversion analysis requires some fixed point or points outside the data on forward and reverse mutations if it is not to remain circular. This is even more true for the second than for the first part. In rare cases, amino acid sequencing of a few mutants may provide a fixed point. In organisms with known nonsense mutations analysis of reversions and interconversions of these mutations may be used for the purpose. Most frequently, the points of reference are taken from chemical data. Unfortunately, these usually leave a choice between various reactions. The manner of choosing depends in part on mutational data, so that

Table 17.2
Classification of rII mutants by genotypic and phenotypic reversion analysis. (After Table I, Champe (57)

Group	*Genotypically reverted by* *BA**	*HA*	*Phenotypically reverted by* *5-FU*	*Mutant base pair*	*Base in reading strand*	*Base in m-RNA*	*Original mutation*
I	+	–	+	AT	A	U	GC → AT
II	+	–	–	AT	T	A	GC → AT
III	+	+	–	GC			AT → GC
IV	–	–	–				not a transition

* Revertibility by one or both of the base analogues 5-BUdR and 2-AP.

the inclusion of chemical information extends the circle rather than providing it with firmly established points of reference. In general, a circular argument gains in persuasiveness with the number of data that can be fitted into the circle without breaking or unduly distorting it. Judged by this criterion, the only system in which the second part of reversion analysis has been very successful is the *in vitro* treatment of bacteriophage. We shall consider this in some detail.

Table 17.2 presents a summary of the results of a reversion analysis that was carried out on a large number of rII mutants in T4. In addition to genotypic reversion by base analogues and HA, phenotypic reversion by 5-fluorouracil (Chapter 4) was scored.

In the analysis of the data, three fixed points were defined by the following assumptions. (1) Base analogues always produce transitions. (2) HA reacts specifically with cytosine; this results in transitions from GC to AT. (3) 5-FU reacts specifically with uracil, being incorporated into m-RNA instead of U. At translation, this results in a certain amount of misreading of A as G, compensating the effects of mutants that carry A on the reading strand. By reference to these points, the mutants were arranged into the four groups shown in the table. The mutants in Group IV, which did not respond to base analogues, were presumed to be due to transversions or to the then newly discovered frameshifts.

For the other three groups, the circle of assumptions and results is well closed. The fact that FU suppressed none of the HA-revertible and about half of the HA-non-revertible mutants confirmed

Table 17.3

Coincidence between revertibility by HA and 5-BUdR or 2-AP. (Data from Table I, Champe (57))

	$BUdR^{rev}$	$BUdR^{non\text{-}rev}$	$2\text{-}AP^{rev}$	$2\text{-}AP^{non\text{-}rev}$
$18HA^{rev}$	16	2	18	0
$51HA^{non\text{-}rev}$	5	46	51	0

rev. = revertible by.

both presumed chemical specificities that had served as basis for the analysis. So did the fact that HA only reverted mutants that, because of their response to base analogues, were presumed to be due to transitions. From the results of this analysis, nucleotide pairs could be allotted to an impressive number of sites in the rII cistrons, and to a proportion of sites in m-RNA.

The data could also be used for deriving the specific effects of the two base analogues 5-BUdR and 2-AP.

Table 17.3 shows that there was a strong correlation between revertibility by 5-BUdR and HA but none between revertibility by 2-AP and HA. The conclusion was drawn that 5-BUdR (or 5-BU) produces mainly transitions from GC to AT, while 2-AP is less specific. This was supported by the data shown in Table 17.4. While HA reverted half of those mutants that had originally been induced by 2-AP, it reverted none of those that had been produced by itself or by 5-BUdR. This would be expected if half of the former but none of the latter had GC at the mutant site. EMS resembled 5-BUdR in producing mainly mutations that had AT at the mutant site; this is in agreement with chemical evidence (Chapter 16). All these conclusions have been confirmed in subsequent work on T4, although there were differences in the degrees of specificity displayed by the base analogues.

In single-stranded DNA phages, reversion analysis has been used to determine the individual nucleotide bases that react with a given mutagen. Table 17.5 shows a reversion analysis on phage S13, treated *in vitro* with HA or EMS.

The reference points for the analysis were again chemical: the specificity of HA for cytosine, and of EMS for guanine. In

Table 17.4

Relationship between origin of mutants and revertibility by HA.

Origin	*n*	HA^{rev}	$HA^{non\text{-}rev}$
HA	9	0	9
5-BUdR	9	0	9
2-AP	8	4	4
EMS	5	1	4

* Data from Table I, Champe (57).

Table 17.5

Mutagen specificity of HA and EMS for the single-stranded DNA phage S13, treated *in vitro*. Forward and reverse mutations at 6 sites ($+^{m1}$ to $+^{m6}$). (After Table 2 in Tessman (54))

		Induced by		*Inferred base change*
		HA	*EMS*	
+	→ m_1	0	+	T → C
m_1	→ +	++	++	C → T
+*	→ m_2	++	++	C → T
m_2	→ +	technically not feasible		
+	→ m_3	0	0	A → G
m_3	→ +	0	++	G → A
+	→ m_4	0	++	G → A
m_4	→ +	0	0	A → G
+	→ m_5	0	++	G → A
m_5	→ +	0	0	A → G
+	→ m_6	0	++	G → A
m_6	→ +	0	+	A → G

* this gene was itself mutant; m_2 was another mutant allele at the same site.

addition, also on chemical grounds, only transitions were presumed to occur. On the basis of these assumptions, the results of forward and reverse mutations at six sites of the genome could be arranged into a consistent pattern, in which HA produces only transitions from C to T, while EMS produces mainly transitions from G to A and C to T, but also acts to a lesser degree in the reverse directions. (see p. 285). Again, the circle of chemical and mutational data remains closed, although it is somewhat strained through the conclusions regarding EMS. A number of mutants for which the base at the mutant site had been inferred with a fair degree of certainty were then used for determining the molecular effects of mutagen treatment inside the host (18,66,87). For base analogues, this is the only possible means of application because incorporation of the analogue is a prerequisite of mutation. NG, too, requires *in vivo* treatment, possibly because of a connection with replication or transcription (Chapter 16). We have already mentioned the results of this analysis: 5-BU as well as NG show different specificities in T4 and S13 (p. 285). More important: HA loses its strict specificity *in vivo*; while still acting preferentially on C, it also produces mutations by reaction with any of the three other bases. Together with the weak mutagenic action of HA on cells, this erosion of its specificity removes the main reference point for the second part of reversion analysis in cellular organisms. While the distinction between frameshifts and base substitutions often can be made with a fair degree of confidence even in these systems, determination of mutant bases requires different techniques such as interconversion of nonsense codons or amino acid sequencing of the gene products. Ten years ago, when the reactions of mutagens with DNA seemed to be the key to the whole of mutagenesis, this would have seemed a tremendous impediment to mutation research. It seems a much less serious drawback nowadays when it has been realized that the relation between the types and frequencies of primary changes in DNA and the types and frequencies of observed mutations is much less direct than had then been assumed. There remains the requirement for sets of mutants that are sufficiently well characterized at the molecular level to serve as testers for potential mutagens in the environment. We shall deal with this in Chapter 23.

References

1. Proceedings of Symposium, (1974), 'Caffeine as an environmental mutagen and the problem of synergistic effects', *Mutation Res.* **26**, (no. 2)51-155.
2. Litman, R. and Pardee, A.B. (1956), 'Production of bacteriophage mutants by disturbance of deoxyribonucleic acid metabolism', *Nature* **178**, 529-531.
3. Benzer, S. and Freese, E. (1958), 'Induction of specific mutations with 5-bromouracil', *Proc. Nat. Acad. Sci. U.S.A.* **44**, 112-119.
4. Freese, E. (1959), 'The specific mutagenic effect of base analogues on phage T4', *J. Mol. Biol.* **1**, 87-105.
5. Rudner, R. (1961), 'Mutation as an error in base pairing. I. The mutagenicity of base analogues and their incorporation into the DNA of *Salmonella typhimurium*', *Zeitschr. Vererb. Lehre.* **92**, 336-360.
6. Strelzoff, E. (1962), 'DNA synthesis and induced mutations in the presence of 5-bromouracil. II Induction of mutations', *Zeitschr. Vererb. Lehre.* **93**, 301-318.
7. Terzaghi, B.E., Streisinger, G. and Stahl, F.W. (1962) 'The mechanism of 5-bromouracil mutagenesis in the bacteriophage T4', *Proc. Nat. Acad. Sci. U.S.A.* **48**, 1519-1524.
8. Litman, R.M. and Pardee, A.B. (1960), 'The induction of mutants of bacteriophage T2 by 5-bromouracil. III. Nutritional and structural evidence regarding mutagenic action', *Biochim. Biophys. Acta* **42**, 117-130.
9. Luzzati, D. (1957), 'Sur les répercussions génétiques de la substitution de la thymine par le 5-iodouracyle dans l'acide desoxyribonucléique de *E.coli* 15T⁻', *C.R. Acad. Sci.* **245**, 1466-1468.
10. Kramer, G., Wittmann, H.G. and Schuster, H. (1964), 'Die Erzeugung von Mutanten des Tabakmosaikvirus durch den Einbau von Fluorouracil in die Virus-nukleinsäure', *Z. Naturf,* **19b**, 46-51.
11. Cooper, P.D. (1964). 'The mutation of poliovirus by 5-fluorouracil', *Virology* **22**, 186-192.
12. Herrington, M.B. and Takahashi, I. (1973), 'Mutagenesis of bacteriophage PBS2', *Mutation Res.* **20**, 275-278.
13. Salganik, R.I., Vasjunina, E.A., Poslovina, A.S. and Andrieva, I.S. (1973), 'Mutagenic action of N^4-hydroxycytidine on *E.coli* B cyt⁻', *Mutation Res.* **20**, 1-5.

14. Litman, R.M. and Pardee, A.B. (1960), 'The induction of mutants of bacteriophage T2 by 5-bromouracil. IV. Kinetics of bromouracil-induced mutagenesis', *Biochim. Biophys. Acta*, **42**, 131-140.
15. Dunn, D.B. and Smith, I.D. (1954), 'Incorporation of halogenated pyrimidines into the deoxyribonucleic acids of bacterium coli and its bacteriophages', *Nature*, **174**, 305-307.
16. Zamenhof, S. and Griboff, G. (1954), '*E.coli* containing 5-bromouracil in its deoxyribonucleic acid', *Nature*, **174**, 307-308.
17. Wacker, A., Kirschfeld, S. and Träger, L. (1960), 'Über den Einbau Purin-analoger Verbindungen in die Bakterien-Nukleinsäure', *J. Mol. Biol.*, **2**, 241-242.
18. Howard, B.D. and Tessman, I. (1964), 'Identification of the altered bases in mutated single-stranded DNA. II. *In vivo* mutagenesis by 5-bromodeoxyuridine and 2-aminopurine', *J. Mol. Biol.*, **9**, 364-371.
19. Eisenstark, A., Eisenstark, R. and Sickle, R. van (1965), 'Mutation of *Salmonella typhimurium* by nitrosoguanidine', *Mutation Res.*, **2**, 1-10.
20. Osborn, M., Person, S., Phillips, S. and Funk, F. (1967), 'A determination of mutagen specificity in bacteria using nonsense mutants of bacteriophage T4', *J. Mol. Biol.*, **26**, 437-447.
21. Sora, S., Panzeri, L. and Magni, G.E. (1973), 'Molecular specificity of 2-aminopurine in *Saccharomyces cerevisiae*', *Mutation Res.*, **20**, 207-213.
22. Hsu, T.C. and Somers, C.E. (1961), 'Effect of 5-bromodeoxyuridine on mammalian chromosomes', *Proc. Nat. Acad. Sci. U.S.A.*, **47**, 396-403.
22a. Hirono, Y. and Smith, H.H. (1969), 'Mutations induced in *Arabidopsis* by DNA nucleoside analogs'. *Genetics*, **61**, 191-199.
22b. Kaufmann, B.P. and Gay, H. (1970), 'Induction by 5-bromodeoxyuridine of sex-linked lethal mutations in spermatogenous cells of *Drosophila melanogaster*', *Mutation Res.* **10**, 591-595.
23. Freese, E. (1959), 'The specific mutagenic effect of base analogues on phage T4', *J. Mol. Biol.*, **1**, 87-105.
24. Koch, R.E. (1971), 'The influence of neighboring base pairs upon base-pair substitution mutation rates', *Proc. Nat. Acad. Sci. U.S.A.*, **68**, 773-776.

25. Drake, J.N. and Greening, E.O. (1970), 'Suppression of chemical mutagenesis in bacteriophage T4 by genetically modified DNA polymerases', *Proc. Nat. Acad. Sci. U.S.A.*, **66**, 823-829.
26. Pietrzykowska, I. (1973), 'On the mechanism of bromouracil-induced mutagenesis', *Mutation Res.*, **19**, 1-9.
26a. Witkin, E.M. and Parisi, E.C.,(1974), 'Bromouracil mutagenesis : mispairing or misrepair', *Mutation Res.*, **25**, 407-409.
27. Fermi, G. and Stent, G. (1962), 'Effects of chloramphenicol and of multiplicity of infection on induced mutation in bacteriophage T4', *Zeitschr. Vererb. Lehre,* **93**, 177-187.
28. Hill, B.T., Tsuboi, A. and Baserga, R. (1974), 'Effect of 5-bromodeoxyuridine on chromatin transcription in confluent fibroblasts', *Proc. Nat. Acad. Sci. U.S.A.*, **71**, 455-459.
29. Witkin, E.M. (1947), 'Mutations in *E.coli* induced by chemical agents', *Cold Spring Harbor Symp. Quant. Biol.*, **12**, 256-267.
30. Clark, A.M. (1953), 'The mutagenic activity of dyes in *Drosophila melanogaster'*, *Amer. Nat.*, **87**, 295-305.
31. Auerbach, C. (1955), 'The mutagenic action of pyronin-B', *Amer. Nat.*, **89**, 241-245.
32. Clark, A.M. (1958), 'The mutagenic action of pyronin in *Drosophila'*, *Zeitschr. indukt. Abst. Vererb. Lehre,* **89**, 123-130.
33. Lerman, L.S. (1961), 'Structural considerations in the interaction of DNA and acridines', *J. Mol. Biol.,* **3**, 18-30.
34. Orgel, A. and Brenner, S. (1961), 'Mutagenesis of bacteriophage T4 by acridines' *J. Mol. Biol.*, **3**, 762-768.
35. Brenner, S., Barnett, L., Crick, F.C.H. and Orgel, A. (1961), 'The theory of mutagenesis', *J. Mol. Biol.*, **3**, 121-124.
36. Hruska, F.E. and Danyluk, S.S. (1968), 'An NMR study of the interactions of purine and pyrimidine derivatives with acridine orange', *Biochim. Biophys. Acta,* **161**, 250-252.
37. Lerman, L.S. (1964), 'Acridine mutagens and DNA structure', *J. Cell. Comp. Physiol.*, **64**, Suppl. 1-18.
38. Lerman, L.S. (1963), 'The structure of the DNA-acridine complex', *Proc. Nat. Acad. Sci. U.S.A.*, **49**, 94-102.
39. Streisinger, G., Okada, Y., Emrich, J., Newton, J., Tsugita, A., Terzaghi, E. and Inouye, M. (1966), 'Frameshift mutations and the genetic code', *Cold Spring Harbor Symp. Quant. Biol.,* **31**, 77-84.
40. Lindstrom, D.M. and Drake, J.W. (1970), 'Mechanics of

frameshift mutagenesis in bacteriophage T4: role of chromosome tips', *Proc. Nat. Acad. Sci. U.S.A.*, **65**, 617-624.

41. Newton, A., Masys, D., Leonard, E. and Wygal, D. (1972). 'Association of induced frameshift mutagenesis and DNA replication in *E.coli', Nature New Biol.*, **236**, 19-22.
42. Sarabhai, A. and Lamfrom, H. (1969), 'Mechanism of proflavin mutagenesis', *Proc. Nat. Acad. Sci. U.S.A.*, **63**, 1196-1197.
43. Magni, G.E., Borstel, R.C. von, and Sora, S. (1964), 'Mutagenic action during meiosis and antimutagenic action during mitosis by 5-aminoacridine in yeast', *Mutation Res.*, **1**, 227-230.
44. Ames, B. and Whitfield, H. Jr. (1966), 'Frameshift mutagenesis in *Salmonella', Cold Spring Harbor Symp.· Quant. Biol.*, **31**, 221-225.
45. Brusick, D.J. and Zeiger, E. (1972), 'A comparison of chemically induced reversion patterns of *Salmonella typhimurium* and *Saccharomyces cerevisiae* mutants, using *in vitro* plate tests', *Mutation Res.*, **14**, 271-275.
46. Malling, H.V. (1967), 'The mutagenicity of the acridine mustard (ICR-170) and the structurally related compounds in *Neurospora', Mutation Res.*, **4**, 265-274.
47. Magni, G.E. and Puglisi, P.P. (1966), 'Mutagenesis of supersuppressors in yeast', *Cold Spring Harbor Symp: Quant.Biol.*, **31**, 699-704.
48. Munz, P. and Leupold, U. (1970), 'Characterization of ICR-170-induced mutations in *Schizosaccharomyces pombe', Mutation Res.*, **9**, 199-212.
49. Werenne, J. and Grosjean, H. (1965), 'Effet de la proflavine sur les *ARN* de transfert', *Arch. Int. Physiol. Biochem.*, **73**, 537-538.
50. Oeschger, N.S. and Hartman, P.E. (1970), 'ICR-induced frameshift mutations in the histidine operon of *Salmonella', J. Bacter.*, **101**, 490-504.
51. Bresler, S.E., Kalinin, V.L. and Perumov, D.A. (1968), 'Inactivation and mutagenesis on isolated DNA. II. Kinetics of mutagenesis and efficiency of different mutagens', *Mutation Res.*, **5**, 1-14.
52. Freese, E. and Strack, H.B. (1962), 'Induction of mutations in transforming DNA by hydroxylamine', *Proc. Nat. Acad. Sci. U.S.A.*, **48**, 1796-1803.

53. Freese, E., Bautz-Freese, E. and Bautz, E. (1961), 'Hydroxylamine as a mutagenic and inactivating agent', *J. Mol. Biol.,* **3**, 133-143.
54. Tessman, I., Poddar, R.K. and Kumar, S. (1964), 'Identification of the altered bases in mutated single-stranded DNA. I. *In vitro* mutagenesis by hydroxylamine, ethyl methanesulfonate and nitrous acid', *J. Mol. Biol.*, **9**, 352-363.
55. Lawley, P.D. (1967), 'Reaction of hydroxylamine at high concentration with deoxycytidine or with polycytidylic acid: evidence that substitution of amino groups in cytosine residues by hydroxylamine is a primary reaction and the possible relevance to hydroxylamine mutagenesis', *J. Mol. Biol.*, **24**, 75-81.
56. Schuster, H. and Wittmann, H.G. (1963), 'The inactivation and mutagenic action of hydroxylamine on tobacco mosaic virus ribonucleic acid', *Virology,* **19**, 421-430.
57. Champe, S.P. and Benzer, S. (1962), 'Reversal of mutant phenotypes by 5-fluorouracil: an approach to nucleotide sequences in messenger-RNA', *Proc. Nat. Acad. Sci. U.S.A.*, **48**, 532-546.
58. Phillips, J., Adman, R., Brown, D. and Grossman, L. (1965), 'The effects of hydroxylamine on polynucleotide templates for RNA-polymerase', *J. Mol. Biol.*, **12**, 816-828.
59. Budowsky, E.I., Krivisky, A.S., Klebanova, L.M., Metlitskaya, A.Z., Turchinsky, M.F. and Savin, F.A. (1974), 'The action of mutagens on MS2 phage and on its infective RNA. V. kinetics of the chemical and functional changes of the genome under hydroxylamine treatment', *Mutation Res.*, **24**, 245-258.
60. Fraenkel-Conrat, H. and Singer, B. (1972), 'The chemical basis for the mutagenicity of hydroxylamine and methoxyamine', *Biochim. Biophys. Acta.*, **262**, 264-270.
61. Freese, E., Bautz, E., Bautz-Freese, E. (1961), 'The chemical and mutagenic specificity of hydroxylamine', *Proc. Nat. Acad. Sci. U.S.A.*, **47**, 845-855.
62. Brenner, S., Barnett, L., Katz, E.R. and Crick, F.H.C. (1967), 'UGA: a third nonsense triplet in the genetic code', *Nature*, **213**, 449-450.
63. Chu, B.C.F., Brown, D.M. and Burdon, M.G. (1973), 'Effect of nitrogen and catalase on hydroxylamine and hydrazine mutagenesis', *Mutation Res.*, **20**, 265-270.

64. Budovsky, E.I., Svertlov, E.D. and Monastyrskaya, G.S. (1971), 'Mechanism of the mutagenic action of hydroxylamine and O-methylhydroxylamine. IV. Reaction of hydroxylamine and O-methylhydroxylamine with the adenine nucleus', *Biochim. Biophys. Acta,* **246**, 320-334.
65. Freese, E., Bautz-Freese, E. and Stuart, G. (1966), 'The oxygen-dependent reaction of hydroxylamine with nucleotides and DNA', *Biochim. Biophys. Acta,* **123**, 17-25.
66. Tessman, I., Ishina, H. and Kumar, S. (1965), 'Mutagenic effects of hydroxylamine *in vivo', Science,* **148**, 507-508.
66a. Budowsky, E.I. (1975), 'The effect of higher structures of macromolecules on the genetic consequences of the action of mutagens', *Mutation Res.,* **27**, 1-6.
67. Levisohn, R. (1970), 'Phenotypic reversion by hydroxylamine: A new group of suppressible phage T4 rII mutants', *Genetics,* **64**, 1-9.
68. Mukai, F. and Troll, W. (1969), 'The mutagenicity and initiating activity of some aromatic amine metabolites', *Ann. N.Y. Acad. Sci.,* **163**, 828-836.
69. Osborn, M., Person, S., Phillips, S. and Funk, F. (1967), 'A determination of mutagen specificity in bacteria using nonsense mutants of bacteriophage T4', *J. Mol. Biol.,* **26**, 437-447.
70. Shukla, T.T. (1972), 'Analysis of mutagen specificity in *Drosophila melanogaster', Mutation Res.,* **16**, 363-371.
71. Guglielminetti, R., Bonatti, St., Loprieno, N. and Abbondandolo, A. (1967), 'Analysis of the mosaicism induced by hydroxylamine and nitrous acid in *Schizosaccharomyces pombe', Mutation Res.,* **4**, 441-447.
72. Malling, H.V. (1966), 'Hydroxylamine as a mutagenic agent for *Neurospora crassa', Mutation Res.,* **3**, 470-476.
73. Putrament, A., Baranowska, H. and Pacheka, J. (1973), 'Mutagenic action of hydroxylamine and methoxyamine on yeast. I. Hydroxylamine', *Molec. Gen. Genetics* , **122**, 61-72.
74. Vries, G. de., and Grosjean, H. (1965), 'Effet de l'hydroxylamine sur la capacité acceptrice des acides ribonucléiques de transfert', *Arch. Int. Physiol.,* **73**, 524-525.
75. Rosenkranz, H.S., and Bendich, A.J. (1964), 'Studies on the bacteriostatic action of hydroxylamine', *Biochim. Biophys. Acta,* **87**, 40-53.

76. Somers, C.E. and Hsu, T.C. (1962), 'Chromosome damage induced by hydroxylamine in mammalian cells', *Proc. Nat. Acad. Sci. U.S.A.*, **48**, 937-943.
77. Borenfreund, E., Krim, M., and Bendich, A. (1964), 'Chromosomal aberrations induced by hyponitrite and hydroxylamine derivatives', *J. Nat. Cancer Inst.*, **32**, 667-679.
78. Lingens, F. (1964), 'Erzeugung von Mangelmutanten von *Escherichia coli* mit Hilfe von Hydrazin und Hydrazinderivaten. *Z. Naturf.*, **19b**, 151-156.
79. Brown, D.M., MacNaught, A.D. and Schell, P.D. (1966), 'The chemical basis of hydrazine mutagenesis', *Biochem. Biophys. Res. Comm.*, **24**, 967-971.
80. Darlington, C.D., and McLeish, J. (1951), 'Action of maleic hydrazide on the cell', *Nature*, **167**, 407-408.
81. Shapiro, R. and Weisgras, J.M. (1970), 'Bisulfite-catalyzed transamination of cytosine and cytidine', *Biochem. Biophys. Res. Comm.*, **40**, 839-843.
82. Shapiro, R., Braverman, B., Louis, J.B. and Servis, R.E. (1973), 'Nucleic acid reactivity and conformation. II. Reaction of cytosine and uracil with sodium bisulfite', *J. Biol. Chem.*, **248**, 4060-4064.
83. Hayatsu, H. and Miura, A. (1970), 'The mutagenic action of sodium bisulfite', *Biochem. Biophys. Res. Comm.*, **39**, 156-160.
84. Summers, I.A. and Drake, J.W. (1971), 'Bisulfite mutagenesis in bacteriophage T4', *Genetics*, **68**, 603-607.
85. Mukai, F., Hawryluk, I. and Shapiro, R. (1970), 'The mutagenic specificity of sodium bisulfite', *Biochem. Biophys. Res. Comm.*, **39**, 983-988.
86. Dorange, J.L. and Dufuy, P. (1972), 'Mise en évidence d'une action mutagène du sulphite de sodium sur la levure', *C.R. Acad. Sci. Paris Ser. D.*, **274**, 2798-2800.
86a. Chambers, R.W., Agyagi, S.Y., Furukawa, Y., Zawadska, H., Bhanet, O.S. (1973), 'Inactivation of valine acceptor activity by a C into U missense change in the anticodon of yeast valine transfer ribonucleic acid', *J. Biol. Chem.*, **248**, 5549-5551.
87. Baker, R. and Tessman, I. (1968), 'Different mutagenic specificities in phages S13 and T4: *in vivo* treatment with N-methyl-N′-nitro-N-nitrosoguanidine', *J. Mol. Biol.*, **35**, 439-448.

CHAPTER EIGHTEEN

Nitrous acid. Other chemical mutagens

In this chapter, I am going to deal with a miscellany of mutagens. During the more than 30 years since the discovery of the first mutagenic chemicals, new ones have been found at an accelerated frequency, so that I shall not be able even to mention all of them. However, most of those that are omitted can be fitted into one or the other of the categories discussed in this chapter and the preceding ones. With a few exceptions, of which nitrous acid is the most important, the substances now to be dealt with play no prominent role in present-day mutation research. Most of them were discovered in the pre-Watson-Crick era, and their mode of action on DNA is not understood. Some certainly, and others probably, act in indirect ways. Yet even these have not been included only for historical reasons. I believe that the time has come to resume the study of at least some of these now neglected mutagens. If in some sections of this chapter, I present what may appear a catalogue of disjointed and incompletely analysed observations, I do this in the hope that some of them may be used as starting points for investigations by modern techniques and on the basis of modern concepts.

Research on directly-acting mutagens has borne rich fruit over two decades. Now the main chemical problems have been solved and the yield in this field is declining. Meanwhile, we have come to realize that mutagenesis, whether induced or spontaneous, is not simply a chemical change in DNA but a highly complex biological process in which enzymes and other proteins, RNA's of various types, membranes and other structural components of the cell are involved. Indirectly acting mutagens may be the best tools for analysing this process and for discovering the homoeostatic devices that normally maintain the fidelity of hereditary transmission. Moreover, some of the substances to be discussed in this chapter may constitute potential environmental hazards and are of interest for this reason (see Chapter 23).

Nitrous acid (HNO_2 ; abbrev. NA)

NA is produced in solutions of $NaNO_2$ or KNO_2 at low pH. It deaminates guanine to xanthine, adenine to hypoxanthine, and cytosine to uracil (1). In RNA genes, the latter change constitutes an immediate mutation by base substitution. In DNA, pairing of uracil with adenine will result in a transition from GC to AT. The pairing properties of hypoxanthine are such that it may produce the opposite transition from AT to GC. On the contrary, xanthine in the place of guanine is not expected to produce mutations. Originally, it was thought that xanthine would pair like guanine and thus restore the *status quo.* Later it was realized that xanthine is a 'nonsense' base that pairs with neither of the two pyrimidines and thus will cause lethality rather than mutation (2).

It was these chemical predictions that prompted the use of nitrous acid as mutagen in the late fifties (for earlier experiments with NA see the article by Auerbach in *Chemical Mutagens,* **3**, Bibliography). The results fully confirmed the expectations. Mutations were produced not only in tobacco mosaic virus but also in its naked RNA (3), used subsequently for infecting tobacco leaves. Soon after these first reports, NA was shown to produce mutations in transforming DNA (4,5), in bacteriophage (6,7), and in bacteria (8). A little later, it was used successfully on fungi (9,10). The pH dependence of its mutagenic action in experiments with viruses paralleled those of the chemical reactions with cytosine and adenine, but not of that with guanine (11); this agreed with the presumed molecular mechanisms of mutagenesis.

Experiments with NA have contributed to the elucidation of the genetic code (12). The proteins of wild-type tobacco mosaic virus and of a large number of NA-induced mutants were subjected to amino acid sequencing. In almost all cases, the mutant protein differed from wild-type in only one amino acid. On the assumption that the mutant codon arose from the wild-type one by a change of either C into U or A into G, already suggested codons could be confirmed and new ones could be suggested. Yet, within the limits of this specificity, base pairs were not mutated in proportion to their frequencies (13). There were deviations from random expectation in regard to the position of the mutant site within its codon, to the site of the mutant codon within the chromosome, and to the particular amino acid change effected. Different scoring systems yielded different spectra of base changes; this suggested that at least a large part of the observed deviations from expectation were due to selection by the method of screening used.

While the specificity of NA for transitions from A to G and C to T may apply also to the two-stranded bacteriophages, it does not hold good for the single-stranded DNA phage S13 (14). Reversion analysis of previously typed base changes showed that NA yields all four types of transition in this phage. This was confirmed when it was found that it can produce conversions of ochre into amber or opal (A into G), of amber or opal into ochre (G into A), reversions of ochre into wild-type (T into C), and mutations of wild-type into ochre (C into T) (15).

In *Neurospora*, reversion and complementation analysis (10,16) led to the conclusion that the majority of NA-induced mutations are base changes. At the *iso*-1-cytochrome *c* locus of yeast, amino acid sequencing revealed a very complex action of NA (17). In at least one site, NA produced primarily transitions from GC to AT. In other sites, about half the mutations occurred by transitions from AT to GC, the other half by various types of transversion. Moreover, the type of change depended on the site of the mutated codon in the cistron and on the site of the mutated base in the codon. We may recall that a site-specific action in this gene has been observed also for UV as mutagen (chapter 14).

In addition to base substitutions, NA induces deletions (18,19). Possibly, their origin is connected with the ability of NA to cross-link the two strands of DNA (20). In eukaryotes, the ability of NA to form cross-links between histone and nucleic acid (21) might conceivably contribute to its mutagenic action.

Dose effect curves for mutation induction by NA are linear after *in vitro* exposure of transforming DNA (22) and bacteriophage (7,23), but more complex after exposure of cellular organisms (8,24). This would be expected from the fact that many cellular conditions such as metabolic state at the time of exposure, temperature and nutrition after exposure, and repair capacities of the cells affect the frequencies of NA-induced mutation (25,26). In *Neurospora* (27), reversions at two loci not only yielded different dose effect curves, but also showed different responses to post-incubation temperature. This suggests specific effects of the treatment on expression; *in vitro*, NA has been shown to change the specificity of transfer-RNA (28).

'4-NQO' (4-nitroquinoline-1-oxide)

'4-NQO' is a powerful carcinogen with mutagenic properties for micro-organisms (29-31). The production of mutations in bacteriophage requires treatment inside the bacterium. The active mutagen thus is a derivative of 4-NQO whose nature is still under debate (32). Classification of 4-NQO-induced mutants at the *iso*-1-cytochrome *c* locus of yeast by Sherman and his collaborators (see Chapter 11, ref. 35) led to the conclusion that all of them were due to transitions or transversions of GC base pairs (32a). Since the primary lesions are not photoreactivable, they are presumed not to be pyrimidine dimers. In spite of this, the mutagenic pathway of 4-NQO, as judged by the effects on repair-deficient bacterial strains, is remarkably similar to that of UV; the lesions in DNA are excisable, mutations to prototrophy are subject to mutation frequency decline (32b), and the production of deletions or mutations requires the *rec*-A function.

Organic peroxides; sodium azide; oxygen

We have seen (Chapter 12) that UV-irradiated medium is mutagenic for micro-organisms, and that this indirect action of UV is mediated by organic peroxides. This suggested that direct treatment with organic peroxides might be mutagenic. Results obtained on *Neurospora* (33), bacteria (34,35), and *Drosophila* (36) confirmed this assumption. In *Neurospora*, hydrogen peroxide by itself was barely or not at all mutagenic but became strongly mutagenic when combined with formaldehyde or acetone, presumably through the formation of a mutagenic peroxide. In *Drosophila*, the addition compound of formaldehyde and hydrogen peroxide, dihydroxydimethylperoxide, produced mutations (37). In bacteria (35), three different organic peroxides yielded very different ratios between four types of mutation; in an adenine-inositol-requiring strain of *Neurospora*, a mixture of formaldehyde and hydrogen peroxide produced adenine-reversions at a higher frequency and with a steeper dose-dependence than inositol-reversions (38). We shall return to these specificities in Chapter 20.

The mutagenic effect of the respiratory inhibitor sodium azide, NaN_3, in *Staphylococcus aureus* has been attributed to the accumulation of metabolically produced mutagenic peroxides (39). In barley, NaN_3 has proved a powerful agent for the production of chlorophyll mutations (40). In *Drosophila*, it is not a mutagen in its own right but enhances the genetical effects of X-rays (Chapter 7). In soy beans, seed treatment with NaN_3 produced leaf spots that were not caused by mutation or somatic crossing-over, but probably by non-disjunction (40a). The suitability of this system for discriminating between treatment effects will be discussed in Chapter 23.

The chemical reactions by which peroxides produce mutations and chromosome breaks are probably related to the mechanism of X-ray mutagenesis in that they proceed via the formation of mutagenic radicals. This may be true also for the production of chromosome breaks (41) and bacterial mutations (42) by pure oxygen. Ethyl alcohol produced aberrations in the chromosomes of *Vicia faba*, and this effect could be inhibited by anoxia (43); probably in this case, too, oxidative processes are involved in mutagenesis.

Formaldehyde

Like caffeine (Chapter 17), formaldehyde is a mutagen that acts differently when applied in different ways. Added to the food of *Drosophila* larvae, it is a powerful mutagen (44). Used in aqueous solution for the injection of *Drosophila* flies or for treatment of micro-organisms, it has at best a weak effect (33,45,46). Although chronologically the first mode of application was used before the second, we shall discuss them in reverse order because the action of formaldehyde in aqueous solution is better understood than its action as food additive.

We have seen that hydrogen peroxide can be turned into an efficient mutagen by mixing it with formaldehyde, and that the addition compound between the two substances is mutagenic for *Drosophila* (33,37). This led to the conclusion that in those cases in which formaldehyde by itself causes mutations it does so via the formation of a mutagenic derivative by combining with small amounts of metabolically produced hydrogen peroxide. This idea was supported by the finding that previous exposure of flies to the catalase-inhibitor KCN approximately doubled the frequency of sex-linked lethals to be obtained from formaldehyde-injection (47,48).

In bacteria, the inability of formaldehyde to produce mutations seems to be due to very efficient removal of potentially mutagenic lesions (49). While the frequency of mutations to tryptophan-independence and streptomycin-resistance was not noticeably increased by formaldehyde-treatment of Hcr^+ bacteria (p. 239), both types of mutation occurred at fairly high frequencies in treated Hcr^- bacteria. When formaldehyde was added to the plating medium of UV-irradiated bacteria, it decreased survival of the Hcr^+ but not of the Hcr^- strain. This suggests that formaldehyde-induced lesions compete with the UV-induced ones for some enzyme in the excision repair system. One possibility for this to happen is that the mutagenically effective lesions in DNA are – or include – dimers between neighbouring purines, and that these are treated like pyrimidine dimers by the repair system. Although this scheme is so far entirely speculative and does not apply to the peroxide-mediated mutagenesis in *Drosophila* and fungi, it is mentioned here because there is evidence for the formation of purine dimers by formaldehyde *in vitro* (50,51) and because dimer formation has also entered into speculations on the alternative mode of formaldehyde mutagenesis, which we shall now discuss.

The strong mutagenic action of formaldehyde food (FF) on *Drosophila* larvae was discovered by the Russian geneticist I.A. Rapoport (44), who was also the first to detect the mutagenic effects of various epoxides, ethylene imines and diazoalkanes. Because of the total eclipse of genetics in Russia during the Lysenko period, these latter results reached the West only after the mutagenicity of these alkylating agents had been discovered independently by Western geneticists. An interesting point to note is that Rapoport's highly successful selection of potential mutagens was based on his belief that the essential part of the gene molecule is its protein moiety; obviously, it sometimes pays to forge ahead on the basis of a wrong theory, guided only by intuition and a spirit of curiosity. In particular, modern models of the molecular action of mutagens would hardly have led to tests with formaldehyde. The results of these tests were reported by Rapoport in the last paper of his that reached the West before the breakdown of communication between Western and Russian geneticists. Subsequently, these findings were confirmed, extended and analysed by Western geneticists (52-55).

The most striking feature about FF-mutagenesis is its extreme stage specificity, which is not matched by any other mutagen. Although, as shown by labelling experiments (56), formaldehyde administered in the food penetrates into all germ cells of adult and larval males and females, mutations occur exclusively during one part of the cell cycle of one germ cell stage in one developmental phase of one sex, namely in the early stage of the primary larval spermatocytes. No mutations are produced in adults of either sex, in female larvae and in larval spermatogonia and late spermatocytes. The sex difference is not attributable to the possession of a Y-chromosome by the male, since XXY females are as recalcitrant to treatment as are XX females (57). Fig. 18.1 gives an outline of the tests by which the sensitivity pattern of the larval testis was established. Males were treated on each of the four days of larval life; the distribution of sensitive cells in the testes of these developmental stages was determined roughly by raising four successive five-day broods from the treated males (brood-pattern analysis, see p. 76). The same procedure was used for female larvae. Subsequent experiments with three-day broods (53) have refined the picture but have not changed it in its essentials. The female germ cells remain unaffected, whatever the age of the larva or the stage of the germ cell. In the testis of newly hatched larvae, the sensitive zone occupies the most posterior part, i.e. it consists of the most advanced germ cell stages. Gradually, as the larva grows, these cells pass the sensitive stage and the sensitive zone moves towards the anterior end of the testis. Cytologically this zone corresponds to the auxocyte stage, i.e. the long growth stage that precedes meiosis; it is during this stage that the chromosomes replicate (58). This suggests that mutagenesis by means of FF is bound up with DNA replication. Indirect evidence supports this assumption (59). It is based on the location of lethals obtained from individual males. Many males have more than 1 lethal but, with very rare exceptions, these are never allelic. Since mutations that occur in spermatogonia will give rise to at least a proportion of clusters, the absence of clusters was one of the pieces of evidence for the

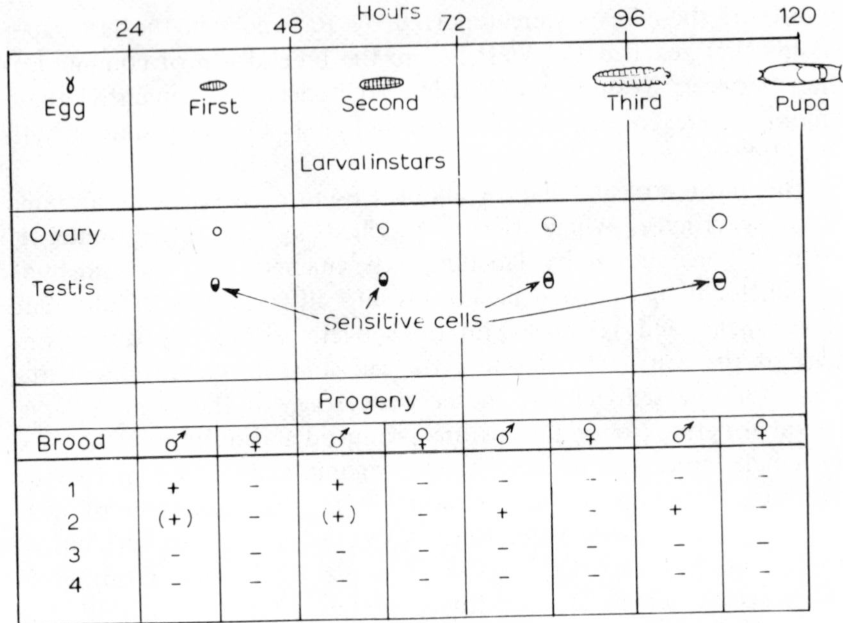

Fig. 18.1. Sensitivity of *Drosophila* germ cells to the mutagenic action of formaldehyde-treated food. (Fig. 2, in Auerbach (52). Courtesy of Mendelian Society of Lund, Sweden.)

refractoriness of spermatogonia to FF. Yet lethals from the same male are not randomly distributed over the chromosome but show a striking topographical correlation. This was found by the mapping of lethals that had occurred in individual males (from 2-6 per male). There were 54 such groups of sex-linked lethals, and 36 of second-chromosome lethals. In both cases, the distribution of lethals within groups deviated very significantly from randomness, but in opposite directions. Groups of sex-linked lethals from the same male often consisted of – or included – 'pseudoclusters' of two or three, rarely more, lethals that were very close to each other, though separable by location or cytological analysis. On the contrary, groups of autosomal lethals from the same male showed a remarkably even distribution, with the distances between them never less than 10 and usually more than 30 cross-over units. One has thus to account for a locational association – positive for the X, negative for

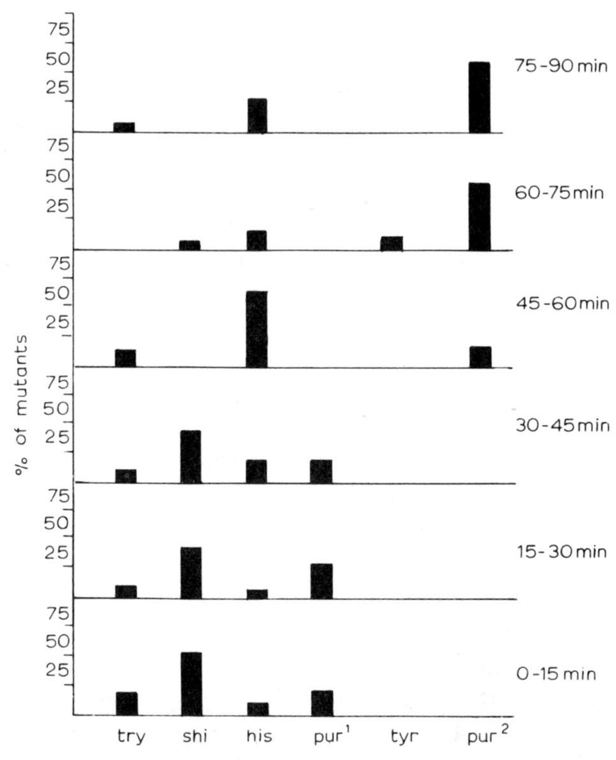

Fig. 18.2. The spectra of auxotrophic mutants of *E.coli* K-12 3080 obtained when formaldehyde ($3.10^{-2}M$)was applied at different time intervals of DNA replication in synchronized bacterial cultures. (Fig. 7, in Salganik (60a). Courtesy Verlag Georg Thieme, Leipzig.

Chromosome II – between lethals produced in the same testis but not as clusters. The only physical association between germ cells in the testis is that between the 16 spermatocytes in the same spermatocytic cyst. These are connected by cytoplasmic bridges (60) and replicate their chromosomes synchronously. A tentative interpretation of the data assumes that the effective mutagen is released and persists for a short time within each cyst and, during this time, acts near a replication point. If these points are numerous in the autosomes, but few in the X, then the chance for two lethals to occur at the same point would be small for the former but high for the latter. Unfortunately, it is not yet possible to determine directly the replication pattern in the male meiosis of *Drosophila*, so that this explanation – although at present the only reasonable one – remains speculative. It is, however, supported by experiments on synchronized bacterial cultures, in which the spectrum of formaldehyde-induced auxotrophs varied with time of treatment and agreed roughly with the map order of the genes (60a) (Fig. 18.2).

Other reasons for assuming that FF acts at replication are derived from speculations on its molecular action which we shall now consider. It seems very unlikely that free formaldehyde should persist in the mixture of warm *Drosophila* food and through ingestion and digestion. Indeed, mutations can be produced by adding to the medium either casein (52) or RNA (61b) that have been treated with formaldehyde and washed free of unreacted or loosely bound traces of it, and a reaction product between CH_2O and amino acids has been found to cause single-strand breaks in bacterial DNA (61a). There must be some mechanism by which free formaldehyde or a mutagenic derivative of it is formed in the cells; the most probable explanation of the stage specificity of FF is that the conditions for this to happen are present only in early spermatocytes. Normal active metabolism is a prerequisite. All conditions that slow down the development of the larvae, including an excessive dose of formaldehyde, decrease the frequency of mutations, and under very poor growth conditions no mutations at all are recovered (55). Alderson, (62) who analysed this dependence on metabolism by adding formaldehyde to a chemically defined medium, found that the presence of adenylic acid or adenosine is indispensable for the production of mutations by FF. Adenine is not; in fact, under certain conditions it acts as antimutagen to FF. Curiously enough, this is exactly the opposite of what has been found for spontaneous and purine-induced mutagenesis in bacteria (Chapter 17), for which adenine is mutagenic and adenosine antimutagenic. Alderson interprets his results as showing that the effective mutagen is an adenosine-dimer which is incorporated into replicating DNA instead of the monomer. Alternatively, one may consider that adenosine plays an essential role in the process by which free formaldehyde is released from its carrier compound, and that it then acts on DNA either by forming a purine dimer (50,51), a reaction product with an amino acid, or in some other way. Since free formaldehyde reacts only with denatured DNA (63), its effect would be expected to be limited to replicating or transcribing regions.

Under the right conditions and in responsive cells, FF is an extremely efficient mutagen. It has produced up to 15% sex-linked lethals ar d to 20% autosomal ones (54). It has also produced visible mutations, small and large deficiencies, inversions and translocations. Translocation frequency was low; but this must at least in part be attributed to the fact that the responding cells are pre-meiotic and that translocations in general are less easily produced in pre-meiotic than in post-meiotic stages. Visible mutations and small deficiencies included a high proportion of mosaics; since one-strand changes in DNA should segregate completely mutant cells at meiosis, this observation points to a delayed action of FF. Even stronger evidence for a delayed effect stems from the cytological analysis of the F_1 salivary glands which revealed a high frequency of complementary deficiencies and duplications (64). These, as we have seen in Chapter 15, require delayed breakage for their formation.

Urethane (ethylcarbamate; $NH_2COOC_2H_5$)

The mutagenic action of urethane was discovered in Germany by Oehlkers (65) simultaneously with that of mustard gas in Scotland. Urethane is, however, a much more 'spotty' mutagen than mustard gas, acting only under certain conditions and in certain organisms. In the meiotic cells of various flowering plants, ethylurethane – and to a lesser extent also methyl, propyl, and butylcarbamate – produced translocations; KCl was a required co-factor (66). In *Drosophila*, a mixture of ethylurethane with KCl or NaCl yielded chromosome rearrangements, recessive lethals and visible mutations (67). The breaks produced by urethane resembled those produced by X-rays to the extent that breaks due to either agent rejoined as freely with each other as did breaks produced by one of them alone (68). The majority of the mutations showed no cytologically detectable changes in the salivary gland chromosomes; at least some of them may have been true intragenic changes (69). So, almost certainly, were tryptophan-requiring fluorescent mutations produced by *in vitro* treatment of transforming DNA (70) (see Chapter 11, p. 191).

Already Oehlkers had assumed that urethane may affect the chromosomes by indirect means. This agreed with his finding that the effect depends markedly on the cytoplasmic environment: the same hybrid genome of *Oenothera* responded differently when treated in reciprocal hybrids (66). *Neurospora* proved entirely recalcitrant to the action of urethane, even when the whole genome was tested for mutations and deletions by the recessive-lethal technique

(Chapter 11) (71). Presumably, conditions for the formation of the ultimate mutagen are not present in *Neurospora*. The nature of this mutagen is not known. As regards chromosome breakage – but not gene mutation – the conclusion has been drawn that the effective agent is hydroxyurethane acting via the formation of hydrogen peroxide and radicals (70). The cogency of this conclusion depends on whether one is prepared to assume that the same mechanism causes breakage of eukaryote chromosomes and inactivation of transforming DNA; for the latter system, the hydroxyurethane-radical mechanism has been made very likely.

Alkaloids

Among the substances that, in Oehlkers' experiments on *Oenothera*, increased the frequency of translocations in meiotic cells were a number of therapeutically used alkaloids (66). Of these, morphine gave a clearly positive result, while the effects of scopolamine, narceine and atropine were on the borderline of significance. It is interesting to note that reciprocal hybrids of *Oenothera*, which differed markedly in their response to the KCl-ethylurethane treatment (p. 345),did not do so in their responses to alkaloids. This does not, of course, exclude the possibility of other strain or species differences, but the early results with these compounds have not been followed up. Perhaps our concern with environmental mutagens will stimulate renewed tests.

Much later, a group of potent mutagens was detected among the pyrrolizidine alkaloids, which are present in many plants (e.g. the ragwort *Senecio*) and produce severe liver damage in grazing animals. In *Aspergillus* they produced forward mutations (103). In *Drosophila* (72-75), some of them produced high frequencies of recessive lethals and low ones of translocations.

There are good reasons for assuming that hepatotoxic and mutagenic pyrrolizidine alkaloids produce their effects via the formation of pyrrolic derivatives with alkylating action (76,77). Whether or not such a substance is mutagenic in a given test system depends therefore not only on the substance itself but also on the metabolism of the organism. The genetic effects of alkaloid narcotica like morphine may well be much more indirect. This was already assumed by Oehlkers, whose vast screening programme on *Oenothera* showed that translocation frequencies can be increased in a great variety of ways, e.g. by simply changing the ionic environment of the meiotic chromosomes. This brings me to the mutagenic action of certain salts.

Inorganic salts

Various salts have been found to favour the production of chromosome rearrangements by other agents or to produce them in untreated cells. Examples of the former action have been mentioned before. Thus the ability of EMS to produce translocations in plants is greatly enhanced by the presence of certain bivalent cations (p. 266), that of ethylurethane by the presence of potassium or sodium chloride (p.345). Fructose, which by itself was at best weakly damaging to *Oenothera* chromosomes, produced translocations when mixed with KCl (65). Acting by themselves, aluminium salts yielded translocations in meiotic cells of *Oenothera* (65), but no chromosome aberrations in mitotic cells of *Vicia* (78).

In micro-organisms, mutagenic, co-mutagenic and antimutagenic effects of salts have been observed. In the bacterium *Proteus mirabilis,* $MnCl_2$ acted as *co-mutagen* to EMS, greatly enhancing both its lethal and mutagenic effects; the increase in mutation frequency was greater than might have been attributed to selective killing of non-mutants (79). $MgCl_2$, on the other hand, reduced the lethal action of EMS without affecting its mutagenic one. In *Penicillium chrysogenum,* the addition of $MnCl_2$ to the plating medium had selective *antimutagenic* effects (80). It suppressed mutations to azaguanine resistance when these had been induced by nitrogen mustard (HN-2), but not when they had been induced by UV or X-rays. HN-2-induced mutations to aza-indole resistance were not suppressed.

The *mutagenic* action of $MnCl_2$ in bacteria was discovered by Demerec (81), who found that mutations – mainly to streptomycin-resistance – were produced in *E.coli* under treatment conditions that allowed high survival. The mutational yield depended strongly on a variety of factors such as temperature, state of growth of the bacteria, composition of the growth medium, the salt mixture used for washing etc. In *Proteus mirabilis,* $MnCl_2$ produced very few mutations in stationary cultures, but many in growing ones (82). In bacteriophage T4, most of the Mn^{2+}-induced mutations were revertible by base analogues and presumably were transitions; some may have been transversions (83).

It is reasonable to assume that Mn ions produce mutagenic and co-mutagenic effects indirectly by affecting those enzyme systems that are involved in replication and stability of chromosomes and genes. A brief discussion of some of the ways in which this might happen is found in the paper by Orgel and Orgel (83). The multiplicity of enzymes whose function depends on Mn^{2+} readily explains why the effects of the treatment depend so strongly on species, cell stage, and treatment conditions. The antimutagenic action of Mn^{2+} in *Penicillium* is more plausibly attributed to interference with enzymes concerned with mutation expression.

Phenols and quinones

These compounds, too, can affect DNA only indirectly either by the production of radicals and peroxides (84) or by acting on enzymes or other proteins. Fragmentation of *Allium* chromosomes by phenols and quinones was observed as early as 1948 (85); the most effective substances, e.g. pyrogallol, produced fragments in up to 60% of the treated cells. The effect was restricted to fragmentation; translocations or other rearrangements were very rare or absent. This is reminiscent of the effect of caffeine on mammalian cells at physiological temperatures (Chapter 17); it would be interesting to test whether phenols, like caffeine, can produce rejoinable fragments under a different set of conditions. Interest in the chromosomal effects of phenols has recently been rekindled by a report that a group of men professionally exposed to ambient benzene had an increased frequency of chromosome aberrations in their peripheral blood (86); the generality of this observation is under investigation.

In *Drosophila* (87), phenol can produce mutations but, like pyronin (Chapter 17), it appears to do so only under very special conditions. While pyronin acts only on certain males in a treated sample, phenol was effective in some experiments and ineffective in others. The causes of this variability could not be established, and the experiments were abandoned. They remain interesting for two reasons: the method of application, and the indication of locus specificity. The method of application was laborious but allowed maximal penetration and may be useful for substances of low penetrating power. It consisted of treating explanted larval ovaries *in vitro* and subsequently re-implanting them into host larvae,

Table 18.1
The distribution of 24 phenol-induced 2nd chromosome lethals and semi-lethals over 8 ovaries. After Hadorn (87), Table 2.)

	Ovaries								
Lethals	*a*	*b*	*c*	*d*	*e*	*f*	*g*	*h*	*i*
A	1								1
B	1	1							2
C	3								3
D		7	3						10
E			1		1	1			3
F				3				1	3
G							1		1
H								1	1
	5	8	4	3	1	1	1	1	24

where a proportion established contact with the oviduct and produced offspring. The mutations scored were recessive lethals on the second chromosome. Table 18.1 shows the distribution of 24 of them over 8 ovaries from the same treated batch. The repeated occurrence of the same lethal in any one ovary can be attributed to cluster formation by a mutant oogonium, but the finding that four out of eight lethals (B,D,E,F) occurred in more than one ovary points to the existence of preferred regions of attack.

In the bacterium *Micrococcus pyogenes,* a naphthaquinone (2-methyl, 1-4 naphthaquinone; 'Menadione') produced mutations to streptomycin- and penicillin-resistance (88). The effect was at first attributed to the affinity of the compound for – SH groups. Other –SH poisons, however, failed to produce mutations in the same system as well as in *Drosophila* (89). Thus, the mutagenic action of this naphthaquinone appears to have a more complex basis.

^{35}S; amino acid analogues; 5-fluorouracil

^{35}S

During what I have called the fourth period of mutation research (Chapter 1), all efforts were directed towards analysing the action of those mutagens that react directly with DNA. Research on the indirectly acting ones, like those discussed in the previous sections of this chapter, was put aside. The more recent realization that the production of mutations by almost any mutagenic treatment depends to a large extent on biochemical processes in the cell as a whole, opened the way for attempts at influencing mutation frequencies indirectly via chemicals that inhibit or deflect these processes. Before discussing recent results in this new area of research, I want to mention earlier work that may find an explanation along these lines. This is the production of mutations by the addition of ^{35}S to the growth medium of *Neurospora* (90). The pitfalls to the interpretation of mutation experiments with radioisotopes have been discussed in Chapter 7. They were, of course, present also in these experiments, but the steps taken to avoid them were such that, on balance, it seems at least possible that a considerable fraction of the mutations were caused by transmutation of ^{35}S into ^{35}Cl (for a critical discussion see *Radiation Biol.* 1954, Chapter 8; Bibliography). The report on these experiments appeared in 1952 and the authors assumed that, if transmutation was indeed involved, it had taken place in sulphur atoms of the gene proteins. While this interpretation is no longer valid, it seems possible that other proteins had been rendered mutagenic by incorporation of ^{35}S. It would be interesting to repeat such tests from this novel point of view.

Amino acid analogues

The first mutagenic amino acid analogue was discovered by accident in experiments in which the growth response of *Chlamydomonas eugametos* to antimetabolites was studied (91). Unexpectedly, one of these, the phenylalanine analogue 2-amino-3phenylbutanoic acid (3-methylphenylalanine) produced high frequencies of mutations conferring resistance to the analogue itself and to two antibiotics, streptomycin and neamine. Fluctuation tests (Chapter 11) showed that the mutations were not due to the selection of preexisting variants but were induced by the treatment. On crossing to sensitives, most of the tested mutants segregated in Mendelian fashion; a few gave patterns of lethal segregation that suggested the presence of a translocation or other chromosome aberration.

A little later, experiments on fungi were undertaken with the express purpose of testing whether the addition of amino acid analogues to the growth medium had mutagenic effects (92). The results were positive. Among a small number of analogues tested, several were clearly mutagenic, with indications of mutagen specificity. In *Ustilago maydis*, *para*-fluoro-phenylalanine (PFPA) and ethionine (ETH) increased the frequencies of methionine-, adenine-, and arginine-reversions, while only the former produced mutations to auxotrophy. In an arginine-requiring strain, canavanine and homoarginine produced reversions to arginine-independence, although only the former seems to be incorporated into proteins instead of arginine. In the same strain, canavanine produced also methionine reversions and forward mutations to amino acid auxotrophy; homoarginine was not tested for these types of mutation. Some analogues, e.g. norleucine and 7-azatryptophan, were ineffective in the systems used. In *Coprinus lagopus* (93) both PFPA and ETH proved strongly mutagenic for suppressor-mutations of the met-1 locus and for dominant reversions of an adenine-requirement. While incorporation into protein seemed to be a prerequisite for mutagenesis by PFPA, there was no correlation between mutagenesis by ETH and its incorporation into proteins. The tentative conclusion was drawn that PFPA produces mutations by acting on enzymes engaged in replication or repair of DNA, whereas ETH does so in some other way, perhaps by donating ethyl groups to DNA. In bacteria, methylated adenine appears to protect the chromosomes against breakage (94). It might do so also in eukaryotes, and ethylation might be less effective. Ethionine has not yet been tested for mutagenic action in prokaryotes. Para-fluorophenylalanine was ineffective (95), and this was tentatively ascribed to differences in the enzymes that mediate replication or repair in the DNA of eukaryotes and prokaryotes. It is of interest in this context that the repair machinery for UV-induced damage in yeast differs between the eukaryotic DNA of the nuclear chromosomes and the, essentially prokaryotic, DNA of the mitochondria (96). Alternatively, the restriction of the mutagenic action of *para*-fluorophenylalanine to eukaryotes might point to an involvement of the chromosome-associated proteins.

Fluorouracil

The experiments on *Ustilago* (92) also included tests with the pyrimidine analogue 5-fluorouracil (5-FU). Growth in the presence of 5-FU yielded forward mutations to auxotrophy as well as reverse mutations at the three loci tested. Again, there was evidence of specificity: while *para*-fluorophenylalanine was more efficient than 5-FU in the production of auxotrophs and of methionine- and adenine-reversions, the opposite was found in tests for arginine-reversions.

We have seen in Chapter 17 that 5-FU, by being incorporated instead of uracil into RNA, can produce mutations in organisms that have RNA as their genetic material. This direct mode of mutagenesis is not possible in organisms with DNA-chromosomes. Here 5-FU must act indirectly. A plausible assumption is that it does so through introducing a tendency for miscoding into messenger-RNA (Chapter 4), and that this in turn may result in malfunctioning of enzymes concerned with replication and stability of DNA.

The mutation tests with analogues of biological macromolecules other than DNA are still in their beginning. It is clear that they have opened up a new and promising path to the study of mutation. In particular, they may throw light on the origin of spontaneous mutations.

DNA and other macromolecules

In 1939, the Russian geneticist Gershenson reported that visible mutations can be induced in *Drosophila* through admixture of calf thymus DNA to the food of the larvae (97). His colleagues, both in Russia and the West, argued that, if these results were correct, the same treatment should also produce sex-linked lethals. The fact that this did not happen was taken as evidence against Gershenson's claim. The war interrupted this work. After the war, Gershenson repeated his claim on the basis of additional data (98), and geneticists in the West experimented with his technique. The discrepancy between the results of the pre-war workers was explained when Mathew (99) found that the X-chromosome, while not wholly insensitive to the effects of the treatment, is much less responsive than the second chromosome. This chromosomal specificity, in turn, is probably a consequence of regional specificity; for lethals in the second chromosome were found to cluster in one particular region, where they formed overlapping deficiencies. Many of the mutations appeared first as mosaics. At least a proportion of the mosaic lethals were caused by replicating instabilities (see Chapter 22).

A special type of small deficiencies, the so-called Minutes, were obtained in the progeny of adult males that had been injected with DNA from *Drosophila*, rat liver, or bacteriophage (100). Minutes are mutants with a characteristic phenotype of small bristles, often associated with other morphological abnormalities and usually with slowed-down larval development. They can occur in many regions of all chromosomes, but those obtained by injection of DNA again showed a regional preference.

There are a number of reasons why these effects of DNA cannot be considered as transformations. First of all, the effective macromolecules need not be homologous to the treated DNA: in the feeding experiments, calf thymus DNA was used; in the injection experiments, heterologous DNA was as effective as DNA extracted from *Drosophila*. Indeed, macromolecules other than DNA (e.g. calf histone) or synthetic macromolecules (e.g. polymethacrylic acid) were at least as effective as DNA, although they tended to prefer different chromosomal regions (100). In other experiments with the injection technique, both deoxyribonuclease and bovine serum albumin yielded the same, rather weak, increase in the frequency of sex-linked lethals (101). Secondly, while transformation is a highly specific effect on one particular gene, for which the transforming DNA must carry a mutant allele, macromolecules used in these experiments produced a variety of more or less unspecific effects, mainly small deletions. Possibly they act by attaching to certain chromosome regions, causing them to under- or misreplicate. In the polytene chromosomes of *Drosophila*, as much as 40% of the chromatin can bind calf thymus histone to DNA (102).

It is only a small step from the production of chromosomal damage by biological macromolecules to the chromosome-breaking and mutagenic effects of certain viruses. Somewhat arbitraily, I shall postpone a discussion of these observations to the chapter on spontaneous mutations (Chapter 21), in which we shall also deal briefly with transformation-like phenomena in *Drosophila* and other eukaryotes.

References

1. Schuster, H. (1960), 'The reaction of nitrous acid with DNA', *Biochem. Biophys. Res. Comm.*, **2**, 320-323.
2. Michelson, A.M. and Monny. C. (1966), 'Polynucleotide analogues. IX Polyxanthic acid', *Biochem. Biophys. Acta.*, **129**, 460-474.
3. Gierer, A. and Mundry, K.W. (1958), 'Production of mutants of tobacco mosaic virus by chemical alteration of its ribonucleic acid *in vitro*', *Nature*, **182**, 1457-1458.

4. Litman, R. and Ephrussi-Taylor, H. (1959), 'Inactivation et mutation des facteurs génétiques de l'acide desoxyribonucléique du Pneumocoque par l'ultraviolet et par l'acide nitreux', *C.R. Acad. Sci.*, **249**, 838-840.
5. Chevallier, M.R. and Greth, M.L. (1971), 'On the mutagenic action of nitrous acid on *H. influenzae* transforming DNA', *Molec. Gen. Genetics* **110**, 27-30.
6. Tessman, I. (1959), 'Mutagenesis in phages ϕX 174 and T4 and properties of the genetic material', *Virology* **9**, 375-385.
7. Bautz-Freese, E. and Freese, E. (1961), 'Induction of reverse mutations and cross-reactivation of nitrous acid-treated phage T4', *Virology* **13**, 19-30.
8. Kaudewitz, F. (1959) 'Production of bacterial mutants with nitrous acid', *Nature* **183**, 1829-1830.
9. Clarke, C.H. (1962), 'A case of mutagen specificity attributable to a plating medium effect', *Zeitschr. indukt. Abst. Vererb. Lehre.* **93**, 435-440.
10. Malling, H.V. (1965), 'Identification of the genetic alterations in nitrous-acid induced *ad*-3 mutants of *Neurospora crassa'*, *Mutation Res.* **2**, 320-327.
11. Vielmetter, W. and Schuster, H. (1960), 'The base specificity of mutation induction by nitrous acid in phage T2', *Biochem. Biophys. Res. Comm.* **2**, 324-328.
12. Wittmann, H.G. and B. Wittmann-Liebold (1966), 'Protein-chemical studies of two RNA viruses and their mutants', *Cold Spring Harbor Symp. Quant. Biol.* **31**, 163-172.
13. Sengbusch, P.V. (1967), 'Influence of protein structure on selection of nitrous acid induced mutants of TMV', *Molec. Gen. Genetics* **99**, 171-180.
14. Tessman, I., Poddar, R.K. and Kumar, S.(1964), 'Identification of the altered bases in mutated single-stranded DNA. I. *In vitro* mutagenesis by hydroxylamine, ethylmethanesulfonate and nitrous acid', *J. Mol. Biol.* **9**, 352-363.
15. Vanderbilt, A.S. and Tessman, I. (1970), 'Identification of the altered bases in mutated single-stranded DNA. IV. Nitrous-acid induction of the transitions guanine to adenine and thymine to cytosine', *Genetics* **66**, 1-10.
16. Malling, H.V. and de Serres, F.J. (1967), 'Relation between complementation patterns and genetic alterations in nitrous acid-induced *ad*-3B mutants of *Neurospora crassa'*, *Mutation Res.* **4**, 425-440.

17. Sherman, F. and Stewart, J.W. (1973), 'Variation of mutagenic action on nonsense mutants at different sites in the *iso*-1-cytochrome *c* gene of yeast', *Genetics* **78**, 97-113.
18. Tessman, I. (1962), 'The induction of large deletions by nitrous acid', *J. Mol. Biol.* **5**, 442-445.
19. Schwartz, D.O. and Beckwith, J.R. (1969), 'Mutagens which cause deletions in *E.coli'*, *Genetics* **61**, 371-376.
20. Geiduschek, P. (1961), 'Reversible' DNA', *Proc. Nat. Acad. Sci. U.S.A.* **47**, 950-955.
21. Potti, N. Dasan and Bello, J. (1971), 'Cross-linking of nucleohistone by nitrous acid', *Mutation Res.* **12**, 113-119.
22. Bresler, S.E., Kalinin, V.L. and Perumov, D.A. (1968), 'Inactivation and mutagenesis on isolated DNA. II. Kinetics of mutagenesis and efficiency of different mutagens', *Mutation Res.* **5**, 1-14.
23. Vielmetter, W. and Wieder, C.M. (1959), 'Mutagene und inaktivierende Wirkung salpetriger Säure auf freie. Partikel des Phagen T2', *Z. Naturf.* **14b**, 312-317.
24. Loprieno, N. and Clarke, C.H. (1965), 'Investigations on reversions to methionine independence induced by mutagens in *Schizosaccharomyces pombe'*, *Mutation Res.* **2**, 312-319.
25. Zimmermann, F.K., Schwaier, R. and von Laer, A. (1966), 'The effect of temperature on the mutation fixation in yeast', *Mutation Res.* **3**, 90-92.
26. Clarke, C.H. (1970), 'Repair systems and nitrous acid mutagenesis in *E.coli B/r'*, *Mutation Res.* **9**, 359-368.
27. Auerbach, C. and Ramsay, D. (1967), 'Differential effect of incubation temperature on nitrous acid-induced reversion frequencies at two loci in *Neurospora'*, *Mutation Res.* **4**, 508-510.
28. Carbon, J. and Curry, J.B. (1968), 'A change in the specificity of transfer-RNA after partial deamination with nitrous acid', *Proc. Nat. Acad. Sci U.S.A.* **59**, 467-474.
29. Ishizawa, M. and Endo, H. (1971), 'Mutagenesis of bacteriophage T4 by a carcinogen, 4-nitroquinoline 1-oxide', *Mutation Res.* **12**, 1-18.
30. Kondo, S., Ishizawa, H., Iwo, K. and Kato, T. (1970), 'Base-change mutagenesis and prophage induction in strains of *Escherichia coli* with different repair capacities', *Genetics* **66**, 187-217.

31. Yamamoto, K. and Ishii, Y. (1974), '4-nitro-quinoline 1-oxide induced deletion mutations in *Escherichia coli* strains with different DNA repair capacities', *Mutation Res.* 22, 81-83.

32. Ishii, Y. and Kondo, S. (1971), 'Differential inactivation of transforming DNA *in vitro* and *in vivo* by 4-hydroxyamino-quinoline 1-oxide', *Mutation Res.* **13**, 193-198.

32a. Prakash, L., Stewart, J.W. and Sherman, F. (1974), 'Specific induction of transitions and transversions of G-C base pairs by 4-nitroquinoline-1-oxide in *iso*-1-cytochrome *c* mutants of yeast', *J. Mol. Biol.* **85**, 51-66.

32b. Williams, P.H. and Clarke, C.H. (1974), 'Mutation frequency decline following chemical mutagenesis of *Salmonella typhimurium'*, *Mutation Res.* **22**, 255-264.

33. Dickey, F.H., Cleland, G.H. and Lotz, C. (1949), 'The role of organic peroxides in the induction of mutations', *Proc. Nat. Acad. Sci. U.S.A.* **35**, 581-586.

34. Luzzati, D. and Chevallier, M.-R. (1957), 'Comparaison entre l'action létale et mutagène d'un peroxyde organique et des radiations sur *E.coli'*, *Ann. Inst. Pasteur* **93**, 365-375.

35. Chevallier, M.-R. and Luzzati, D. (1960), 'Action mutagène spécifique de trois peroxydes organiques sur les mutations reverses de deux loci de *E.coli* 15T-9-13', *C.R. Acad. Sci.* **250**, 1572-1574.

36. Altenburg, L.S. (1954), 'The production of mutations in *Drosophila* by tertiary-butyl hydroperoxide', *Proc. Nat. Acad. Sci. U.S.A.* **40**, 1037-1040.

37. Sobels, F.H. (1956), Mutagenicity of dihydroxydimethyl peroxide and the mutagenic effects of formaldehyde', *Nature* **177** 979-980.

38. Auerbach, C. and Ramsay, D. (1968). 'Analysis of a case of mutagen specificity in *Neurospora crassa.* 1. Dose-response', *Molec. Gen. Genetics* **103**, 72-104.

39. Wyss, O., Clark, J.B., Haas, F. and Stone, W.S. (1948), 'The role of peroxides in the biological effects of irradiated broth', *Proc. Nat. Acad. Sci., U.S.A.* **56**, 51-57.

40. Nilan, R.A., Sideris, E., Kleinhof, A., Sander, C. and Konzak, C.F. (1973), 'Azide – a potent mutagen' *Mutation Res,* **17**, 142-144.

40a. Vig, K.B. (1973), 'Somatic crossing-over in *Glycine max* (L) Merrill: mutagenicity of sodium azide and lack of synergistic effect with caffeine and mitomycin C', *Genetics* **75**, 265-277.

41. Conger, A.D. (1952), 'Breakage of chromosomes by oxygen', *Proc. Nat. Acad. Sci., U.S.A.* **38**, 289-299.
42. Gifford, G.D. (1968), 'Mutation of an auxotrophic strain of *Escherichia coli* by high pressure oxygen', *Biochem. Biophys. Res. Comm.* **33**, 294-298.
43. Rieger, R. und Michaelis, A. (1960), Über die radiomimetische Wirkung von Athylalkohol bei *Vicia faba', Abh. Deutsche Akad. Wiss. Berlin. Kl. Medizin. Jg.* **1960**, 54-65.
44. Rapoport, I.A. (1946), 'Carbonyl compounds and the chemical mechanism of mutations', *C.R. Acad. Sci. U.S.S.R.* (Russian with English summary) **54**, 65-67.
45. Jensen, K.A., Kirk, I., Kølmark, G. and Westergaard, M. (1951), 'Chemically induced mutations in *Neurospora', Cold Spring Harbor Symp. Quant. Biol.* **16**, 245-261.
46. Auerbach, C. (1952), 'Mutation tests on *Drosophila melanogaster* with aqueous solutions of formaldehyde', *Amer. Nat.* **86**, 330-332.
47. Sobels, F.H. and Simons, J.W.I.M. (1956), 'Studies on the mutagenic action of formaldehyde in *Drosophila.* I. The effect of pretreatment with cyanide on the mutagenicity of formaldehyde and of formaldehyde-hydrogen peroxide mixtures in males', *Zeitschr. indukt. Abst. Vererb. Lehre.* **87**, 735-742.
48. Sobels, F.H. (1956), 'Studies on the mutagenic action of formaldehyde in *Drosophila.* II. The production of mutations in females and the induction of crossing-over', *Zeitschr. indukt. Abst. Vererb. Lehre.* **87**, 743-752.
49. Nishioka, H. (1973), 'Lethal and mutagenic action of formaldehyde in Hcr^+ and Hcr^- strains of *Escherichia coli', Mutation Res.* **17**, 261-265.
50. Feldman, M.Ya. (1967), 'Reaction of formaldehyde with nucleotides and ribonucleic acid', *Biochim. Biophys. Acta* **149**, 20-34.
51. Collins, C.J. and Guild, W.R. (1968), 'Irreversible effects of formaldehyde on DNA', *Biochim. Biophys. Acta* **157**, 107-113.
52. Auerbach, C. (1951), 'Some recent results with chemical mutagens', *Hereditas* **37**, 1-16.
53. Auerbach, C. and Moser, H. (1953), 'An analysis of the mutagenic action of formaldehyde food. I. Sensitivity of *Drosophila* germ cells', *Zeitschr. indukt. Abst. Vererb. Lehre.* **85**, 479-504.

54. Auerbach, C. and Moser, H. (1953), 'Analysis of the mutagenic action of formaldehyde food. II. The mutagenic potentialities of the treatment', *Zeitschr. indukt. Abst. Vererb. Lehre* **85**, 547-563.

55. Auerbach, C. (1956), 'Analysis of the mutagenic action of formaldehyde food. III. Conditions influencing the effectiveness of the treatment', *Zeitschr. indukt. Abst. Vererb Lehre* **87**, 627-647.

56. Kaplan, W.D. and Pelc, S.R. (1950), 'Autoradiographic studies of *Drosophila* gonads following the feeding of C^{14} labelled formaldehyde', *Zeitschr. indukt. Abst. Vererb Lehre* **87**, 356-364.

57. Ratnayake, W. (1968), 'Tests for an effect of the Y-chromosome on the mutagenic action of formaldehyde and X-rays in *Drosophila melanogaster', Genet. Res., Camb.* **12**, 65-69.

58. Kaplan, W.D. and Siskin, J.E. (1960), 'Genetic and autoradiographic studies of tritiated thymidine in testes of *Drosophila melanogaster', Experientia* **16**, 67-69.

59. Naafei, H. and Auerbach, C. (1964), 'Mutagenesis by formaldehyde food in relation to DNA replication in *Drosophila* spermatocytes', *Zeitschr. indukt. Abst. Vererb. Lehre* **95**,351-367.

60. Meyer, G.F. (1961), 'Interzelluläre Brücken (Fusome) im Hoden und im Ei-Nährzellverband von *Drosophila melanogaster', Z. Zellforsch.* **54**, 238-251.

60a. Salganik, R.J. (1972), 'Some possibilities of mutation control concerned with local increase of DNA sensitivity to chemical mutagens', *Biol. Zbl.* **91**, 49-59.

61a. Poverenny, A.M., Siomin, Yu.A., Saenko, A.S. and Sinzinis, B.J. (1975), 'Possible mechanisms of lethal and mutagenic action of formaldehyde', *Mutation Res.* **27**, 123-126.

61b. Alderson, T. (1964), 'The mechanism of formaldehyde-induced mutagenesis. The monohydroxymethylation reaction of formaldehyde with adenylic acid as the necessary and sufficient condition for the mediation of the mutagenic activity of formaldehyde', *Mutation Res.* **1**, 77-85.

62. Alderson, T. (1961), 'Mechanism of mutagenesis induced by formaldehyde. The essential role of the 6-amino group of adenylic acid (or adenosine) in the mediation of the mutagenic activity of formaldehyde', *Nature* **191**, 251-253.

63. Tikchonenko, T.I. and Dobrov, E.N. (1969), 'Peculiarities of the secondary structure of bacteriophage DNA in situ. II. Reaction with formaldehyde', *J. Mol. Biol.* **42**, 119-132.
64. Slizynska, H. (1957), 'Cytological analysis of formaldehyde-induced chromosomal changes in *Drosophila melanogaster*', *Proc. Roy. Soc. Edinb.* **66**, 288-304.
65. Oehlkers, F. (1943), 'Die Auslösung von Chromosomenmutationen in der Meiosis durch Einwirkung von Chemikalien', *Zeitschr. indukt. Abst. Vererb. Lehere* **81**, 313-341.
66. Oehlkers, F. and Linnert, G. (1951), 'Weitere Untersuchungen über die Wirkungsweise von Chemikalien bei der Auslösung von Chromosomenmutationen', *Zeitschr. indukt. Abst. Vererb. Lehre* **83**, 429-438.
67. Vogt, M. (1948), 'Mutationsauslösung bei *Drosophila* durch Äthylurethan', *Experientia* **4**, 68-69.
68. Oster, I.I. (1958), 'Interaction between ionizing radiation and chemical mutagens', *Zeitschr. indukt. Abst. Vererb. Lehre* **89**, 1-6.
69. Vogt, M. (1950), 'Urethane-induced mutations in *Drosophila*', *Publ. Stazione Zool. Napoli* **22** (suppl), 114-124.
70. Bautz-Freese, E. (1967), 'Der Chemismus der urethaninduzierten Mutationen 25 Jahre nach ihrer Entdeckung', *Molec. Gen. Genetics* **100**, 150-158.
71. Auerbach, C. (1966), 'Chemical induction of recessive lethals in *Neurospora crassa*', *Microb. Genet. Bull.* **17**, 5.
72. Clark, A.M. (1959), 'Mutagenic activity of the alkaloid heliotrine in *Drosophila*', *Nature* **183**, 731-732.
73. Clark, A.M. (1960), 'The mutagenic activity of some pyrrolizidine alkaloids in *Drosophila*', *Zeitschr. indukt. Abst. Lehre* **91**, 74-80.
74. Brink, N.G. (1966). 'The mutagenic activity of heliotrine in *Drosophila*. I. Complete and mosaic sex-linked lethals', *Mutation Res.* **3**, 66-72.
75. Brink, N.G. (1969), 'The mutagenic activity of the pyrrolizidine alkaloids in *Drosophila melanogaster*. II. Chromosome rearrangements', *Mutation Res.* **8**, 139-146.
76. Culvenor, C.C.J., Downing, D.T. Edgar, J.A. and Jago, M.V. (1969), 'Pyrrolizidine alkaloids as alkylating and antimitotic agents', *Ann. N.Y. Acad. Sci.* **163**, 837-847.
77. Mattocks, A.R. and White, I.N.H. (1971), 'Pyrrolic metabolites from non-toxic pyrrolizidine alkaloids', *Nature New Biology* **231**, 114-115.

78. Michaelis, A. und Rieger, R. (1959), 'Einige Bemerkungen zur 'mutagenen' Wirkung von Aluminium-chlorid bei *Vicia faba*', *Monatsber. Deutsch. Akad. Wiss. Berlin* **1**, 596-599.

79. Böhme, H. (1962), 'Die Beeinflussung der inaktivierenden und mutagenen Wirkung von Äthyl-Methansulfonat durch Mangan- und Magnesium-chlorid', *Biol. Zbl.* **81**, 269-276.

80. Arditti, R.R. and Sermonti, G. (1962), 'Modification by manganous chloride of the frequency of mutation induced by nitrogen mustard', *Genetics* **47**, 761-768.

81. Demerec, M. and Hanson, J. (1951), 'Mutagenic action of manganous chloride', *Cold Spring Harbor Symp. Quant. Biol.* **16**, 215-227.

82. Böhme, H. (1961), 'Streptomycin-abhängige Mutanten von *Proteus mirabilis* und ihre Verwendung in Mutationsversuchen mit Manganchlorid', *Biol. Zbl.* **80**, 5-32.

83. Orgel, A. and Orgel, L.E. (1965), 'Induction of mutations in bacteriophage T4 with divalent manganese', *J. Mol. Biol.* **14**, 453-457.

84. Walling, C. (1957), 'Free radicals in solution', John Wiley and Sons, New York.

85. Levan, A. and Hin Tjio, J. (1948), 'Induction of chromosome fragmentation by phenols', *Hereditas* **34**, 453-484.

86. Tough, I.M., Smith, P.G. Court Brown, W.M. and Harnden, D.G. (1970), 'Chromosome studies on workers exposed to atmospheric benzene: the possible influence of age', *Europ. J. Cancer* **6**, 49-55.

87. Hadorn, E., Bertani, G. and Rosin, S. (1949), 'Ergebnisse der Mutationsversuche mit chemischer Behandlung von *Drosophila* Ovarien *in vitro*', *Proc. 8th Int. Cong. Genetics; Suppl. Hereditas* 256-266.

88. Clark, J.B. Wyss, O. and Stone, W.S. (1950), 'Induction of mutation in *Micrococcus pyogenes* by chemical inactivation of sulphydryl groups', *Nature* **166**, 340.

89. Auerbach, C. (1950), 'SH-poisoning and mutation', *Experientia* **6**, 17.

90. Hungate, F.P. and Mannell, T. (1952), 'Sulfur-35 as a mutagenic agent in *Neurospora*', *Genetics* **37**, 709-719.

91. McBride, A.C. and Gowans, C.S. (1969), 'The induction of gene mutation and chromosome aberration in *Chlamydomonas eugametos*', *Genet. Res. Camb.* **14**, 121-126.

92. Lewis, C.M. and Tarrant, G.M. (1971), 'Induction of mutation by fluorouracil and amino acid analogues in *Ustilago maydis'*, *Mutation Res.* **12**, 349-356.
93. Talmud, P.J. and Lewis, D. (1974), 'The mutagenicity of amino acid analogues in *Coprinus lagopus'*, *Genet. Res., Camb.* **23**, 47-61.
94. Marinus, M.G. and Morris, N.R. (1974), 'Biological function of methyladenine residues in the DNA of *Escherichia coli* K12', *J. Mol. Biol.* **85**, 309-322.
95. Talmud, P. and Lewis, D. (1974), 'Mutagenicity of amino acid analogues in eukaryotes', *Nature* **249**, 563-564.
96. Moustacchi, E. and Enteric, S. (1970), 'Differential "liquid holding recovery" for the lethal effect and cytoplasmic "petite" induction by ultraviolet light in *Saccharomyces cerevisiae'*, *Mol. Gen. Genetics* **109**, 69-83.
97. Gershenson, S. (1939), 'Induction of directed mutations in *Drosophila'*, *Dokl. Akad. Nauk. SSSR* **25**, 224-227.
98. Gershenson, S. (1965), 'Induction of lethal mutations in *Drosophila melanogaster* by DNA', *Genet. Res., Camb.* **6**, 157-162.
99. Mathew, C. (1965), 'The production of recessive lethals by calf-thymus DNA in *Drosophila'*, *Genet. Res., Camb.* **6**,163-174.
100. Fahmy, O.G. and Fahmy, M.J. (1965), 'Differential induction of chromosome deletions by natural and synthetic macromolecules in *Drosophila melanogaster'*, *Genetics* **52**, 861-873.
101. Kaufmann, B.P., Gray, H., Buchanan, J., Weingart, A., Maruyama, and Akey, A. (1961), *'The nature of the materials of heredity'*, Carnegie Inst. Wash. Year Book, **60**, 476-493.
102. Khesin, R.B. (1973), 'Binding of thymus histone F1 and *E.coli* RNA polymerase to DNA of polytene chromosomes of *Drosophila'*, *Chromosoma* **44**, 255-264.
103. Alderson, T. and Clark, A.M. (1966), 'Interlocus specificity for chemical mutagens in *Aspergillus nidulans'*, *Nature* **210**, 593-595.

CHAPTER NINETEEN

The relation between induced crossovers, rearrangements and mutations. The origin of complete (non-mosaic) mutations

In this chapter, we shall deal with two problems that are common to the production of mutations by all agents, physical and chemical. In a slightly different form, these problems also play a role in spontaneous mutagenesis and will be taken up again in Chapter 21. Two further problems of general importance for mutagenesis – mutagen specificity and unstable mutations – will be discussed in Chapters 20 and 22.

The relation between crossing-over, rearrangement formation and mutation

Most mutagens produce the whole triad of these effects, although in various proportions. We shall see in Chapter 23 that reliance on this correlation plays a role in the choice of test methods for environmental mutagens. In particular, it seems that most or all mutagens induce crossing-over (1) and gene conversion (2); this makes it very likely that these processes have one or more steps in common with mutagenesis.

A more specialized hypothesis which was put forward in the early days of mutation research is still under consideration. It assumes that the processes by which mutagens produce rearrangements on the one hand, crossovers on the other, are identical *throughout*, either because rearrangements are due to 'illegitimate crossing-over' between non-homologous chromosomes or, conversely, because crossovers are translocations between homologous chromosomes. It is this hypothesis in its two versions that we shall now discuss.

Evidence that X-rays do not produce rearrangements by illegitimate crossing-over was derived from the near-quadratic dose-effect curve for rearrangements (Chapter 6). For mustard gas, the quadratic dose-effect curve for translocations (Chapter 16) leads to the same conclusion. Further evidence was obtained in experiments with chemical mutagens. In *Oenothera,* Oehlkers and Linnert searched for a correlation between the frequencies of translocations and chiasmata after treatment of meiotic cells with a variety of mutagens (see Chapter 18). While most chemicals produced both effects, there was no correlation whatever between their efficiencies in this regard. In a total of 118 experiments with urethane, salts, and alkaloids, the correlation coefficient between the frequencies of chiasmata and translocations was −0.2, with a probability of 10% of being due to chance (3).

In *Drosophila*, where cytological comparisons are not possible, much the same situation was found in genetical tests. If chemically induced crossing-over in the male is considered to be true crossing-over – and we shall give reasons below for justifying this assumption – then most or all mutagens are found capable of inducing crossing-over, but do so with efficiencies that are not correlated with their efficiencies in the production of rearrangements. Thus, at doses yielding similar frequencies of sex-linked lethals, formaldehyde food was much more effective than mustard gas in producing crossovers (4); yet mustard gas produced more translocations than formaldehyde food. Mustard gas itself yielded more crossovers in pupae than in adults at doses that produced approximately the same frequencies of sex-linked lethals in both developmental stages (5). Altogether, there is little or no support for the idea that translocations arise from illegitimate crossing-over. Neither, as we have seen in Chapter 15, is there a reason for assuming that chemicals produce small deletions and repeats by 'non-homologous' crossing-over.

The opposite hypothesis, which considers induced crossovers a special kind of translocation, has stood up much better to experimental scrutiny. Indeed, it would be expected that any treatment which produces rearrangements will yield *some* translocations between homologous chromosomes; these will result in apparent recombinants when the breakpoint falls between two marker genes. The question, therefore, is not whether recombinants *can* be due to translocations but whether all of them have this origin. For

X-ray-induced crossing-over in *Drosophila* males, the problem has been studied extensively. The methods used and the, in part, contradictory results have been reviewed in reference 6 and, more fully, in reference 7.

One method is based on the consideration that true crossovers should be one-hit phenomena, while translocations require two independent hits (Chapter 6). The most recent experiments by this method (6) gave results that supported the translocation hypothesis: the dose exponent for X-ray induced crossovers in late spermatogonia was 1.8; fractionation of the dose significantly reduced the frequency of crossovers and this effect disappeared when the flies were kept in nitrogen between the fractions; the oxygen enhancement ratio (Chapter 7) was that expected for a two-hit process. It must be concluded that, in *late* spermatogonia, X-rays produce crossing-over mainly by translocations between homologous chromosomes.

Different conclusions were drawn from a method that goes back to the early days of X-ray genetics. Friesen, who was the first to induce crossovers in *Drosophila* spermatogonia, was also the first to analyse their nature (8). He argued that, since translocations between homologous chromosomes will rarely, if ever, have both breakpoints in the same position, the resulting chromosomes must carry complementary deficiencies and duplications. While the latter may be viable, the former will very rarely be so. If, then, genetical and cytological analysis should show that most crossover chromosomes, and particularly the two complementary products of the same crossover event, are viable and normal, this is good evidence for their origin through true crossing-over. Friesen analysed 37 chromosomes from irradiated *early* spermatogonia, which often yield bundles of the two complementary classes. Only three of them were lethal in homozygous condition and, since the viable complementary chromosome carried no cytologically detectable duplication, may have carried an adventitious lethal. Spermatogonial crossing-over is mitotic crossing-over and, as such, occurs preferentially in the heterochromatic region near the centromere, where cytological analysis is difficult and deficiencies may be viable. It is therefore of particular importance that 15 out of 16 chromosomes with exchanges in euchromatin were viable in homozygous condition. It may be concluded that X-rays produce true crossing-over in *early* spermatogonia, in which they produce very few translocations. Discrepancies obtained in later experiments with the same method may have been due to differences in the age of the irradiated spermatogonia. Thus Ives, who in early experiments (quoted in 6) found that the frequencies of lethals on irradiated chromosomes were the same whether or not the chromosomes had been involved in a crossover event, subsequently obtained excess lethality among crossover chromosomes when slightly older spermatogonia had been irradiated (9).

In contrast to X-rays, formaldehyde food (FF) produces very few translocations (Chapter 18). It is therefore more likely that the recombinant chromosomes found after FF-treatment arise through crossing-over. This has been confirmed by two methods, both of which dissociate the ability of FF to induce mutations from its ability to induce crossing-over. It may be recalled that FF exercises its mutagenic action exclusively on early spermatocytes, and that it does so only in the presence of adenosine. Neither of these limitations applies to the induction of crossing-over. Ratnayake (1) showed that FF induces crossing-over in spermatogonia, including very early ones. Alderson (10) showed that it induces crossing-over in the absence of adenosine. Moreover, both workers found that the majority of recombinant chromosomes were viable in homozygous condition. Particularly relevant is the finding that this was true for both recombinant chromosomes in 11 out of 12 complementary clusters from early spermatogonia (1).

In summary, it appears that X-rays and FF, and probably most or all mutagens, can induce true crossing-over, and that they do this by a mechanism that is distinct from aberration formation by breakage and reunion, although the two mechanisms probably have an early step in common. Whether, as Ratnayake (1) suggests, this is the production of single-strand breaks, will need to be tested by further experiments. A very similar situation obtains for the relation between recombination and intragenic mutations. In Chapters 14 and 16 we have discussed the role of recombination in the production of mutations by UV and chemicals. Here, too, some step or steps appear to be common to both processes, but the connection is far from simple.

The origin of 'complete' mutants

The problem

We have seen (Chapters 1, 10, 15) that the tendency of UV and chemicals to produce mosaic mutations posed a problem that was solved only with the advent of the double-helix model of DNA. Solution of this old problem created a new one which is still under debate. How, it is asked, can mutagenic changes in one strand of DNA yield non-mosaic (complete) mutants? We have seen in the preceding chapters that a large majority of premutational lesions affect only one strand of DNA. On the other hand, there is no doubt that most mutagens can produce complete mutations. In *Drosophila* work, this faculty can be attributed *ipso facto* to any mutagen that yields sex-linked lethals in F_2 (Chapter 5). In microorganisms, completely mutant clones have been produced not only in one-stranded bacteriophage (11) but also in all tested organisms with double-stranded DNA.

In the phages T4 and T2, nitrous acid yielded mixtures of completely mutant and mosaic (mottled) plaques (11,12), while 5-bromodeoxyuridine produced a prevalence of mosaics (13). In *E.coli*, on the contrary, most of the 5-BU induced mutations from *lac*$^-$ to *lac*$^+$ were completes (14). Other experiments on bacteria, too, show a prevalence of completes. X-rays produced mainly completes among prototrophic revertants (15), and all of several hundred nitrous acid-induced auxotrophs were completes (16). No mosaics at all were found after treatment of bacteria in the chemostat with such different agents as photodynamically acting dyes, caffeine, and 2-aminopurine (17). Similar results were obtained with fungi and algae. The majority of UV-induced mutations in the fungus *Ustilago maydis* (18) and in the algae *Chlamydomonas Reinhardi* (19) and *Ulva mirabilis* (20) were completes. So were nitrous acid-induced mutations to pyrimidine dependence in *Neurospora* (21). In fission yeast, however, UV and various chemical mutagens produced both complete and mosaic mutations (22). Four main hypotheses have been put forward to explain the origin of completes (22). One of them assumes different origins for these two mutational types. The other three postulate that all mutations start as potential mosaics but that some of them are transformed into completes by replication, repair, or lethality. We shall consider them in turn, keeping in mind that they are not mutually exclusive.

The dual-action hypothesis

This assumes that mosaics and completes result from different primary effects of the same mutagen, e.g. base changes *versus* cross-linkage, or one-strand breakage *versus* two-strand breakage. It is conceivable that X-rays produce complete lethals and deficiencies by two-strand breakage; it is unlikely that they produce prototrophic revertants (15) in this way. The complete auxotrophs in nitrous acid-treated *Neurospora* have been atrributed to cross-linkage (21). While there is no direct proof for the dual-action hypothesis, it may account for some cases of completes. Others, like the complete mutants produced by base analogue-treatment of bacteria (14,17), cannot be plausibly interpreted in this way. For their explanation we shall have to turn to one or more of the remaining hypotheses.

The master-strand hypothesis (17)

This assumes that only one strand of DNA, the master-strand, serves as template for replication, its daughter strand serving as template for the second strand. Semi-conservative replication is preserved if this second strand pairs with its complement in the parental double helix. Mutational changes in the master-strand give rise to complete mutants; those arising in the other strand are not expressed. If this hypothesis is adopted, it reverses the historical sequence by re-opening the old problem: how do mosaics arise? A facile way of answering this question is to assume that all mosaics trace back to genes that were already duplicated at the time of treatment. This was shown not to be true for *Schizosaccharomyces pombe,* in which nitrous acid produced about 50% mosaics even after exposure of a pure G1 (p. 372) population of cells (23). In any case, the master-strand hypothesis has meanwhile been proved wrong for the replication of phage DNA (24); there is no reason to assume its validity for other organisms.

The lethal-hit hypothesis

This assumes that potential mosaics are transformed into completes by lethal hits on the complementary strand of DNA. It provides a plausible explanation for the origin of completes in experiments in which high doses were used. Thus, the doses of nitrous acid that yielded exclusively complete auxotrophs in *E.coli* (16) gave survival values between 10^{-3} and 10^{-5}. In other cases, however (e.g. 14,15,21), completes arose at high survival levels. A direct means of

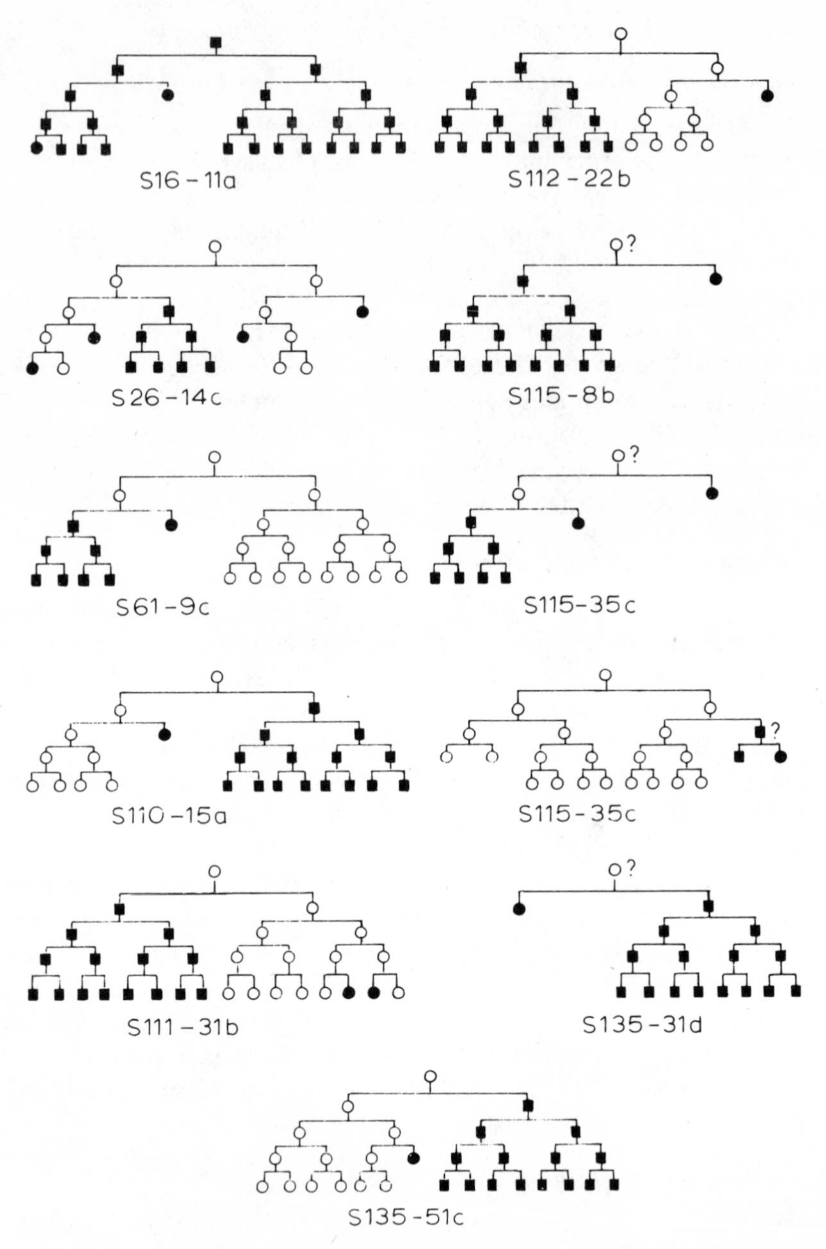

Fig. 19.1. Pedigrees of UV-treated fission yeast, containing both lethal and mutant sectors. ○ = colony-forming, unmutated; ■ = colony-forming, mutated; ● = non-colony forming, lethal; ? = cell whose mutational status cannot be determined (Fig. I in Haefner (27). Courtesy Genetics Foundation.)

showing the existence of lethal sectors and estimating their frequencies is pedigree analysis of treated cells by micromanipulation. This has been done for bacteria and fission yeast after treatment with X-rays (25), UV (26-29), nitrous acid (29,30), and other chemicals (31). In every case, lethal sectors have been found in excess of those occurring in the controls. It is obvious that a mutant cell will form a pure clone if its non-mutant sister dies without forming a colony. This will happen independently of whether lethality is due to a 'lethal hit' on one of the strands of DNA or whether it has a cytoplasmic origin (25). The pedigrees confirmed this but showed at the same time that many pure mutant clones cannot be accounted for by lethal sectoring. Fig. 19.1 shows pedigrees of UV-irradiated cells of *Schizosaccharomyces pombe*; only pedigrees with both mutant and lethal progeny are included. It will be seen that most of the pedigrees were mosaic for mutant and non-mutant progeny; since a proportion of the cells had already duplicated genomes at the time of treatment, the number of mosaics that are due to one-strand mutations is overestimated. Examples for the transformation of mosaic into complete mutants by lethal sectoring are pedigrees S115-8, S115-31c, and S115-31d. Pedigree S16-11a was included because it formed a lethal sector in the second generation; it is, however, likely to have arisen as a complete mutant, with an adventitious lethal in one of its descendants. Not included in Fig. 19.1 are 5 out of the 19 analysed pedigrees that were pure mutant clones without lethal sectors. These must have arisen by a different mechanism. The same applies to UV-induced morphological mutations in the unicellular desmid alga *Cosmarium*, 13 of which out of a total of 49 affected both daughter cells (32).

Under special conditions, it may be the *repair of lethal lesions* rather than the lesions themselves that transforms mosaic into complete mutations. This appears to be the explanation of results obtained with phage T4 (33). This phage is not subject to host cell reactivation (Chapter 14) but carries its own gene (v^+) for excision repair. Phage particles were treated with a high dose of hydroxylamine (HA), known to produce mainly mutations and few inactivations (Chapter 17). This was followed by UV-irradiation *in vitro* under conditions known to produce inactivations but no mutations. (For conditions leading to UV-mutagenesis *in vitro* see 34.) The majority of mutant *r* plaques were mottled but, with increasing UV-dose, there was an increasing proportion of completes among them. This might be interpreted by the conventional lethal-hit hypothesis, were it not for the finding that transformation of mottled into complete plaques required the function of the v^+ gene; in v^- strains, no such transformation took place. On the lethal-hit hypothesis, exactly the opposite result would have been expected. The following mechanism was postulated to explain these results. Excision repair of the lethal lesion often removes a sufficiently long piece of DNA to uncover the mutant base on the complementary strand; when this happens, this base will serve as template for re-synthesis and the resultant mutation will affect both strands. It should be noted, however, that lethal hits as such appear to contribute to the formation of pure clones after treatment with hydroxylamine, for at a lower dose of HA, known to produce inactivations as well as mutations, pure clones arose in both v^+ and v^- strains without concomitant UV-treatment (33,35).

The repair hypothesis

This assumes the existence of a repair system that recognizes mismatched bases and corrects them before the first replication. Depending on which strand suffers correction, repair will transform a one-strand mutation into a pure mutant clone or a pure wild-type clone. Uncorrected heteroduplex molecules will give rise to a mixed clone (mosaic). The scheme is shown in Fig. 19.2.

This figure is taken from a paper on transformation (36). It is known that transformation starts with the formation of a hybrid duplex (heteroduplex) between one strand of donor DNA and the complementary strand of host DNA; thus the origin of pure transformed clones raises the same question as does the origin of pure mutant clones. In this particular investigation, clonal analysis of transformed cells gave strong support to the repair model.

For a direct test of the hypothesis, artifically constructed heteroduplex molecules of phage DNA were used. These were obtained by annealing complementary strands of wild-type DNA and of 21 plaque type mutants. The heteroduplex molecules were then introduced into host bacteria by transfection, i.e. by infection with naked phage DNA. Without correction, the single bursts should yield only mixed clones; with correction, there should be a proportion of pure mutant and pure wild-type clones. The results (37,38) proved the latter prediction true: independently of whether the mutations had been induced by hydroxylamine, nitrous acid,

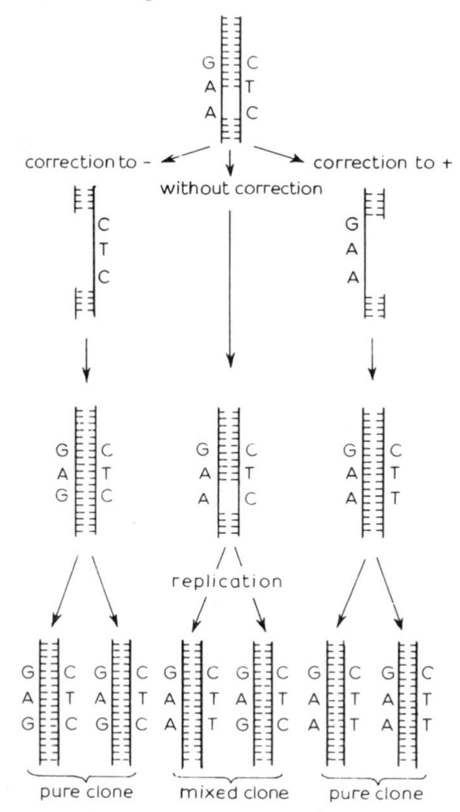

Fig. 19.2. Correction of a molecular heterozygote. Wild-type (+): $\frac{\text{GAA}}{\text{CTT}}$, mutant (–): $\frac{\text{GAG}}{\text{CTC}}$, heterozygote: $\frac{\text{GAA}}{\text{CTC}}$, (Fig. 1, in Bresler (36). Courtesy Berlin-Heidelberg-New-York: Springer.

nitrosoguanidine or acridine, more than half of them appeared as pure clones. The ability of the host to perform excision- or recombination-repair made no difference; thus, bacterial genes for UV-repair were not involved in this process of 'conversion' at the molecular level. Interestingly, the ratio between pure mutant and pure wild-type clones, i.e. between correction affecting the wild-type or mutant strand, varied markedly between alleles. The reason for this specificity is not known except that it is not simply a preference for either a purine or a pyrimidine as target of conversion. It would be very interesting to test whether a similar specificity also exists for repair of mutational heterozygotes: if it did, the distribution of alleles of a given mutant gene should differ between complete and mosaic mutants.

Dose-effect curves for the ratio between mosaic and complete mutants

One of the main differences between the predictions of the repair hypothesis and the lethal-hit hypothesis is the shape of the expected dose-effect curve for the proportion of mosaics among all mutants. The lethal-hit hypothesis predicts that this ratio will decrease with dose for, whatever the nature of the lethal hit, higher doses will produce more lethal hits and, concomitantly, transform more mosaics into completes. The repair hypothesis predicts that the proportion of mosaics will be independent of dose or, if repair should become saturated or inhibited at high doses, will increase with dose. Early experiments gave contradictory results. Sometimes, the proportion of mosaics decreased with dose (30,35); sometimes it remained unchanged (39,40); sometimes it increased with dose (41). This was difficult to understand at a time when lethal hits were generally supposed to be the sole or at least the prevalent source of completes. If, instead, one admits that lethal hits as well as repair may be implicated, the dose effect curve will be composite and its overall shape will depend on the relative contributions of these two components. This was indeed found to be so in more recent experiments with fission yeast (23,29).

In order to avoid complications from already replicated genes, Abbondandolo and his collaborators used a pure G1 population and treated it with increasing doses of nitrous acid. Table 19.1 shows that, within the dose range used, the proportion of mosaics among mutants increased steadily and significantly with dose. While this is no proof for the repair hypothesis, it is incompatible with the lethal-hit hypothesis. The frequency of complete mutants per survivors changed very little from the first to the second dose and not at all from the second to the third, while that of mosaic mutants increased steeply from dose to dose; this suggests that saturation rather than inhibition of repair was involved.

At lower levels of survival, the creation of completes by lethal hits would be expected to counteract the increase in the proportion of mosaics, and this was indeed found by the same group of workers in experiments with UV, shown in Table 19.2. It will be seen that the proportion of mosaics reached its peak at a survival level that corresponded to that of the highest dose of nitrous acid in Table 19.1. At still lower survival, it started to decline. Concomitant pedigree analysis of samples from the irradiated population (last column of

Table 19.1

The production of mosaic and complete red *ad*-6 or *ad*-7 mutants by nitrous acid in a pure G1 population of *Schizosaccharomyces pombe*. (After Table III, Abbondandolo (23).

Survival (%)	*No. of expts.*	*No. of colonies scored*	*No. of mutants*		*Mutants per 10^4 survivors*			*Percent mosaics among mutants*
			mosaics	*completes*	*overall*	*mosaics*	*completes*	
84,94	3	9600	37	31	70.8	38.5	32.3	54.4 (I)
44,47	2	8219	93	40	161.8	113.1	48.7	69.9 (II)
18,22	2	4183	115	20	322.7	274.9	47.8	85.2 (III)

II *versus* I $P < 0.05$
III *versus* II $P < 0.01$

Table 19.2

The production of mosaic and complete red mutants by UV in a pure G1 population of *Schizosaccharomyces pombe* (4-9 experiments for each dose). (From Tables II and III in Abbondandolo (29)

UV dose ($ergs/mm^2$)	*Survival range (%)*	*No. of colonies scored*	*Mutation frequency per 10^4 survivors*	*Number*		*Percent mosaics among all mutants*	*Percent pedigrees with lethal sector in first generation*
				mosaics	*completes*		
780	72 – 100	230 001	1.6	8	30	21.0	3.4
2340	30 – 53	115 083	7.2	28	55	33.7	13.9
4160	5 – 21	85 256	13.4	58	56	50.9	36.0
7020	0.5 – 5	105 920	14.3	65	86	43.0	

Table 19.2) showed an increase in the frequency of lethal sectors with dose; extrapolation to the highest dose (for which pedigree analysis proved prohibitively laborious) made it likely that, at this low survival, all completes were due to lethal sectoring.

Mosaics in repair-deficient strains

On the repair hypothesis, it might be expected that the ratio of mosaic to complete mutants would be higher in strains with impaired repair efficiencies than in normal strains. Two considerations make this prediction doubtful. One is the difficulty of knowing which repair enzymes are involved in the conversion of a heteroduplex to a homoduplex; we have seen that, in *Bacillus subtilis*, UV-repair enzymes are not among them (37). Moreover, even in a strain with reduced efficiency for conversion, there will be an increase not only in the frequency of mutational mosaics but also of lethal sectors and, as happens in the dose-dependence of mosaicism, these two effects will partially cancel each other. It is therefore not surprising that tests of repair-deficient strains have not given spectacular results, although some definitely support the repair hypothesis (42-44).

In summary, we may say that lethal hits and repair are the main sources of complete mutants, the former being more effective at high doses, the latter, at low ones. In addition, some completes may arise from mutagenic mechanisms that simultaneously affect both strands of DNA.

References

1. Ratnayake, W.E. (1970), 'Studies on the relationship between induced crossing-over and mutation in *Drosophila melanogaster*', *Mutation Res.* **9**, 71-83.
2. Zimmermann, F.K. (1971), 'Induction of mitotic gene conversion by mutagens', *Mutation Res.* **11**, 327-337.
3. Linnert, G. (1953), 'Der Einfluss von Chemikalien auf Chiasmenbildung und Mutationsauslösung bei *Oenothera*', *Chromosoma* **5**, 428-453.
4. Sobels, F.H. and Steenis, van H. (1957), 'Chemical induction of crossing-over in *Drosophila* males', *Nature* **179**, 29-31.
5. Auerbach, C. and Sonbati, E.M. (1960), 'Sensitivity of the *Drosophila* testis to the mutagenic action of mustard gas', *Zeitschr. indukt. Abst. Vererb. Lehre* **91**, 237-252.

6. Olivieri, G. and Olivieri, A. (1964), 'Evidence for the two-hit nature of X-ray induced crossing-over in the centromeric region of *Drosophila* males', *Mutation Res.* **1**, 279-295.
7. Hannah-Alava, A. (1968), 'Induced crossing-over in the pre-sterile broods of *Drosophila melanogaster', Genetics* **39**, 94-152.
8. Friesen, H. (1936), 'Spermatogoniales crossing-over bei *Drosophila', Zeitschr. indukt. Abst. Vererb. Lehre* **71**, 502-526.
9. Ives, P.T. (1972), 'Residual lethality in induced rearrangements and the chromosomal distribution of male crossovers', *Drosophila Information Service* **49**, 94-95.
10. Alderson, T. (1967), 'Induction of genetically recombinant chromosomes in the absence of induced mutation', *Nature* **215**, 1281-1283.
11. Tessman, I. (1959), 'Mutagenesis in phages ϕX174 and T4 and properties of the genetic material', *Virology* **9**, 375-385.
12. Vielmetter, W. and Wieder, G.M. (1959), 'Mutagene und inaktivierende Wirkung salpetriger Säure auf freie Partikel des Phagen T2', *Z. Naturf.* **14B**, 312-317.
13. Pratt, D. and Stent, G. (1959), 'Mutational heterozygotes in bacteriophage', *Proc. Nat. Acad. Sci. U.S.A.* **45**, 1507-1515.
14. Witkin, E.M. and Sicurella, C. (1964), 'Pure clones of lactose-negative mutants obtained in *Escherichia coli* after treatment with 5-bromouracil', *J. Mol. Biol.* **8**, 610-613.
15. Munson, R.J. and Bridges, B.A. (1964), 'Segregation of radiation-induced mutations in *Escherichia coli', Nature* **203**, 270-272.
16. Kaudewitz, F. (1959), 'Inaktivierende und mutagene Wirkung salpetriger Säure auf Zellen von *Escherichia coli', Z. Naturf.* **14**, 528-537.
17. Kubitschek, H.E. (1964), 'Mutation without segregation in bacteria with reduced dark repair ability', *Proc. Nat. Acad. Sci. U.S.A.* **55**, 269-274.
18. Holliday, R. (1962), 'Mutation and replication in *Ustilago maydis', Genet. Res. Camb.,* **3**, 472-486.
19. Gillham, N.W. and Levine, R.P. (1962), 'Pure mutant clones induced by ultraviolet light in the green alga, *Chlamydomonas Reinhardi', Nature* **194**, 1165-1166.
20. Fjeld, A. (1970), 'Mosaic mutants: absence in a eukaryotic organism', *Science* **168**, 843-844.

21. Reissig, J. (1963), 'Induction of forward mutants in the *pyr*-3 region of *Neurospora*', *J. Gen. Microbiol.* **30**, 317-325.
22. Nasim, A. and Auerbach, C. (1967), 'The origin of complete and mosaic mutants from mutagenic treatment of single cells', *Mutation Res.* **4**, 1-14.
23. Abbondandolo, A. and Bonatti, S. (1970), 'The production, by nitrous acid, of complete and mosaic mutations during defined nuclear stages in cells of *Schizosaccharomyces pombe*', *Mutation Res.* **9**, 59-69.
24. Russo, V.E., Stahl, A.M.M. and Stahl, F.W. (1970), 'On the transfer of information from old to new chains of DNA duplexes in phage λ: destruction of heterozygotes', *Proc. Nat. Acad. Sci. U.S.A.* **65**, 363-365.
25. James, A.P. and Werner, M.M. (1966), 'Radiation-induced lethal sectoring in yeast', *Radiation Res.* **29**, 523-536.
26. Swann, M.M. (1962), 'Gene replication, ultraviolet sensitivity and the cell cycle', *Nature* **193**, 1222-1227.
27. Haefner, K. (1967), 'Concerning the mechanism of ultraviolet mutagenesis. A micromanipulatory pedigree analysis in *Schizosaccharomyces pombe*', *Genetics* **57**, 169-178.
28. Haefner, K. and Striebeck, U. (1967), 'Radiation-induced lethal sectoring in *Escherichia coli B/r* and B_{s-1}', *Mutation Res.* **4**, 399-407.
29. Abbondandolo, A. and Simi, S. (1971), 'Mosaicism and lethal sectoring in G1 cells of *Schizosaccharomyces pombe*', *Mutation Res.* **12**, 143-150.
30. Guglielminetti, R. (1968), 'The role of lethal sectoring in the origin of complete mutations in *Schizosaccharomyces pombe*', *Mutation Res.* **5**, 225-229.
31. Nasim, A. and James, A.P. (1970), 'Inactivation and the induction of lethal sectoring by chemical mutagens in yeast', *Canad. J. Genet. Cytol.* **12**, 10-14.
32. Korn, R.W. (1970), 'Induction and inheritance of morphological mutations in *Cosmarium turpinii. Brel.*', *Genetics* **65**, 41-49.
33. Bautz-Freese, E. and Freese, E. (1966), 'Induction of pure mutant clones by repair of inactivating DNA alterations in phage T4', *Genetics* **54**, 1055-1067.
34. Drake, J.W. (1966), 'Ultraviolet mutagenesis in bacteriophage T4. I. Irradiation of extracellular phage particles', *J. Bacter.* **91**, 1775-1780.

35. Freese, E., Bautz-Freese, E. and Bautz, E. (1961), 'Hydroxylamine as a mutagenic and inactivating agent', *J. Mol. Biol.* **3**, 133-143.
36. Bresler, S.E., Kreneva, R.A., and Kushev, V.V. (1968), 'Correction of molecular heterozygotes in the course of transformation', *Molec. Gen. Genetics* **102**, 257-268.
37. Spatz, H. Ch. and Trautner, T.A. (1970), 'One way to do experiments on gene conversion?', *Molec. Gen. Genetics* **109**, 84-106.
38. Schlaeger, E.J. and Spatz, H.Ch. (1974), 'The extent of repair synthesis in SPPI transfection of *B. subtilis', Molec. Gen. Genetics* **130**, 165-175.
39. Schuster, H. and Vielmetter, W. (1961), 'Studies on the inactivating and mutagenic effects of nitrous acid and hydroxylamine on viruses', *J. Chim. Phys.* **58**, 1005-1010.
40. Nasim, A. and Clarke, C.H. (1965), 'Nitrous acid-induced mosaicism in *Schizosaccharomyces pombe', Mutation Res.* **2**, 395-402.
41. Tazima, Y. and Onimaru, K. (1969), 'Frequency pattern of mosaic and whole-body mutants induced by ionizing radiations in post-meiotic cells of the male silkworm', *Mutation Res.* **8**, 177-190.
42. Nasim, A. (1968), 'Repair mechanisms and radiation-induced mutations in fission yeast', *Genetics* **59**, 327-333.
43. Resnick, M.A. (1969), 'Induction of mutations in *Saccharomyces cerevisiae* by ultraviolet light', *Mutation Res.* **7**, 315-332.
44. Kaplan, R.W. and Stoye, H. (1970), 'Influence of host-cell reactivation and UV-dose on formation of pure and mixed clones of *c* mutants in phage κ', *Mutation Res.* **10**, 257-267.

CHAPTER TWENTY

Mutagen specificity

In the narrow sense, this term has been applied to the types of molecular change that different mutagens produce in DNA (1). In its widest sense, the term covers all cases in which mutagens vary from each other or from spontaneous mutability in the proportions of the effects they produce, e.g. in the ratios between dominant lethals and translocations, between chromosome breaks in different chromosomes or chromosomal regions, between deletions and gene mutations, between true revertants and suppressors, between forward mutations at different loci or to different alleles at the same locus, or between reverse mutations of genes in the same cell. Mutagen specificity in the narrow sense has been dealt with in previous chapters in relation to the most important mutagens. It is, however, well to bear in mind that there are only very few systems in which specificity at this level can be analysed reliably, and that extrapolation to other systems are fraught with difficulties; we have seen examples of this. In this chapter I am going to deal with specificity in the widest sense. We shall see that only a minority of observed specificities can be attributed unambiguously to reactions between mutagens and DNA, and that often this explanation can be excluded. I shall restrict myself to examples of specificity for gene mutations; more examples will be found in the review by Auerbach and Kilbey (Bibliography). Examples for specificities involving chromosome breakage will be found in the same review and in previous chapters of this book.

Early results

From the very beginning of modern genetics, mutagen specificity has been an intriguing possibility. Already de Vries (Chapter 1) had dreamt of the induction of 'directed mutations' that would provide man with 'unlimited power over nature' and this dream has been a major spur in the search for chemical mutagens. Fulfilment still eludes us and probably will always elude us; but partial success is possible and some steps towards it have already been made. With these we shall deal in the last chapter. In the present chapter we shall discuss possible causes of mutagen specificity, and consider examples in which one or more of them are likely, or proven, to be implicated. I shall try to show that the analysis of mutagen specificity in the wider sense is a prime tool for studying mutagenesis as a process that includes, but is not restricted to, the primary reaction between mutagen and DNA and that, in addition, it may provide information on cellular processes such as repair, translation, or enzyme formation that transcend the interests of mutation research.

The first clear cases of mutagen specificity were reported in 1953. Several of them had occurred in bacteria, in which auxotrophic genes in the same cell reverted in different proportions depending on the mutagen used (2). An example is shown in Table 20.1; particularly striking is the fact that the ratio between leucine- and phenylalanine-revertants was between 50 and 60 after treatment with $MnCl_2$, but less than 1 after treatment with UV.

The second organism to give evidence of mutagen specificity was a strain of *Neurospora,* in which the ratio between adenine- and inositol-reversions depended markedly on the mutagen (3). I shall deal more fully with this case at the end of the chapter.

Historically, it was ironical that both these reports appeared in the same year in which the discovery that all genes are made up of the same four nucleotides rendered it very unlikely that any mutagen should single out any gene for preferential action – an idea that had appeared much less far-fetched when the essential part of the gene was thought of as a complex protein molecule. What saved the situation for the moment was the consideration that one was dealing with *reversions* of auxotrophic defects, each presumably due to a specific kind of lesion in DNA and revertible by a specific kind of molecular event. When differences between the effects of mutagens on DNA became established, it seemed easy to consider

Table 20.1

Frequencies of mutations from leucine- and phenylalanine-deficiencies in a doubly mutant strain of *Escherichia coli* after treatment with three mutagens. (After Table 3, Demerec (2).)

Mutation	*Frequency per 10^8*		
	$MnCl_2$	*UV*	*β-propiolactone*
$leu^- \rightarrow leu^+$	594	57	28
$ph\text{-}al^- \rightarrow ph\text{-}al^+$	11	100	3

these as the cause of the observed mutagen specificities.

Yet already at that time evidence was available which showed that this could not be the whole explanation. Thus, in the experiments on *Neurospora*, diepoxybutane had yielded about 40 times as many adenine- as inositol-reversions at a low dose, but 500 times as many at the highest. It was difficult to account for this on purely chemical grounds. Even more difficult to explain was the finding that, in *E.coli*, a gene that was wholly recalcitrant to the action of diepoxybutane became highly mutable by the same treatment after having been introduced into an arginine-requiring strain (4). A little later, Witkin (5) made a similar observation: in a strain of *E.coli* that had become streptomycin-dependent, UV completely ceased to produce mutations from tryptophan-auxotrophy to prototrophy. In bacteriophage, the existence of 'hot spots' in the rII gene created difficulties of interpretation (6). Other instances of mutagen specificity for forward mutations had already been reported *prior* to 1953 and continued to turn up. Since there seemed no reasonable way to account for them in the framework of the Watson-Crick model, they were not followed up and were, indeed, almost forgotten. Yet, some were striking. Thus, among auxotrophs produced in the fungus *Ophiostoma multiannulatum*, there were less than 5% inositol-requirers after UV-irradiation, but more than 40% after treatment with purine derivatives (7). In bacteria growing in the chemostat, purine derivatives (Chapter 17) produced more resistance mutations to phage T5 than to phage T6, while X-rays and UV did the opposite (8). Recently, it has been reported that radiation on the one hand, chemicals on the other, produce different ratios of forward mutations in the *ad*3A and *ad*3B loci of *Neurospora* (1). Even closely related mutagens, which presumably act by very similar

Table 20.2
Relative efficiency of UV and three organic peroxides in the production of different mutations in *E.coli*. (After Chevallier (9).)

Mutagen	*Mutation scored*			
	sm-s → *sm-r*	T_1*-s* → T_1*-r*	*pro*$^-$ → *pro*$^+$	*try*$^-$ → *try*$^+$
UV	++	++	++	+
Succinic peroxide	(+)	0	++	(+)
Cumene peroxide	0	+	0	+
Thymine peroxide	–	–	+	+

sm-s, sm-r sensitive or resistant to streptomycin
T_1-s, T_1-r sensitive or resitant to phage T_1
pro proline
try tryptophane
(+) weakly mutagenic
0 not mutagenic
– not tested

mechanisms, may yield specific effects on forward and reverse mutations, as shown in Table 20.2 (9).

The mutagenic pathway (Fig. 20.1)

Chevallier, from whose work these data are taken, was one of the first to stress that such specificities do not necessarily arise from the reactions between mutagen and DNA but may be due to treatment effects on the fixation or expression of potential mutations. It is now clear that this is often true. Many cases of mutagen specificity arise at later levels in the mutagenic pathway. As a guideline to discussion, Fig. 20.1 shows a simplified model of the mutation process. It is useful to consider the primary changes in DNA as a population of potential mutations and the following steps in the pathway as sieves that allow only a proportion of them to proceed towards the final product of a mutant clone. Usually only a small fraction of the initial changes reaches this stage. How many and which types manage to do so may depend on the very same mutagenic treatment that produced them. It should always be kept in mind that most mutagens do not single out DNA for their action. In the spermatozoa of mice that had been injected with mutagenically

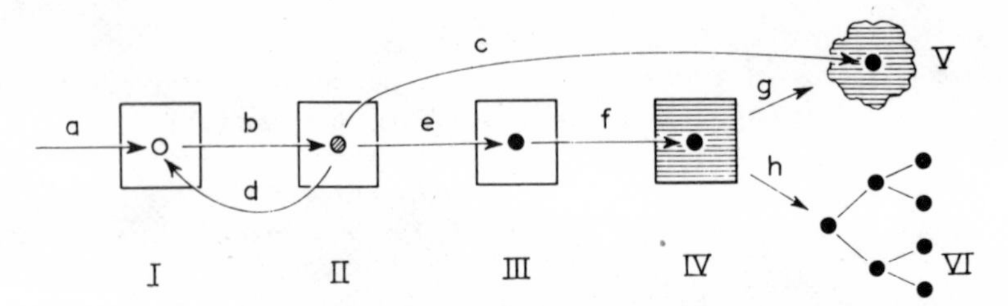

Fig. 20.1. Mutagenic pathway for induced mutations. I. Non-mutated gene in non-mutated cell. II. Gene with premutational lesion. III. Mutated gene in non-mutated cell. IV. Mutated gene in mutated cell. V. Dead mutant cell. VI. Mutant clone (only genes are shown). *a*. Penetration of mutagen to DNA of gene. *b*. Production of a premutational lesion. *c*. Death from unrepaired damage. *d*. Repair of lesion, restoring normal gene. *e*. Fixation of premutational lesion as mutated gene. *f*. Expression: formation of mutant cell. *g*. Death of mutant cell. *h*. Formation of clone of mutant cells.

effective doses of ethylmethane-sulphonate, at least 82% of the ethylations had occurred outside DNA (10).

Penetration (a)

Penetration of the mutagen to DNA involves more than entry into the cell. A mutagen may be trapped or modified by extrachromosomal elements. In eukaryotes, it has to penetrate the chromosome-associated protein and, in resting cells, the nuclear membrane. Kihlman (11) has shown that, among a large array of purine derivatives, only those able to penetrate the perinuclear lipid membrane produced chromosome breaks in non-dividing cells. Without tests on dividing cells, the chromosome breaking ability of these substances would have been missed. On the whole, it is unlikely that many gene-specific effects arise at this step in mutagenesis. There might, however, be regional differences of chromosome structure that make penetration to certain genes easier than to others. Conceivably, such differences in accessibility might account for the specificities among mutations produced in fungi by amino acid analogues and fluorouracil (Chapter 18).

A peculiar type of cell-cycle-related specificity may be created if a mutagen acts preferentially on genes that have opened up for replication. We have seen that this applies to nitrosoguanidine (Chapter 16) and formaldehyde (Chapter 18). By treating bacteria at different times during the replication cycle, mutations can be

Table 20.3

The effect of incubation time on the type of mutant produced in aerosols of semi-dried *Escherichia coli* cells (strain HfrH) by 320-400 nm radiation. (Table 6 in Webb (12). Courtesy Pergamon Press.)

Incubation time (min)	*Number of colonies isolated (1) and independent experiments (2)*		*Number of mutants found*	*Mutant characters and their number ()*		*Conjugation transfer time of the genes mutated (min)**
0	(1)	(2)				
	654	(11)	413	Threo⁻	(402)	0
				Threo-Leu⁻	(9)	0
				Serine⁻	(2)	1,57
10	743	(11)	702	Lac⁻	(506)	10
				Pro⁻	(42)	7-10
				Arg⁻	(21)	–
				Mucoid	(133)	11
20	604	(6)	460	Tryp⁻	(316)	26,65
				Glut⁻	(84)	20
				Cysteine⁻	(43)	25,65
				Unknown	(17)	–
30	622	(8)	419	Hist⁻	(419)	39
40	503	(9)	406	Gly⁻	(102)	49
				Pur⁻	(24)	–
				Unknown	(208)	–
50	417	(8)	208	Lys⁻	(94)	55
				Lys-Pur⁻	(63)	–
				Serine⁻	(44)	1,57
				Arg⁻	(7)	–
60	686	(12)	440	Str	(406)	64
				Cysteine⁻	(12)	25,65
				Tryp⁻	(8)	26,65
				Unknown	(14)	–
70	714	(7)	385	Met⁻	(322)	75-78
				Isoleu-Val⁻	(12)	74
				Unknown	(51)	–

*Taken from Hayes [19] and Taylor and Trotter [30]. Where two times are mentioned they represent the times at which different genes involved in the metabolism of the same metabolite are transferred.

induced in those genes that are replicating at the time. A similar situation has been found for mutations induced by 'black light' (320-400nm) in aerosols of semi-dried bacteria (12). When synchronized cells were dried and irradiated at different times during the replication cycle, the shifting spectrum of auxotrophs could be correlated with the order of the genes on the chromosome as known from conjugation studies (Table 20.3). As a tentative explanation, the author suggests that, in semi-dried cells, a complex is formed between the membrane, the replication point of DNA and a component of the cytochrome chain that absorbs black light.

Brock (13) assumed that the preferential response of replicating genes to certain mutagens is due to the strand separation which exposes the base to chemical attack, and he reasoned that the same should apply to genes that have opened up for transcription. He tested this by comparing the reversion rates of point mutations in the β-galactosidase gene of *Escherichia coli* under conditions when the operon was induced or non-induced. He found, indeed, that various alkylating agents, but not base analogues or X-rays, produced higher mutation frequencies in induced than in non-induced genes (Table 20.4).

An intriguing correlation of the opposite kind has been reported by Webb (14). He grew *E.coli* in minimal medium or in media that had been supplemented with a variety of growth factors. After 24 hours, cells from these various media were harvested, washed and semi-dried in an atomizer to 40% relative humidity. At various intervals after atomization, the cells were exposed to ultraviolet light and plated for estimates of survival and mutation frequency to auxotrophy. Few mutants occurred in cells grown in minimal medium, but when the medium had been enriched with amino acids, nucleotide bases, or vitamins, more mutations appeared and these were preferentially auxotrophs for that class of chemical that had been added as supplement to the growth medium. Webb attributes these specificities to differences in the amount and types of DNA-associated proteins in the semi-dried cells. Alternatively, one might suppose that messenger-RNA had remained attached to the derepressed genes in the semi-dried cells and had protected them from the action of UV. The repressed genes, i.e those whose action had been made redundant by supplements, would then show a preferential response to irradiation.

Table 20.4

Frequency of *lac*$^+$ revertants of mutant U120 by 0.0435M DES (1h at 25^0) and 100μg/ml MNG, 10^{-2}M AP and 10^{-2}M BUdR when applied for 1h at 37^0 to induced and non-induced cells. Mutagen removed by washing. Cells plated to 0.2% lactose agar. (Table III, Brock (13). Courtesy Elsevier Scientif. Publ. Cy.)

Mutagen	*Lac revertants per 10^6 viable cells*	
	induced	*non-induced*
DES	3.13 ± 0.29	0.42 ± 0.07
MNG	1.45 ± 0.18	0.43 ± 0.08
AP	0.12 ± 0.07	0.07 ± 0.05
BUdR	0.13 ± 0.04	0.11 ± 0.13
Control	0.08 ± 0.03	0.07 ± 0.03

± standard error

Reaction with DNA (b)

This is the stage at which mutagen specificity in the narrow sense is created by the reactions between mutagens and DNA. Account has to be taken of the fact that a mutagen may act in different ways under different experimental conditions. We have discussed this for caffeine and hydroxylamine in Chapter 17, and for formaldehyde in Chapter 18. It is likely to apply to many other mutagens. Possibly the finding that, in *Aspergillus* (15) and bacteriophage (16), ionizing radiation produced different mutational spectra in the presence and absence of oxygen, can be explained along these lines.

Superimposed on these primary specificities are others of a higher order, which presumably reflect differences in the nucleotide sequences surrounding the site of primary damage. A case in question is the origin of frameshifts. These seem to arise preferentially in runs of identical bases (17) through mispairing followed by faulty replication (18). (Fig. 21.1). Acridines might promote frameshift mutagenesis by stabilizing the mispaired regions, thus increasing the time during which faulty resynthesis can occur (Chapter 17). (For a fuller discussion see Drake, The molecular basis of mutation; (Bibliography).

In the production of base changes, 'neighbourhood' effects from adjoining nucleotides have been shown to play a role. Fig. 17.3, e.g., illustrates the influence of neighbouring bases on the efficiency with which 2-aminopurine produces the transition from AT to GC. In Chapters 14 and 18 we have seen that the position of a base within the *iso*-1-cytochrome *c* gene of yeast determines its response to UV and to nitrous acid. It is difficult to see how the purely chemical reactions between DNA and mutagen can be controlled to such a striking degree by the nucleotide sequences surrounding the reacting base. The realization – to be discussed more fully in the next chapter – that DNA polymerase takes an active part in replication, makes it easier to account for these specificities; for recognition of specific sites in the substrate is a property of all enzymes. An example of the mutagen specificity that may arise from this cause has been described for phage T4; a mutant DNA-polymerase strongly suppresses mutations induced by base analogues, moderately suppresses mutagenesis by EMS at GC sites and by nitrous acid at AT sites, and has no effect on mutagenesis by hydroxylamine, or by nitrous acid at GC sites (19). The authors of this paper come to the conclusion that 'mutagenic specificities of chemical agents depend as much upon the characteristics of the enzymatic apparatus of DNA replication as they do upon the chemistry of primary mutational lesions'. I should like to extend this statement to include later steps in mutagenesis as equally important for the creation of mutagen specificity. My reasons for doing this will be given in the remaining sections of this chapter.

Repair and its alternatives (c,d,e)

A lesion in DNA opens up three divergent paths: (1) death from unrepaired damage such as a double-strand break; (2) restoration of the original state through error-proof repair, e.g. photorepair, with concomitant loss of a potential mutation; (3) fixation of a premutational lesion through replication, recombination, or error-prone repair. Specificities may arise when the relative frequencies with which these pathways are taken differ between sites, genes and mutagens.

In *E.coli*, mutation frequency decline (Chapter 14) creates specificity through an error-proof repair mechanism which removes suppressor mutations but not true reversions. Witkin in her article in Annual Review of Genetics (Bibliography) suggests that conditions promoting MFD trigger repression of the *t*-RNA genes that are the source of suppressor mutations, and that repressed genes become more accessible to excision repair.

In other cases, error-prone repair may be a source of specificity. It is probable that the site specificities within the *iso*-1-cytochrome *c* gene, mentioned in the previous section, arise during repair rather than during replication. They persisted unchanged in an excision-deficient strain, but were greatly diminished in a strain that probably lacks an error-prone repair system (20). In a strain of *Aspergillus nidulans* judged to be deficient in an error-prone repair mechanism, UV induced much higher ratios of methionine- to adenine-reversions than in the parent strain (21).

Photorepair, too, may create specificity. We have seen in Chapter 13 (Fig. 13.5) that, in *Neurospora*, one particular allele at the inositol-locus responds very little to photorepair. The reason for this is not known. Most probably, it is some peculiarity of the structure of DNA or of the nucleotide sequence surrounding this particular site. Harm and Rupert (22) found differences of photoreparability between sites in transforming DNA and attribute them to neighbourhood effects on the stability of the complexes formed between the photoreactivating enzyme and its substrate, the cyclobutane dimer (Chapter 13). Whatever its cause, this recalcitrance of the inositol-allele to photorepair leads to an increase, by photorepair, in the ratio between inositol- and adenine-reversions in the same cell; in other words, photorepair makes UV more specific for the production of inositol-reversions.

Expression (*f*)

This term covers a wide variety of processes, depending on the type of mutation concerned. Most mutations from prototrophy to auxotrophy are expressed as soon as the cell has lost a substance produced by the normal gene; starvation before exposure may shorten the time required for expression. Some auxotrophs require loss or change of a structural component of the normal cells; thus inositol-dependence affects membrane structure. Conversely, revertants to inositol-dependence have to acquire normal membrane structure for expression. For most revertants from auxotrophy to prototrophy, expression means synthesis of a new protein, according to the well known sequence of transcription, translation, polypeptide formation, and assembly of polypeptides into a functional enzyme. This is true also for many suppressor mutations. Super-suppressors, however, require formation of a new transfer-RNA. The expression of resistance may depend on synthesis of a new membrane structure, on formation or loss of a permease, on the production of a new enzyme able to deal with the noxious agent, or on the production of a new kind of ribosome. If we keep in mind that most of these cellular processes are affected by the mutagen used, and that often these effects are selective (see Chapter 12 for UV, Chapters 16, 17, 18 for chemical mutagens), we are led to expect specificities to arise at this stage. Indeed, there are by now a fair number of cases where this is true. Unfortunately, biochemical analysis has lagged far behind genetical research, and for none of these cases has the underlying chain of events been laid bare. Yet, it seems likely that their analysis would not only contribute to a better understanding of the mutational process but that it also would be a means for spotting concatenations between biochemical events that are not readily detected and analysed outside mutation studies. Some examples will make this clear.

Effect of plating medium

It is well known, and has been mentioned in Chapter 11, that it is often necessary to add a small amount of the required supplement to the medium on which revertants to prototrophy are scored. Not so well known is the possibility that, in multiply auxotrophic strains, an extra large amount of one of the other supplements may be required to give late appearing revertants the opportunity of competing successfully with background growth of the parent strain. Neglect of this precaution may lead to the spurious appearance of mutational stability. A strain of *Bacillus subtilis* that requires threonine and histidine, yielded no histidine-revertants unless the amount of threonine in the plating medium had been greatly increased; once expressed, histidine revertants grew well on the normal amount of threonine (23).

In fission yeast, a striking case of mutagen specificity was found to be due to an inhibitory effect of the plating medium on a particular type of reversion (24). In a strain that was auxotrophic for methionine and adenine, nitrous acid produced about half as many adenine- as methionine-reversions, while UV produced many methionine-reversions but hardly any adenine-reversions (Table 20.5).

It would have been tempting to attribute this specificity to the different ways in which UV and nitrous acid react with the lesions in DNA that characterize the two auxotrophs. Analysis established that this was not so. Fig. 20.2 shows what happened when methionine

Table 20.5

Mutagen specificity in fission yeast. Adenine-and methionine-reversions in a doubly auxotrophic strain. (Data from Clarke (24).)

Mutagen	*Survival in % of controls*	*Mutations per 10^7 survivors*		*Ratio*
		ad^+	met^+	met^+/ad^+
Nitrous acid	92	6.0	12.2	2
UV	86	0.3	99	330

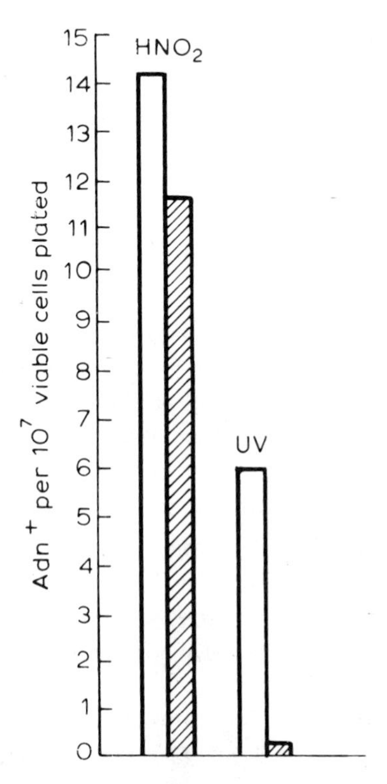

Fig. 20.2. The influence of methionine in the plating medium on the production of adenine-reversions in an *ad*⁻ strain of fission yeast, treated with nitrous acid or UV. Unshaded columns: minimal medium. Shaded columns: minimal medium plus 40 μg/ml methionine. (Fig. 1, Clarke (24). Courtesy of Springer-Verlag)

was added gratuitously to a strain that required only adenine. There was a slight reduction in the frequency of nitrous acid-induced adenine-revertants; after UV, hardly any revertants appeared. Thus, methionine in the plating medium inhibits the expression of adenine-reversions. Once expression has occurred, methionine is without effect. A hypothesis to account for the inhibition is that the formation of S-adenosylmethionine from methionine and ATP may inhibit the protein synthesis that is required for expression (25). Whatever the cause of the methionine inhibition, it obviously acted much more strongly in UV-irradiated cells than in those treated with nitrous acid, and this produced the mutagen specificity. Somewhat surprisingly, different alleles of the adenine-locus responded quite differently to methionine inhibition after the two types of treatment (26); in doubly auxotrophic strains, this must result in allele-specific mutagen effects. These are usually taken as evidence that specificity occurs at the level of reaction between DNA and mutagens (Path *b* in our diagram); it is well to keep in mind that this is not necessarily true.

Specificity for forward mutations may also be created by the plating medium. In the ascomycete *Ophiostoma multiannulatum,* Zetterberg (27) found no histidine-requirers among thousands of UV-induced auxotrophs; yet, nine such occurred among the first 65 auxotrophs that had been induced by nitrosomethylurethane (NMU). Realizing that the complete medium contains components which may inhibit growth of the revertants (Chapter 11), Zetterberg plated the treated cells on minimal medium that had been supplemented only with histidine. Under these conditions, UV was as effective as NMU in the production of histidine-requirers. It is, however, not clear whether this specificity arose at the expression stage or later during colony formation (Paths *g* or *h*): established histidine prototrophs seemed to grow more poorly on complete medium when they had been induced by UV than when they had been induced by NMU.

This is not true in a case of mutagen specificity that has already been mentioned in Chapter 18 (28). After treatment of *Penicillium chrysogenum* with nitrogen mustard, mutations to 8-azaguanine resistance were inhibited by $MnCl_2$ in the medium, while mutations to aza-indole resistance were unaffected. Radiation-induced mutations to 8-azaguanine resistance were not subject to inhibition by $MnCl_2$. Thus, on $MnCl_2$-supplemented medium, UV would produce both types of resistance mutations, while nitrogen mustard would be specific for mutations to aza-indole resistance. When nitrogen mustard-treated cells were kept in complete medium before plating, the response to $MnCl_2$ gradually disappeared. Whether it occurred at the level of repair or expression could not be decided; its restriction to one type of mutation suggests the latter.

Most of these specific plating medium effects have not been analysed at the biochemical level. For one of them, a plausible model has been offered (29). Revertants were scored in a strain of yeast that carried ochre mutations in five different genes, including one in the histidine pathway and one in the adenine pathway. Spontaneous and UV-induced mutations were scored on minimal medium and on 30 media that were supplemented with from one to four of the required nutrilites. Theoretically, reversion frequencies should be the same on all these media; in practice, there were large differences. The most striking of these was the complete absence of suppressor mutations – but not of true revertants – on all media that contained histidine but lacked adenine. A hypothesis that accounts satisfactorily for all features of these experiments was based on the interaction of adenine and histidine with the pathway leading to the formation of a new *t*-RNA; it would take us too far to discuss it in detail.

Residual genotype

The cell is the internal plating medium for its genes. It is therefore not surprising that a change in genotype may create mutational specificities which resemble those due to a change in external plating medium. In fact, since a changed genotype often requires a changed plating medium, it may not be known which of the two is responsible. This applies, e.g., to the cases described on p. 380, where mutability of a resident gene was profoundly changed by introduction of a second mutant gene into the strain. In some recent cases, however, the genotype was clearly shown to be responsible. In *E. coli*, a normally revertible gene for tryptophan dependence (*tryp*$^-$) completely ceased to mutate when an adenine-requirement arose in the strain. Yet, the strain had not become generally stable towards mutation, for reversions to adenine-independence occurred at the normal frequency. In all revertants, the *tryp*$^-$ gene had regained its original mutability, and this was not affected by the addition of gratuitous adenine to the plating medium (30). In yeast (31), three different alleles to tryptophan requirement were shown to suppress mutations to 'rough' phenotype; here, too, it could be shown that it was the *tryp*$^-$ genotype and not the required tryptophan supplement in the medium that produced the change in

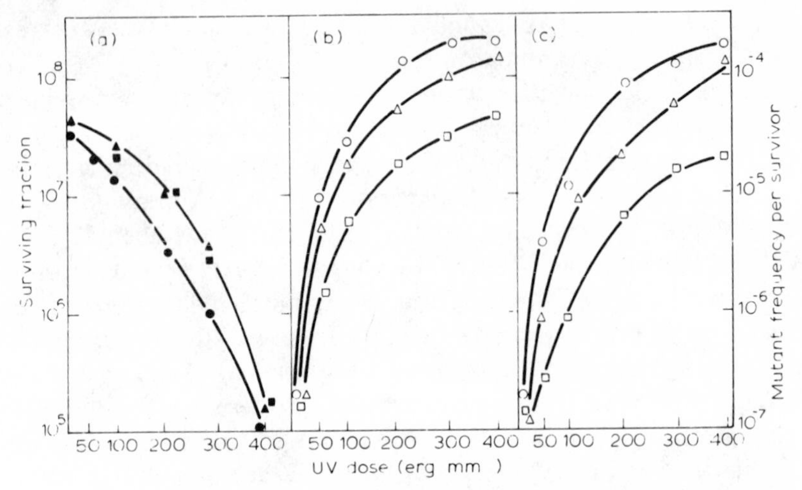

Fig. 20.3. UV-induced reversions to prototrophy in a streptomycin-sensitive and a streptomycin-resistant strain of *E.coli*; the strains were isogenic apart from their response to streptomycin. (*a*) survival; (*b*) reversion to tryptophan-independence; (*c*) reversion to threonine independence. □ strep^r strain ○ strep^s strain △ strep^r strain on medium containing streptomycin (Fig. 1, in Skavronskaya (33). Courtesy of Elsevier Publications.

mutability. In *Neurospora* (32), a striking change in mutagen specificity was produced by the cytoplasmic background (see p. 398).

In none of these cases has it been shown which stage of the expression process is influenced by the residual genotype. There are, however, a few instances in which the stage can be identified with fair assurance as translation. In Chapter 4, we have discussed the rôle played by ribosomes in the fidelity of translation and, through this, in the efficiency of supersuppression. We have seen that ribosomal function can be modified phenotypically by antibiotics and genotypically by mutations to antibiotic resistance. Results that fit into this framework have been obtained in two laboratories (33,34). Fig. 20.3 shows data from experiments in which the frequencies of UV-induced tryptophan- and threonine-reversions were measured in two strains of *E.coli* that were isogenic apart from a gene for streptomycin-resistance carried by one of them. It will be seen that both reversions occurred much more frequently in the streptomycin-sensitive than in the streptomycin-resistant strain. Addition of streptomycin to the plating medium of the resistant strain raised mutation frequencies to intermediate levels; the effect was the same whether streptomycin was present in the medium from the start or was added up to 24 hours later. In addition, the mutational spectra differed between the two strains. The ratio of suppressors to true revertants was higher in the sensitive than in the resistant strain, and there were differences in the proportion of mutations that suppressed both amber and ochre mutations.

Clone formation (g,h)

Once a mutant gene has achieved expression, its fate depends on the kind of mutation, the type of cell in which it occurred, and the conditions of the experiments. In diploid cells, few mutants will follow path *g* that leads to death; in haploid cells, many will do so. The plating medium often plays a decisive rôle. A whole class of previously unknown arginine-requirers was detected in *Chlamydomonas* when the ammonium-ion to which these mutants are sensitive was omitted from the medium for the scoring of auxotrophs (35). Sometimes, pleiotropic effects of a mutation create new requirements, so that mutants are missed when scoring is not carried out on complete medium. In the bacterium *Proteus mirabilis,* a majority of spontaneously arisen mutants from streptomycin-dependence to streptomycin-independence had become auxotrophic, mainly for isoleucine and valine (36). Conversely, in *E.coli* isoleucine-requirers were found among mutants from streptomycin-sensitivity to streptomycin-resistance or –dependence (37). In *Serratia marcescens,* some suppressor mutations created auxotrophy for various new requirements (38). Probably, in all these cases the primary mutation consisted of a change in the translation machinery, which secondarily produced the new requirement. It is an intriguing problem why, in two different bacterial species, mutations concerned with response to streptomycin preferentially produced a requirement for isoleucine.

Success or failure of a mutant during clone formation would not be expected to depend on the kind of treatment by which it had been produced. Where this seems to be the case, it is likely that selection at this late stage does not *create* specificity but *reveals* a specificity that arose earlier in the mutagenic pathway. If, as mentioned on p.390, mutagen specificity for histidine auxotrophs in *Ophiostoma* did indeed arise at the stage of clone formation, then the cause of this specificity must lie at some earlier stage, possibly as far back as the reaction of the mutagens with DNA. In the experiments on *Proteus* (36), mutants to streptomycin-independence contained a much smaller proportion of auxotrophs when they had been induced by $MnCl_2$ than when they had arisen spontaneously, The effect of Mn^{++} ions on translation suggests that this specificity originated at the translation stage.

The situation is quite different when expression is completed already during treatment, for then the treatment itself may select for or against mutants. In micro-organisms, tests for this source of specificity can be made through reconstruction experiments (Chapter 11). In higher organisms, where the mutagen usually cannot be removed after treatment, reconstruction experiments are not feasible, and observed mutagen specificities might well be due to selection for or against mutant cells by the same chemical that had produced the mutations. This is especially likely when – as in tests for sex-linked mutations in male germ cells of *Drosophila* – the mutant genes are effectively haploid, so that any recessive pleiotropic action on germ cell development can find full expression.

A striking example occurred in experiments on *Drosophila* (39), in which visible sex-linked mutations were scored among the progeny of males that had been given chemically treated food as larvae. Analysis has so far been restricted to two mutagens, hydroxylamine (HA) and hydrazine (HZ), and to two mutations, vermilion (*v*) and miniature (*m*) (40). At doses yielding similar frequencies of sex-linked lethals, both mutagens produced similar frequencies of *v*. In addition, HA yielded approximately as many *m* as *v* mutants. In contrast, not a single *m* mutant was found among 37 000 HZ-treated chromosomes that had yielded 90 *v* mutants. It is *a priori* extremely unlikely that recalcitrance of a whole locus to one particular mutagen should be due to the selective action of this mutagen on nucleotides. It is true that Lifschytz and Falk (41) have put forward an attractive hypothesis for locus specificities of mutagens. They assume that genes differ in the number of 'essential sites' (Chapter 3, p. 40) in their code message. Mutagens that produce base changes will yield more mutations in genes with many than in genes with few essential sites; mutagens that produce frameshifts or deletions will be less selective. This hypothesis, however, does not apply to the case in question. Both HA and HZ produce mainly or exclusively base changes (Chapter 17). They appear to do this also in *Drosophila*, where the ratio of lethals to visibles was very low, and no deletions were found among visibles scored as heterozygotes. The argument against attributing mutagen specificity to the reaction between HZ and DNA was clinched when it was found that HZ does, in fact, produce *m* mutations in cells in which the locus is effectively diploid, i.e. in females, and in males carrying a duplication for this region of the chromosome. Evidently, by some pleiotropic effect of the *m* gene on germ cells, the development of *m* spermatogonia into spermatozoa is inhibited by an environment that has been treated with HZ. Inhibition results in delay of development rather than in death; for a few *m* mutants were found when sampling of offspring was continued until several weeks after treatment.

In summary, we conclude that mutagen specificity may arise at any one of the steps in the mutagenic pathway, and that it may be very difficult if not impossible to identify the step or steps involved by presently available techniques. A sustained attempt to do so has been made for one of the earliest detected cases of mutagen specificity (3). I shall finish the chapter by briefly summarizing the results that have so far been obtained; they provide a good illustration of the complex processes underlying a clear case of mutagen specificity which, without further analysis, might have been attributed to the primary reactions between mutagens and DNA.

Analysis of a case of mutagen specificity in *Neurospora*

In a doubly auxotrophic strain of *Neurospora, ad*3A 38701 *inos* 37401, the ratio between adenine-and inositol-reversion was found to depend strongly on the mutagen. In particular, diepoxybutane (DEB) produced a striking preponderance of adenine- over inositol-reversions, while UV produced from two to three times as many inositol- as adenine-reversions. Meanwhile, other mutagens have been tested in this strain, and all have been found to act specifically on one or the other gene. Here we shall limit ourselves to a consideration of the first discovered specificities, which have been analysed in greater depth than the others. These specificities turned up at a time when the Watson-Crick model of the gene was not yet established. Nevertheless, speculations about their origin centred at once around the primary reactions between mutagens and gene. The molecular approach to mutagenesis, by showing differences between the primary reactions of UV and of alkylating agents, gave substance to these speculations, and the *Neurospora* case has often

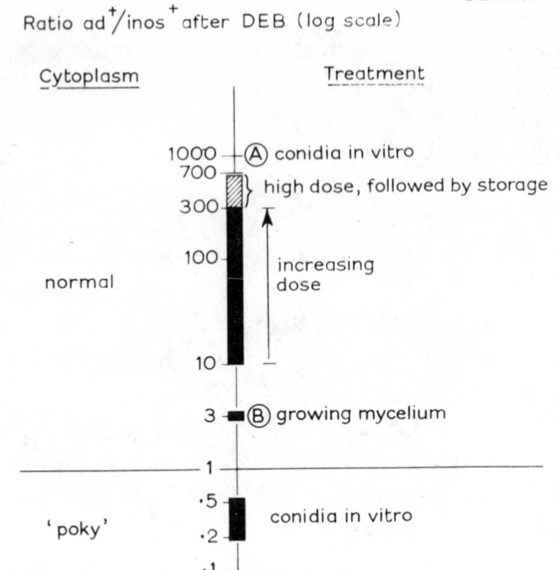

Fig. 20.4. Specificity of DEB for reversions in the *ad*38701 *inos*37401 strain of *Neurospora* (log scale). Specificity is measured as the ratio between adenine- and inositol-reversions ($ad^+/inos^+$). The upper part of the diagram is based on data by Auerbach and Ramsay (for references see Auerbach in 'Mutation as Cellular Process', Bibliography); the lower part is based on data by Paterson (32).

been quoted in evidence of the role of DNA lesions in mutagen specificity. Yet, analysis showed that cellular processes at the repair and expression level can account for a large part of the specificity and may account for all of it. Some of the results have already been discussed in Chapter 16; they will be mentioned here only summarily. The relevant references in Chapter 16 are 72-80; literature up to 1969 is found in the articles by Kilbey and by Auerbach in 'Mutation as cellular process' (Bibliography).

The common feature of all experiments was a search for conditions that modify specificities, followed by attempts to locate the step in the mutagenic pathway at which the modification took place. Figs. 20.4 and 20.5 summarize the main findings for DEB and UV.

DEB

Specificity is measured by the ratio of adenine- to inositol-reversions. Values above 1 represent adenine-specificity; values below 1, inositol-specificity. The ratio varies over almost three orders of magnitude; the diagram has therefore been drawn to a logarithmic scale.

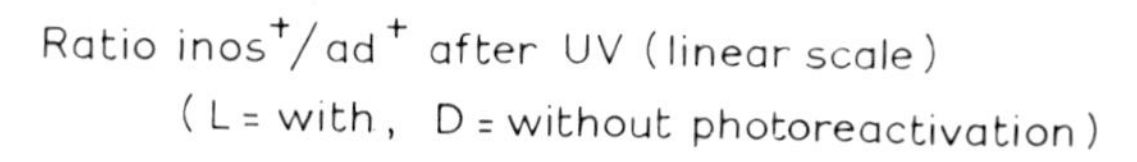

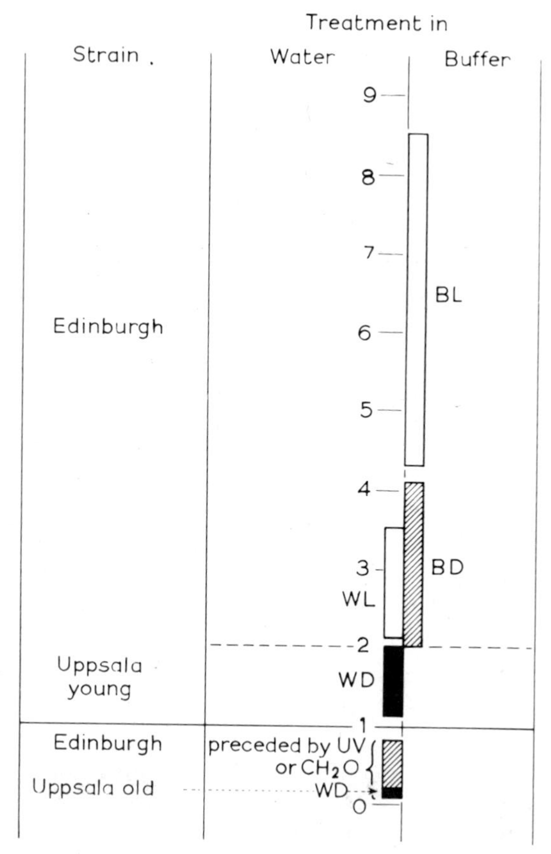

Fig. 20.5. Specificity of UV for reversions in the *ad*38701 *inos*37401 strain of *Neurospora* (linear scale). Specificity is measured as the ratio between inositol- and adenine-reversions (*inos*$^+$/*ad*$^+$). The upper part of the diagram is based on data by Allison (43), the lower part on data by Kilbey (44).

The main factors that determine specificity are the following.

Dose of treatment

The dose-effect curve for adenine-reversions is very steep, that for inositol-reversions is quite flat (Chapter 16). As a result, adenine-specificity increases quickly with dose.

Dose rate

In the usual experiments, conidia are exposed to fairly high concentrations of DEB for periods of up to 1 hour, usually less (high dose rate). Instead, they can be exposed to very low concentrations of DEB for periods up to 12 hours (low dose rate). Under low dose rate conditions, the dose-effect curve for adenine-reversions becomes linear (42). Since that for inositol-reversions is less than linear (Chapter 16), adenine specificity persists, but it is much weaker. When protein synthesis is prevented during exposure at low dose rate, the dose effect curve for adenine-reversions is steep like that obtained at high dose rate, and adenine-specificity is strong.

Storage

When treated washed conidia are stored in water before plating, the frequency of adenine-reversions increases several fold, while that of inositol-reversions remains unchanged. Thus storage increases adenine-specificity. The storage effect is caused by sensitization of the treated spores to the very small doses of DEB that are not removed by centrifuge washing. It can be imitated by exposing treated washed spores to a very low dose of DEB, which is hardly mutagenic for previously untreated cells. The state of sensitization decays gradually; this can be prevented by inhibiting protein synthesis.

Type of treatment

When DEB is mixed into the growth medium, adenine-specificity is greatly diminished.

Cytoplasmic background

In the background of 'poky', a cytoplasmically conditioned respiratory mutant, the adenine-specificity of DEB is transformed into inositol-specificity (32). When the two genes are transferred back into normal cytoplasm, their original specific responses are fully restored.

In Chapter 16, we have given reasons for attributing the specificity of DEB to two different cellular effects. One is the inhibition of a repair mechanism; this leads to the steep dose-effect curve for adenine-reversions and the sensitization of the cells to further adenine-reversions. The other is inhibition of some step in the expression of inositol-reversions, which results in their very flat dose-effect curve and their recalcitrance to sensitization.

UV

Originally, UV was considered a specific mutagen for inositol-reversions and, indeed, it usually produces more inositol- than adenine-reversions. In Fig. 20.5, the specificity of UV has been expressed as the ratio of inositol- to adenine-reversions. Compared with DEB, UV yields only a small range of specificity values. They extend over approximately two orders of magnitude and have therefore been entered on an arithmetic scale. Within this range, specificity can be modified by a variety of factors. In order not to overload the diagram, not all of them are shown.

Dose

The dose effect curves are strain-dependent. In the most frequently used strain, they are divergent, that for inositol-reversions lying above that for adenine-reversions. As a result, inositol-specificity increases with dose. At very low doses, the curves may intersect, so that specificity becomes reversed. In other strains, the dose effect curves run parallel or converge, with concomitant results on specificity.

Temperature during treatment

Conidia irradiated at 30^0 yield only half as many adenine-reversions as conidia irradiated at 2^0; the frequency of inositol-reversions is the same at both temperatures. As a result, inositol-specificity is more pronounced at high than at low temperature.

Photorepair

The recalcitrance of the inositol-allele to photorepair has been mentioned before (p.229). Since photorepair reduces the number of adenine-reversions but not that of inositol-reversions, inositol-specificity is increased by photorepair (WD and WL in Fig. 20.5).

Ionic environment during treatment (43)

Irradiation in phosphate buffer increases the frequency of inositol-reversions, while leaving that of adenine-reversions unchanged. As a result, inositol-specificity is increased (WD and BD in Fig. 20.5). Since photorepair and phosphate buffer increase specificity by different means, their effects are synergistic (WD and BL).

Age of conidia

Like the shape of the dose effect curve, this effect is strain dependent. While most strains used so far have not shown much age dependence of specificity, a very striking age effect was obtained in one of them. In this strain (Uppsala), conidia collected during the first week after onset of conidation showed the usual inositol-specificity. With increasing age of the spores, specificity decreased and finally was reversed to adenine-specificity (44). This was due, not to an effect on the frequency of inositol-reversions, but to an increasing frequency of adenine-reversions. If, as seems possible, this aging effect arises from the weakening of a repair mechanism, it would form an interesting parallel to the case of the progeroid, i.e. prematurely aging children, whose fibroblasts are reported to be deficient in the ability to rejoin radiation-induced breaks in DNA (45).

Chemical pre- or post-treatment

In Chapter 16, we have seen that pre-treatment of conidia with even very low and barely mutagenic doses of DEB reduces the inositol-specificity of UV or reverses it into adenine-specificity (Fig. 16.4). A similar, though weaker, effect is obtained by post-treatment with DEB. It is plausible to attribute this modification of UV-specificity to the same cellular effects of DEB that create its own adenine-specificity. The fact that DEB is a strong mutagen in its own right creates difficulties of interpretation. It is, therefore, of special interest that a reversal of UV-specificity has been obtained also by pre- or post-treatment with formaldehyde which, in aqueous solutions, is at best marginally mutagenic in this system (46). Table 20.6 illustrates this for two out of a large number of experiments, all of which yielded similar results.

Five other mutagens (X-rays, nitrous acid, nitrosomethylurethane, nitrosoethylurethane, hydrogen peroxide mixed with formaldehyde) have been tested for specificity in this system. All of them showed specificity of one kind or another, and in every case specificity could be modified by various means. Although the analyses were less thorough than those carried out with DEB and UV, their results make it quite clear that these mutagens, too, produce at least part, possibly all, of their specific effects through action on cellular processes.

These results illustrate the danger of attributing observed but not further analysed mutagen specificities to specific reactions

Table 20.6

Reversal of the inositol-specificity of UV by pre- or post-treatment with formaldehyde (CH_2O). (Data from Malling *et al* . (46).)

Treatment	*Survival in % of controls*	*Number of colonies**		*Specificity*
		inos$^+$	*ad*$^+$	*inos*$^+$/*ad*$^+$
CH_2O	64	1	0	
UV	60	262	93	2.8
CH_2O + UV	35	49	100	0.5
UV	83	532	180	2.9
CH_2O	50	1	6	
UV + CH_2O	27	142	226	0.6

* corrected for controls

between mutagens and DNA. The only cases in which this can be done with some confidence are those in which the amino acid sequences of the normal and mutant protein regions have been identified. Since these cases form only a small fraction of all observed ones, it seems unrealistic to restrict the term 'mutagen specificity' to them, as has often been suggested. If, instead, the term is used to cover all *observed* specificities, it can then be subdivided into narrower categories as causal analysis progresses. This is the usual procedure in the development of scientific terminology. It has, e.g., been followed for the terms 'suppressor' and 'dark repair', which first were used to describe observed phenomena and still are used in this sense in unanalysed cases, while in analysed ones there have been subdivisions into causally defined categories such as 'intracistronic suppressors' or 'excision repair'.

References

1. De Serres, F.J., Brockmann, H.E., Barnett, W.E. and Kølmark, H.G. (1971), 'Mutagen specificity in *Neurospora crassa'*, *Mutation Res.* **12**, 129-142.
2. Demerec, M. (1953), 'Reaction of genes of *Escherichia coli* to certain mutagens', *Symp. Soc. Exp. Biol.* 7, 43-54.
3. Kølmark, H.G. (1953), 'Differential response to mutagens as studied by the *Neurospora* reverse mutation test', *Hereditas* **39**, 270-276.

4. Glover, S.W. (1956), 'A comparative study of induced reversions in *Escherichia coli'*, 'Genetic studies with bacteria', *Carnegie Inst. Wash. Pub.* **612**, 121-136.
5. Witkin, E.M. and Theil, E.C. (1960), 'The effect of post-treatment with chloramphenicol on various ultraviolet-induced mutations in *Escherichia coli'*, *Proc. Nat. Acad. Sci. U.S.A.* **46**, 226-231.
6. Benzer, S. (1961), 'Genetic fine structure', *Harvey Lectures* **56**, 1-21.
7. Fries, N. and Kihlman, B. (1948), 'Fungal mutations obtained with methylxanthines', *Nature* **162**, 573.
8. Novick, A. (1955), 'Mutagens and antimutagens', *Brookhaven Symp. Biol.* **8**, 201-215.
9. Chevallier, M.R. (1964), 'Contribution a l'étude de l'action mutagène du peroxyde succinique chez *Escherichia coli'*, Thèse pour le titre de Docteur Es-Sciences, University of Strasbourg.
10. Sega, G.A., Cumming, R.B. and Walton, M.F. (1974), 'Dosimetry studies on the ethylation of mouse sperm DNA after in vivo exposure to (^{3}H) ethylmethane-sulfonate', *Mutation Res.* **24**, 317-333.
11. Kihlman, B. (1952), 'A survey of purine derivatives as inducers of chromosome changes', *Hereditas* **38**, 115-127.
12. Webb, S.J. and Tai, C.C. (1970), 'Differential lethal and mutagenic action of 254nm and 320-400nm radiation on semi-dried bacteria', *Photochem. Photobiol.* **12**, 119-143.
13. Brock, R.D. (1971), 'Differential mutation of the β-galactosidase gene of *Escherichia coli'*, *Mutation Res.* **11**, 181-186.
14. Webb, S.J. (1967), 'The influence of growth media on proteins bound to DNA and their possible role in the response of *Escherichia coli B* to ultraviolet light', *Canad. J. Microbiol.* **13**, 57-68.
15. Alderson, T. and Scott, B.R. (1971), 'Induction of mutation by γ-irradiation in the presence of oxygen or nitrogen', *Nature New Biology*, **230**, 45-48.
16. Bleichrodt, J.F. and Verhey, W.S.D. (1974), 'Influence of oxygen on the induction of mutations in bacteriophage φX174 by ionizing radiation', *Int. J. Rad. Biol.* **25**, 505-512.
17. Okada, Y., Streisinger, G., Owen, J., Newton, J., Tsugita, A. and Inouye, M. (1972), 'Molecular basis of a mutational hot spot in the lysozyme gene of bacteriophage T4', *Nature New Biology*, **236**, 338-341.

18. Streisinger, G., Okada, Y., Emrich, J., Newton, J., Tsugita, A., Terzaghi, E. and Inouye, M. (1966), 'Frameshift mutations and the genetic code', *Cold Spring Harbor Symp. Quant. Biol.* **31**, 77-84.
19. Drake, J.W. and Greening, E.O. (1970), 'Suppression of chemical mutagenesis in bacteriophage T4 by genetically modified polymerases', *Proc. Nat. Acad. Sci. U.S.A.* **66**, 823-829.
20. Lawrence, C.W., Stewart, J.W., Sherman, F. and Christensen, R. (1974), 'Specificity and frequency of ultraviolet-induced reversion of an *iso*-1-cytochrome *c* ochre mutant in radiation-sensitive strains of yeast', *J. Mol. Biol.* **85**, 137-162.
21. Chang, T.L., Lennox, J.E. and Tuveson, R.W. (1968), 'Induced mutation in UV-sensitive mutants of *Aspergillus nidulans* and *Neurospora crassa'*, *Mutation Res.* **5**, 217-224.
22. Harm, H. and Rupert, C.S. (1968), 'Analysis of photoenzymatic repair of UV lesions in DNA by single light flashes. I. In vitro studies with *Haemophilus influenzae* transforming DNA and yeast photoreactivating enzyme', *Mutation Res.* **6**, 355-370.
23. Corran, J. (1969), 'Analysis of an apparent case of 'gene-controlled mutational stability': the auxotrophic preemption of a specific growth requirement', *Mutation Res.* **7**, 287-295.
24. Clarke, C.H. (1962), 'A case of mutagen specificity attributable to a plating medium effect', *Zeitschr. Indukt. Abst. Vererb. Lehre* **93**, 435-440.
25. Clarke, C.H. (1965), 'Methionine as an antimutagen in *Schizosaccharomyces pombe'*, *J. Gen. Microbiol.* **39**, 21-31.
26. Clarke, C.H. (1963), 'Suppression by methionine of reversions to adenine independence in *Schizosaccharomyces pombe'*, *J. Gen. Microbiol.* **31**, 353-363.
27. Zetterberg, G. (1962), 'On the specific mutagenic effect of N-nitro-N-nitroso-methylurethan in *Ophiostoma'*, *Hereditas* **48**, 371-389.
28. Arditti, R.R. and Sermonti, G. (1962), 'Modification by manganous chloride of the frequency of mutations induced by nitrogen mustard', *Genetics* **47**, 761-768.
29. Queiroz, C. (1973), 'The effect of the plating medium on the recovery of nonsense suppressors in *Saccharomyces cerevisiae'*, *Biochem. Genet.* **8**, 85-100.

30. Chopra, V.L. (1967), 'Gene-controlled change in mutational stability of a tryptophanless mutant of *Escherichia coli* WP2', *Mutation Res.* **4**, 382-384.
31. Šilhankova, L. (1969), 'Suppression of rough phenotype in *Saccharomyces cerevisiae', Antonie van Leeuwenhoek* **35** (Suppl.), C11-12.
32. Paterson, H.F. (1974), 'Investigations into a reversal of di-epoxybutane specificity in *Neurospora crassa', Mutation Res.* **25**, 411-413.
33. Skavronskaya, A.G., Aleshkin, G.J. and Likchoded, L.V. (1973), 'The dependence of UV-induced reversions to prototrophy on the streptomycin resistance allele in *Escherichia coli', Mutation Res.* **19**, 49-56.
34. Clarke, C.H. (1973), 'Influence of streptomycin on UV-induced $tryp^+$ reversions in a streptomycin-resistant strain of *Escherichia coli B/R', Mutation Res.* **19**, 43-47.
35. Loppes, R. (1969), 'A new class of arginine-requiring mutants in *Chlamydomonas reinhardi', Molec. Gen. Genetics* **104**, 172-177.
36. Böhme, H. (1961), 'Über Rückmutationen und Suppressor-mutationen bei *Proteus mirabilis', Zeitschr. Vererb. Lehre* **92**, 197-205.
37. Charkabarti, S.L. and Maitra, P.K. (1974), 'Development of auxotrophy by streptomycin-resistant mutations', *J. Bacter.* **118**, 1179-1180.
38. Kaplan, R.W. (1961), 'Spontane Mutation von einer Mono-auxotrophie zu einer Anderen in einem Schritt (Auxotrophiesprungmutation)', Zeitschr. Vererb. Lehre **92**, 21-27.
39. Jain, H.K. and Shukla, P.T. (1972), 'Locus specificity of mutagens in *Drosophila', Mutation Res.* **14**, 440-442.
40. Shukla, P.T. (1972), 'Analysis of mutagen specificity in *Drosophila melanogaster', Mutation Res.* **16**, 363-371.
41. Lifschytz, E. and Falk, R. (1969), 'Fine structure analysis of a chromosome segment in *Drosophila melanogaster.* Analysis of ethylmethanesulphonate-induced lethals', *Mutation Res.* **8**, 147-155.
42. Kilbey, B.J. (1974), 'The analysis of a dose-rate effect found with a mutagenic chemical', *Mutation Res.* **26**, 249-256.
43. Allison, M.J. (1972), 'The effect of phosphate buffer on the

differential response of two genes in *Neurospora crassa* to UV', *Mutation Res.* **16**, 225-234.

44. Kilbey, B.J. and Purdom, S.M. (1974), 'The modifying effect of age and strain on the mutagenic specificity of ultraviolet light in *Neurospora crassa', Molec. Gen. Genetics.* **135**, 295-308.
45. Epstein, J., Williams, J.R. and Little, J.B. (1973), 'Deficient DNA repair in human progeroid cells'. *Proc. Nat. Acad. Sci. U.S.A.* **70**, 977-981.
46. Malling, H., Miltenburger, H., Westergaard, M. and Zimmer, K.G. (1959), 'Differential response of a double mutant – adenineless, inositolless – in *Neurospora crassa* to combined treatment by ultraviolet radiation and chemicals', *Int. J. Rad. Biol.* **1**, 328-343.

CHAPTER TWENTY-ONE

Spontaneous mutations

It is not possible to draw a hard and fast line between induced and spontaneous mutations, since agents with which we induce mutations in the laboratory, e.g. X-rays or certain chemicals, are present in the environment of living organisms or are produced during metabolism. In this chapter, I shall discuss mutations that arise without consciously applied radiation or chemical treatment.

Although spontaneous mutations were found, and their rate of occurrence was measured (Chapter 1), years before means for inducing mutations artificially had become available, our knowledge of the origin and nature of induced mutations far outstrips that relating to spontaneous ones. This is not surprising. It is always easier to analyse processes that one can manipulate experimentally than naturally occurring ones. Certain questions concerning spontaneous mutation could, however, be asked and tested without knowledge of the nature of mutation and, indeed, of the gene. One of these questions, which is still under discussion, concerns the dependence of mutation on gene replication. There is now good evidence that mutations do occur in non-replicating genes. I shall present this evidence before discussing the nature and causes of spontaneous mutations.

Mutation without replication

The extraordinary fidelity of gene replication has been one of the most intriguing problems of genetics from its earliest days. It seemed reasonable to assume that fidelity was not perfect and that mutations could arise through 'copy errors'. There remained, however, the question whether spontaneous mutations *always* are produced in this manner, or whether they can occur also in the 'existing' i.e. non-replicating genes. For induced mutations, we have dealt with this question in relation to the most important mutagens. For spontaneous mutations, it was not resolved during the early periods of mutation research. Experiments with the chemostat confused rather than clarified the issue by showing that, in a continuously growing culture of bacteria, the frequency of spontaneous mutations increased with time, independent of how many generations had occurred during a given interval (1). The situation in the chemostat is still not clearly understood, and we shall not discuss it here. A recent review of the older data in the light of new ones is to be found in reference (2).

A more straightforward situation obtains when non-replicating cultures are monitored for the occurrence of new mutations. Surprisingly, these seemingly simple experiments are beset with difficulties that have taken years to resolve. One is the low frequency at which spontaneous mutations occur. As long as *Drosophila* was the only suitable object for this research, it was difficult to obtain sufficiently large numbers for statistical evaluation. When microorganisms had become available for mutation experiments, a new difficulty arose from the fact that it is not easy to ensure *complete* absence of replication in stationary cultures; this makes it possible to attribute any observed mutations to the small fraction of replicating genomes. It is true that these mutations would give rise to mosaic clones, containing both mutant and non-mutant cells, but tests for mosaicism are not always feasible. This applies in particular to the screening of prototrophs among an auxotrophic parent population, or of resistant cells among a sensitive parent population; for in these experiments, the parental cells cannot grow and all mutant clones are pure. In addition, realization of the double-stranded nature of the gene has deprived mosaicism of much of its diagnostic value for the recognition of mutations that arise during replication; for mosaics may also derive from one-strand mutations in non-replicating genes. Yet, while mosaics are no proof *for* mutation by replication errors, completes are proof *against* it; in this sense, tests for mosaicism have retained their usefulness. Moreover, the occurrence of complete mutants from resting cells provides evidence against a final possibility that has to be considered: the possibility that genes in the stationary stage become in some way mutation-prone, but that the actual mutations occur only after the resumption of replication. With these criteria in mind, let us look at the experimental results.

In *Drosophila*, the frequency of sex-linked lethals in spermatozoa was approximately doubled by three weeks' storage in the seminal receptacles of the female (3). It is highly unlikely that genes in stored spermatozoa replicate. Moreover, mutations that arise as replication errors will give rise to mosaics, while the test for recessive sex-linked lethals in F_2 detects only complete mutants, with the exception of those rare mosaics in which the whole of the gonads but not all of the soma carries a lethal. Mosaic lethals are detected in F_3 (Fig. 5.3), and their frequency was not increased by storage.

In *Neurospora,* too, the test for recessive lethals detects only complete mutations (Fig. 11.2; mosaics would have two methionineless genomes, only one of which carries a lethal). When dried *Neurospora* conidia were stored at different temperatures, the frequency of recessive lethals increased steadily at a weekly rate of 0.3% at 30°C, and less steeply at 4° (4).

In *bacteria,* early experiments on reverse mutations to prototrophy during stationary phase provided suggestive but not conclusive evidence for mutation without replication. Since completes and mosaics could not be distinguished (p. 407), the possibility of cryptic replication during storage was not excluded (5). This stricture does not apply to other experiments on bacteria, where the storage conditions clearly made replication impossible. Thus, high frequencies of auxotrophs were obtained from dried bacterial cells or from spores that had been exposed *in vacuo* for 16 minutes to temperatures ranging from 135° to 155°C (6), and from semi-dried cells of *E. coli* that had been stored for one hour in an aerosol at 25° (7). In dry, but not in wet, spores of *Bacillus subtilis,* mutations to azide-resistance accumulated over many months of storage (8). Finally, it is very probable that conditions of 'thymineless death' produce mutations in non-replicating bacterial genes; we shall come back to this on p. 416.

When *bacteriophage* is stored in broth, complete absence of replication can be confidently presumed. Yet, under these conditions forward and reverse mutational changes in the rII gene of phage T4 accumulated at approximately linear rates over several years (9). As in *Neurospora*, the rate of accumulation was higher at high than at low temperature (20° *versus* 0°). A majority of the mutant plaques was mottled, as would be expected from one-strand changes in DNA, but there was also a sizeable proportion of completely mutant plaques. These may have originated from one-strand changes by one of the means described in Chapter 19.

Taken together, the experimental evidence leaves no doubt that mutations occur in non-replicating genomes of both pro- and eukaryotes. Yet, wherever it has been possible to compare the frequencies of mutation with and without replication in the same system, the former has always been higher, usually considerably so. Obviously, replication opens up sources of mutagenesis that are not available to non-replicating genomes. We shall now turn to a consideration of how mutations arise in these two situations.

The nature and origin of spontaneous mutations

For reasons set out at the beginning of this chapter, we still know little about the causes of spontaneous mutations, but we can be certain that they are manifold and that they result in a variety of genetical damage, from base changes to chromosome rearrangements. In this section, we shall deal only with mutational mechanisms that can produce intragenic changes of the base substitution or frameshift types. Grosser types of mutational damage will be discussed in later sections and in the next chapter.

Replication-independent causes of mutation

Let us start by looking for mechanisms that might produce mutations in non-replicating genomes. Doubtless, some of these will affect also replicating ones. Ionizing radiation and, in transparent cells growing in sunlight, UV and visible light will certainly contribute their quota of damage to DNA; but it will usually be small. This has been calculated a long time ago for *Drosophila* and plants (Chapter 1) and is likely to hold good in general. The mere fact that the spontaneous mutation rate has a positive temperature coefficient (4,9,10) supports this conclusion, for radiation-induced mutation frequencies do not increase with temperature during exposure. Mutagenic chemicals that have been produced during metabolism or have been introduced from outside may continue to produce mutations for some time after replication has ceased, but accumulation of mutations over long periods (4,7,8,9) requires processes that can go on continuously in stationary cultures. Depurination is a possible candidate (Chapter 16). Heat has been shown to depurinate DNA *in vitro* and may do so *in vivo* (11). Low pH, too, acts as depurinating agent (12); in bacteriophage kept *in vitro*, it produces mutations, mainly transitions or transversions at GC sites (13). Whether sufficient depurination occurs at physiological temperatures to account for the mutations in stored phage particles is doubtful. The mutational spectrum (14) is compatible with the assumption but does not prove it. Most or all of the mutations were transitions or transversions at GC sites. There were no frameshifts, and the two large hot spots for spontaneous mutations in replicating phage (15) were missing.

Desiccation may stabilize chemical or configurational changes in DNA that otherwise would decay or revert. In addition, it may itself produce structural changes that result in deletions; conceivably, many or most of the mutations from dry *Neurospora* conidia or dry bacteria were deletions. Inositol, which protects dried bacterial cells against death, also protects their chromosomes against mutation (7).

Replication-dependent causes of mutation. Mutators and antimutators

When Watson and Crick proposed the double-helix model of the gene ((5) Chapter 1), they also suggested a possible origin for spontaneous mutation. It was based on the assumption that the nucleotide bases will occasionally undergo transitory tautomeric shifts (Fig. 21.1). When one of these short-lived shifts happens to coincide with the moment of replication, it may lead to mutation *via* an error in base-pairing; these mutations will be base substitutions. Subsequently, frameshifts were recognized as an additional class of mutation, probably produced by 'strand-slippage' at replication (Fig. 21.2). While frameshift mutations obviously require some kind of enzyme action, base changes were originally thought to arise from purely chemical accidents, perhaps – as the early mutation workers had surmised – through randomly increased temperature oscillations near a gene. A fundamental change of outlook took place when it was realized that DNA polymerase plays an active role in the selection of bases at replication. This realization grew out of the analysis of 'mutator genes'.

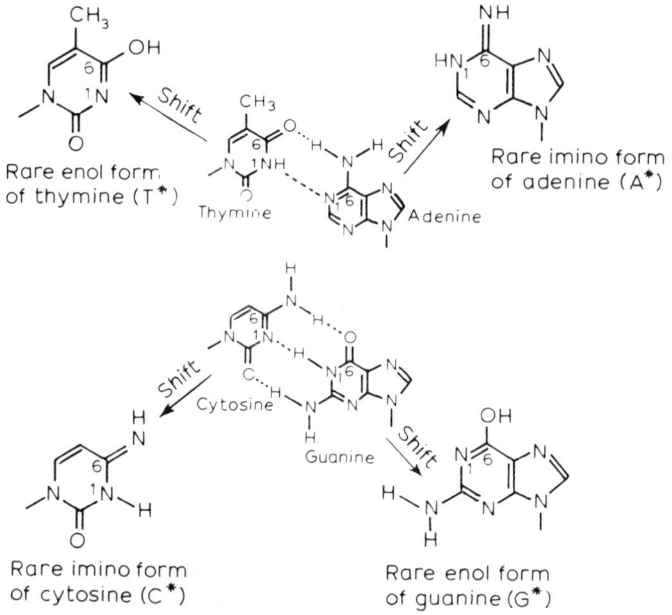

Fig. 21.1. The tautomeric forms of pyrimidines and purines. In the centre are shown the stable forms of thymine, adenine, cytosine and guanine as they participate in AT and GC pairing. The tautomeric form of each molecule is indicated at arrows. (Fig. 5.7 in *Genetics* by Goodenough and Levine, 1974 Courtesy Holt, Rinehart and Winston Inc.)

Bacteriophage. The role of DNA-replication

In the original model for semi-conservative replication of the double helix, the role of DNA-polymerase was simply the linking together of nucleotides that had been aligned opposite the parental strand on principles of chemical complementarity. The possibility that polymerase might play a more active and selective role in replication was considered by a number of scientists and was finally put to the test by Speyer and his collaborators (16). Basing themselves on the fact that gene 43 of phage T4 codes for DNA-polymerase, they screened temperature-sensitive mutants of this gene for mutator activity. Since a functional DNA-polymerase is essential for survival, these mutants are lethal at the non-permissive temperature but produce a functioning, albeit abnormal, polymerase at the permissive one. A proportion of them were, indeed, found to act as mutators, and this property could be shown to be connected with their role in replication (17). Subsequently, Drake and his group (18)

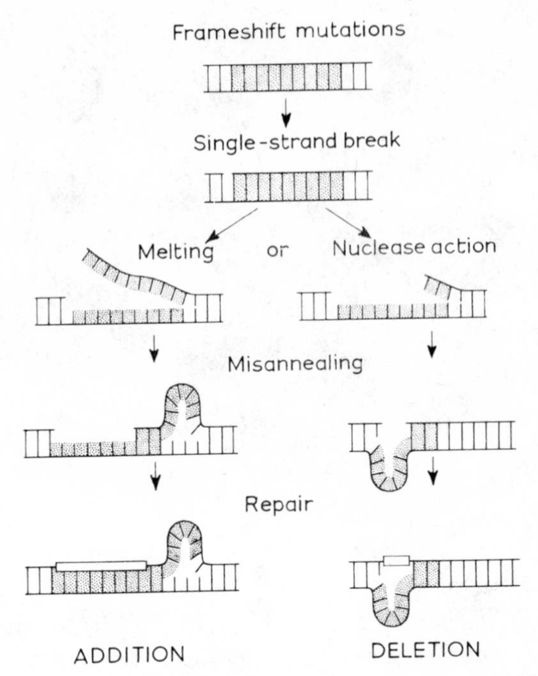

Fig. 21.2. Proposed mechanism for the origin of frameshifts. The heavy vertical bars represent identical base pairs. The open horizontal bars represent newly synthesized DNA. Modified from Fig. 12.8 in Drake, The molecular basis of mutation (Bibliography).

found that some mutants of gene 43 act as 'antimutators', i.e. they reduce mutation frequencies below that of the wild-type. Mutant polymerases can affect the mutation rates of all types of mutation: transitions, transversions, frameshifts (19,20,21). Allelic differences in response are presumably due to 'neighbourhood effects' (Chapter 20). Here, as in induced mutagenesis, specificity at the level of DNA is governed by base sequences rather than by individual base pairs. The involvement of an enzymatic process makes this readily understandable.

In vitro studies (22,23) have shown that, in the synthesis of polynucleotides on templates of natural DNA or synthetic polynucleotides, mutator DNA polymerases are more 'error-prone' than the normal enzyme. The gene 43 enzyme has a dual function, acting both as polymerase and exonuclease. In the latter capacity, it appears to have an 'editing' role, removing mismatched bases both before and during synthesis (Fig. 21.3). Whether a mutant enzyme has mutator or antimutator properties may in part depend on the

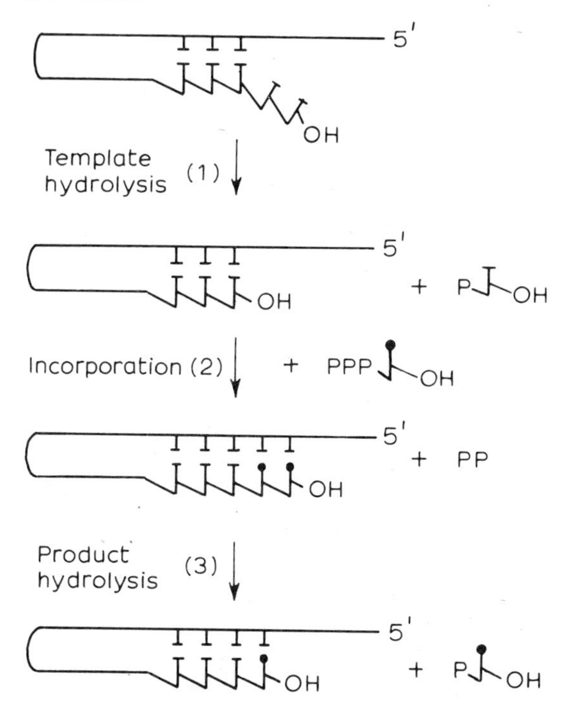

Fig. 21.3. Schematic drawing of reactions catalysed by T4 DNA polymerase in the presence of complementary deoxynucleoside triphosphates *in vitro.* The preferred template is single-stranded DNA whose 3′ end is looped around to hydrogen bond with a complementary section of the same strand. Reaction 1: non-complementary residues near the 3′ terminus of the primer DNA are removed prior to synthesis. Reaction 2: New residues are added to the 3′OH end of the DNA. Reaction 3: Exonuclease activity of the polymerase during DNA synthesis. (Fig. 1, in Hershfield and Nossal (22) Courtesy Genetics Business Office.

ratio between its two functions.

Among other T4 genes concerned with DNA-replication, gene 32 has given clear effects on mutation rates (18,24). The product of this gene is required for replication; probably it helps in the unwinding of the double helix. Mutants were found to act as mutators for some but not all transitions from AT to GC, as antimutators for some transitions from GC to AT, and as mutators or antimutators for different frameshifts.

Still other genes increase mutation frequencies by interfering with the metabolism of the two pyrimidines in T4-DNA: thymine and hydroxymethylcytosine (HMC). Thymine deprivation can be achieved in two ways: phenotypically, through inhibition of the synthesis of thymidylic acid with fluorodeoxyuridine; genotypically, through a combination of mutant genes in phage and host. Both conditions resulted in greatly increased frequencies of transitions from AT to GC, but did not affect frameshifts or transitions from GC to AT. Conversely, when the synthesis of HMC was inhibited by mutations of the relevant gene 42, the frequency of transitions from GC to AT was specifically increased (18).

Bacteria. The role of repair

Mutator genes have been discovered in many bacterial strains (for review see 25). They produce different mutational spectra. The most striking specificity is displayed by the 'Treffers' mutator gene (*mut T*) of *E.coli*, which was discovered by Treffers *et al.* in 1954 (26); it produces exclusively transversions from AT to CG (27). Other mutator genes are less specific (28). In some strains spontaneous deletions arise at high frequencies (29,30).

In parallel with what had been found for T4, it was expected that some at least of these mutator genes would act at replication *via* an error-prone DNA-polymerase. *E.coli* and *Bacillus subtilis* and, perhaps, bacteria in general have three different DNA-polymerases: I, II and III (31,32). DNA polymerase III is the replication enzyme; in *E.coli*, the gene coding for it (dnaE) has been located. Temperature-sensitive mutants of this gene, and mutants of *Bacillus subtilis* with a temperature-sensitive polymerase III were shown to have mutator properties (33,34). In *E.coli*, these mutator genes enhanced the frequencies of several transversion events but did not affect the frequencies of transitions or frameshifts. This makes it plausible – although not compelling – to assume that the *mut T* gene acts *via* a mutant DNA-polymerase with an even higher degree of specificity.

Mutator genes, like mutagens, show specificities both in regard to the sites in DNA that they attack and to the conditions under which they are most effective. Thus in the tryptophan synthetase gene of *E.coli*, *mut S3* produces transitions in one AT pair but fails to do so in another less than 50 nucleotides away (35).

In *E.coli*, a 'conditional mutator gene', *mut D5,* increases mutation frequencies moderately on minimal medium, but up to 100 times as much on rich medium (36). The component of the rich medium that acts as 'effector' has been identified as thymidine. Like mutants of gene 43 of T4, *mut D5* promotes all types of base changes as well as frameshifts. This catholicity of action has been tentatively explained by a model first proposed by Fresco and Alberts (see Fig. 3a in (36), according to which mutations may arise when the parental strand at the replication fork loops out by mistake so that the incoming base pairs incorrectly with the neighbouring parental one.

The conclusion that repair is involved in spontaneous mutagenesis was drawn from experiments on bacterial strains lacking DNA-polymerase I, the enzyme concerned with repair of damage to DNA. Strains of *E.coli* that lack the enzyme (*polA*$^-$) are viable, but they are very sensitive to UV, X-rays and other mutagens (37). These strains also yield increased frequencies of spontaneous mutations and deletions (29,30,38). A mutant gene that resembles *polA*$^-$ in producing increased UV sensitivity and increased spontaneous mutability is *uvr502* (39); it increases the frequency of transitions but not that of transversions. The double mutant *polA*$^-$*uvr502* is inviable (40). The tentative conclusion has been drawn that the normal function of one or the other of these two genes is required not only for the rejoining of artificially induced strand breaks in DNA but also for the mending of discontinuities that arise during normal replication (41). Excision repair and recombination repair do not appear to influence spontaneous mutation frequencies in bacteria (38,44). Yet, there have been a number of cases in which UV-sensitive strains exhibited increased spontaneous mutability (e.g. 42, 43), and *recA* or *exrA* strains appear to have lower spontaneous mutation rates than wild-type (43a). *Vice versa,* spontaneous mutation frequencies were reduced in two strains of *E.coli* that were resistant to the lethal and mutagenic action of nitrosoguanidine (45).

Genes with antimutator activity have not so far been identified in bacteria. Certain chemicals, however, act as 'antimutators' or, as they also reduce induced mutation frequencies, as 'antimutagens'. The means by which they achieve this effect are not yet known; they probably differ between substances. Thus, the antimutator action of purine ribosides on spontaneous mutation in the chemostat (46) is obviously related to mutagenesis in the course of normal purine metabolism, since these same ribosides totally abolished the response of the cells to otherwise mutagenic purines (Chapter 17). On the other hand, spermine and other polyamines are presumed to inhibit the formation of mutational lesions by combining with

DNA (47, 47a). One might also consider the possibility that, if spermine is not completely removed from the cells before plating, its influence on the fidelity of translation (48) could affect mutation expression.

Thymine deprivation has been reported to produce bacterial mutations under two quite different sets of conditions and, presumably, by two quite different mechanisms (49). When thymine-requiring bacteria were grown on limiting supplies of thymine, which allowed some vegetative replication to take place, spontaneous mutation frequency was increased by several orders of magnitude. Deficiencies for the *recA* or *exrA* functions did not prevent this effect, indicating that it does not depend on an error-prone repair process, at least not on that required for UV-mutagenesis (Chapter 14). The simplest interpretation is to assume that errors accumulate during slow replication under conditions of precursor scarcity. Much smaller increases of mutation frequency have been reported for experiments in which thymine requiring bacteria were kept for some time under conditions of thymine deprivation, leading to 'thymineless death' (50,51). Since in these experiments there was no absolute increase in the number of mutations, their interpretation is open to question. Mutagenesis by FUdR, which produces a phenocopy of genetic thymine requirement, has been clearly demonstrated in exr^+ strains, but does not occur in exr^- ones (52). This suggests that thymine deprication may cause a low level of mutagenesis by means of the same error-prone repair system that is believed to produce mutations from UV damage to DNA (Chapter 14).

Fungi. The role of recombination

Tests for spontaneous mutation frequencies have been mainly carried out on yeast. Both in *Saccharomyces cerevisiae* (53) and *Schizosaccharomyces pombe* (54), mutants with enhanced spontaneous mutation frequencies have been found among radiation-sensitive strains. In *Saccharomyces,* von Borstel and his collaborators have developed an accurate method for detecting strains with increased spontaneous mutation frequencies (53). They isolated a large number of such strains and identified some mutator loci, each with several alleles. Only a few of these strains were radiation-sensitive. Mutation to the respiratory-deficient 'petite' type, whether due to a nuclear or cytoplasmic change, specifically increased mutation frequency of one particular gene. 'Petite' is the physiological homologue of 'poky' in *Neurospora.* It is of interest that poky, too, affects mutation frequencies specifically (Chapter 20). It seems likely that, in both organisms, the effect occurs at the repair- or expression-stage rather than at the stage of primary changes in DNA.

The most interesting result to emerge from the study of spontaneous mutability in eukaryotes is evidence that recombination may be a source of mutation. For bacteria, this possibility had been considered by Demerec (55), who found protrotrophic revertants in excess of controls among 'selfers', i.e. the progeny of auxotrophs that had been transduced with phage from the same auxotrophic strain. He suggested that the revertants might arise through unequal crossing-over in the homologous segments of phage and bacterium. This interpretation, however, had to be abandoned when it was found that the frequency of prototrophs remained the same when the donor strain carried a deletion for the locus concerned. We shall come back a little later to the 'selfing' phenomenon.

In eukaryotes, we have the novel and important process of meiotic recombination, and it is here that mutations may arise. The clearest evidence comes from the work by Magni and his collaborators on yeast (56,57) (also see his article in 'Mutation as Cellular Process', Bibliography). In diploid *Saccharomyces cerevisiae,* many more mutations to auxotrophy occur during meiosis than during mitosis. This has been called the 'meiotic effect'. It has been obtained also for super-suppressors and some reverse mutations. These latter were studied more in detail, and two observations showed that the meiotic effect was due to recombination. (1) Many of the meiotic revertants were associated with nearby crossovers and (2) the meiotic effect was abolished when the homologous chromosome carried a deletion for the locus in question. When mutants that did or did not show a meiotic effect were classified by reversion analysis and by the usual criteria for mis-sense *versus* nonsense and frameshift mutations (Chapter 3), it appeared that the responding ones were frameshifts. This led to the conclusion that mutations are produced at meiosis by intragenic recombination resulting in frameshifts. The fact that supersuppressor mutations show a meiotic effect was taken to mean that frameshifts in *t*-RNA genes can produce suppressors (Chapter 4). While some of the details of this picture are still hypothetical, the contribution of meiotic recombination to spontaneous mutation in

Saccharomyces cannot be doubted.

Attempts to extend these findings to fission yeast yielded only partial confirmation (58). Mutation frequencies were indeed higher in meiosis than in mitosis, but the excess was small. This is probably not due to the difference in the species used but in the scoring systems employed. In fission yeast, forward mutations at five loci in the adenine pathway were scored by colour (red-into-white; Chapter 11). Mitotic mutations arose at a frequency of about 4×10^{-6}; all those that could be classified were considered to be base substitutions. This agrees with the classification of mutations from white to red (59). In *Saccharomyces*, in contrast, forward mutations to auxotrophy were scored by their inability to grow on minimal medium; this would prevent the detection of even mildly leaky base substitutions. No mitotic mutations were found among over 400 000 tested cells. The effect of meiotically produced frameshifts will be amplified by the stringency of this system; conversely, in the less stringent colour test it will be swamped by base substitutions. The spectrum of 118 mitotic and 59 meiotic white mutants in fission yeast fits well into this interpretation. While base changes were only moderately less frequent in the meiotic than in the mitotic sample, all of seven frameshifts were found in the former.

No other fungus has been subjected to a systematic comparison between mitotic and meiotic mutation frequencies. Some results of experiments carried out for different purposes have been attributed to a mutagenic effect of meiotic recombination (e.g. 60); others argue against such an effect (61). A very high meiotic reversion frequency of a nicotinic-acid requiring gene in *Coprinus* was not due to recombination (61a).

Drosophila

In *Drosophila*, the first mutator gene was discovered as early as 1937 (62). Subsequently, a second mutator gene turned up in a wild population of *Drosophila melanogaster* (63). It may have been allelic with the first, which meanwhile had been lost. It differed from it not only by a much greater efficiency but also by a different spectrum of visible mutations. More recently, a mutator gene has been detected that acts only in females and produces lethals that consist to about 50% of deficiencies (64). The mode of action of mutator genes in *Drosophila* is as yet only a matter for speculation. An exception is the mutator property of a gene that, in many strains from a natural population in Texas, consistently yields small numbers of recombinants in male germ cells, (65). Tests for sex-linked and autosomal recessive lethals showed that their frequencies were increased in lines yielding male recombination, and that the increase paralleled the degree of male recombination. The resemblance to Magni's findings on yeast is interesting; but it may be fortuitous since recombination in *Drosophila* males appears to be mainly premeiotic (65a).

'Mutable genes' of *Drosophila* which, in contrast to mutator genes, have themselves a high mutation frequency but do not impart it to other genes, will be mentioned in the next chapter.

Summarizing the results on the origin of mutations in replicating genomes, we may say that replication, repair and recombination all play a rôle but that their relative importance varies with organism, type of mutation and system used for scoring.

Mutation rates

Geneticists have long been familiar with the idea that spontaneous mutation rates are objects as well as tools of evolution and open to modification by mutation and selection. The mathematics of this process has recently been discussed by Kimura (66) (see also the section on evolution in 'The genetic control of mutation', Bibliography). Insight into the enzymatic processes that play a role in spontaneous mutagenesis has made it possible to visualize the means by which an organism can control its mutation frequency (67).

A curious fact emerges when spontaneous mutation rates are compared in organisms differing as widely from each other as bacteriophage and *Drosophila* (Table 21.1).

The estimates of mutation rates in Table 21.1 are only very approximate and may be out by factors of 5 or more. All the same they show clearly that, both in replicating and non-replicating genomes, overall mutation rates vary very much less between organisms than do the contents of DNA. As a consequence, the frequency of mutations per base pair per replication decreases from about 10^{-8} in bacteriophage to 10^{-11} or less in eukaryotes. Obviously, selection has militated against large differences in overall spontaneous mutability. When bacteriophage λ is integrated into the bacterial genome as prophage, it comes under the control of the bacterial system; for under these conditions its mutation rate per base per replication

Table 21.1

Estimates of spontaneous mutation rates in non-replicating and replicating genomes of various organisms

Organism	*Base pairs per genome*	*Mutation rate*	
		per replication	*without replication per week*
Bacteriophage λ	4.8×10^4	1.2×10^{-3}	
Bacteriophage T4	1.8×10^5	3.0×10^{-3}	2×10^{-3}
Salmonella typhimurium	4.5×10^6	0.9×10^{-3}	
Escherichia coli	4.5×10^6	0.9×10^{-3}	
Neurospora crassa	4.5×10^7	2.9×10^{-4}	3×10^{-3}
Drosophila melanogaster	2.0×10^8	$> 10^{-4}$	3×10^{-3}

The second and third columns have been taken from Drake (67), except the mutation rate in *Drosophila*, which has been roughly estimated from data on sex-linked lethals. The last column is based on data from references 3, 4 and 9; for the calculations see Auerbach and Kilbey, *Ann. Rev. Genetics*, bibliogr.

decreases to that of the bacterial genes. It is of interest to compare these results with the approximate invariance of X-ray induced mutation frequencies per base pair (Fig. 9.10). Evidently, while there is selection for spontaneous mutability, there is none for that induced by X-rays. This would be expected from the fact that ionizing radiation contributes very little to spontaneous mutability.

Selection for a mutator gene has been shown to occur in *E.coli* where, in competition experiments in the chemostat (68), a mutator strain consistently outgrew the strain without mutator. The reason for this selective advantage can be found in much older experiments by Novick and Szilard (1) and by Atwood *et al.* (69) on bacterial populations kept growing in the chemostat or through serial transfer. Mutants with improved growth ability arose periodically under these conditions and replaced the population in which they had arisen. Since these beneficial mutations are likely to occur among the large proportion of non-mutant cells rather than among the much smaller ones of spontaneous mutants of neutral or near-neutral value, a side-effect of this periodic selection was a periodic cleansing of the population from selectively unimportant mutations. In a similar way, the mutagenic DNA polymerase which has recently been detected in human leukaemic cells (69a) may help these cells to get established by providing a constant source of new mutations to quicker growth rate.

A number of geneticists, notably Goldschmidt (70), have not been content with the idea that evolution proceeds wholly through selection of small-step mutations. They have postulated the existence of an additional type of process, called 'macromutation', which results in a remodelling of the whole genotype. In a novel guise, this idea has very recently been taken up by American workers who used frogs and mammals for a comparison between the relative rates of protein evolution as measured by immunological differences between albumins, and of chromosome evolution as measured by changes in chromosome number (71). They found that protein evolution, which presumably proceeds *via* base pair changes, has taken place at the same rate in both groups. Chromosome evolution, on the contrary, has occurred much more rapidly in mammals. They relate this to the anatomical diversification, which is much more rapid in mammals than in frogs, and they speculate that the remodelling of the karyotype by loss, gain or rearrangements may speed up anatomical changes, perhaps *via* the regulation of gene expression.

Not surprisingly, spontaneous mutation rates and aberration frequencies depend on physiological, metabolic and nutritional conditions. For plants, this field has been reviewed in reference 72. A newer report (73) deals specifically with the effect of age on mutation frequencies in plant seeds. In a modern theory of ageing, increased mutation frequencies during senescence are predicted as a result of an accumulation of errors in the synthesis of enzymes concerned with replication and repair of DNA (74).

Mutagenesis by transduction and transformation

Demerec, as mentioned on p. 417, found an increased frequency of prototrophic revertants among 'selfed' transductants of *Salmonella typhimurium.* It is less well known that Demerec also observed excess mutations among the products of transformation in *Salmonella* (75): about 50% of the transformants to streptomycin-resistance carried a new mutation in the closely linked thiamine gene. Recently, transformation-induced mutagenesis has been observed and studied in *Bacillus subtilis* (76) and in the blue-green alga *Anacystis nidulans* (77). In *Bacillus subtilis*, the frequencies of both forward and reverse mutations were greatly increased in transformants. In experiments with well-marked pieces of donor DNA, there was evidence for linkage between the transferred donor gene and the new mutations. The many additional mutations that occurred outside this region could be plausibly attributed to transformation with unmarked pieces of DNA. About 30% of the new mutations were leaky and temperature-sensitive, suggesting that they were base changes of the mis-sense kind. In *Anacystis,* the frequency of auxotrophic mutations in transformants was extremely high; in recombinants for the markers used, it was 100%. This and other observations strongly suggest that the mutations arose during recombination. If this should apply also to *Bacillus subtilis,* recombination during transformation does not appear to result exclusively in frameshifts, as has been concluded for yeast (57).

Instabilities produced in the course of transformation will be discussed in Chapter 22.

Mutagenic effects of extrachromosomal elements, bacteriophages and viruses. Insertion mutations

In 1963, Taylor (78) noticed that insertion of the temperate phage Mu into the chromosome of *E.coli* produced both forward and reverse mutations at sites that were closely linked to that of the prophage. In the *lac*-operon it increased the frequency of *lac*$^-$ mutations about 100-fold when the gene was repressed, and less when it was induced and, presumably, protected by RNA. The majority of 180 *lac*$^-$ mutations in 11 segments of the gene had occurred very close to the site at which the prophage had inserted itself; it seems that insertion in some way produces the mutations. A small percentage of the mutations were deletions of the whole gene. These were not linked to the site of the prophage and their frequency was not reduced by induction of the genes; they are supposed to have arisen in some different way (79). While Mu can insert anywhere in the genome, phage λ has specific attachment sites. When it inserts at abnormal attachment sites, it generates mutations in or close to the genes at these sites (80).

Mutations may occur also when a prophage changes into the free state. Strains of *E.coli* carrying the prophage P2 in close linkage to the histidine locus spontaneously segregated his$^-$ cells which were due to deletions covering the histidine locus (81). This deviation from the normal, non-mutagenic prophage excision has been called 'eduction'.

Both in bacteria and phage, mutations have been shown to occur after insertion into the chromosome of DNA-sequences of unknown origin (82-84). Some of these inserted pieces of DNA are very long, between 1000 and 2000 nucleotides; others consist of a few hundred nucleotides. Almost all insertion mutations, whether due to phage or foreign DNA-sequences, are characterized by extreme polarity (Chapter 3). This polarity effect is so strong that it has been used for ordering genes within an operon (85). Its basis is not yet fully understood. Recent experiments (86) suggest that insertions may act as transcriptional termination signals within the gene. In the presence of the termination factor ρ, they act as polar mutations also *in vitro* by binding ρ at concentrations at which it does not bind to the termination signal at the end of the gene.

Extrachromosomal elements have been shown to produce mutations and chromosome breakage also in eukaryotic organisms, but it is not known whether any of them act by insertion. In *Drosophila,* an extrachromosomal element, δ, which was first discovered by its distorting action on the segregation of certain chromosome pairs, was later found to produce recessive lethals (87,88). Their distribution showed an interesting similarity to those produced by calf thymus DNA (Chapter 18). While the frequency of lethals on the second chromosome was clearly in excess of the controls, that of sex-linked lethals was only slightly increased. Moreover, lethals induced in the second chromosome showed a non-random distribution, with the greatest number occurring in the same region in which Gershenson had located most of his DNA-induced lethals. In *Drosophila robusta,* a cytoplasmically transmitted agent produced large numbers of chromosome rearrangements; unfortunately it was lost before it could be more closely identified. It had the remarkable properties of acting only in the eggs and only on paternally derived chromosomes (89).

Virus infection has been found to produce chromosome breakage in mammalian cells (90,91), chlorophyll mutations in maize (92), and mutations and rearrangements in *Drosophila* (93,94). The mechanism of this action is not known. The suggestion that it is due to the release of deoxyribonuclease from lysosomes was not borne out by experiment (95).

The production of unstable mutations by viruses or virus-like particles will be discussed in the next chapter.

Genotrophs

In 1963, Durrant (96) reported a very remarkable finding. By growing plants of a flax variety under different fertilizer regimes, he obtained large (L) and small (S) lines, which subsequently bred true to type independent of the fertilizer treatment. On crossing, the lines behaved as genetically distinct, and Durrant assumed that some nuclear change was involved. He called these lines 'genotrophs'. A cytological study of tetraploid pollen mother cells showed that differentiation of the two genotrophs had resulted in some loss of homology between their chromosomes. In a tetraploid, the ratio of quadrivalents to bivalents at meiosis is a measure of the degree of homology between the chromosomes of the parental lines: the higher this ratio, the greater the degree of homology. On this criterion, homology was significantly less in crosses between L and S lines than in crosses between two L and two S lines (97). In addition, the L plants contained 16% more nuclear DNA than the S plants. Analysis of the DNA by ultracentrifugation and renaturation indicated that the extra DNA consisted of a highly multiplied fraction of some of the unique sequences (98). This makes it appear that genotrophs do not arise through mutation proper but through some process that is related to gene amplification, such as has been found in, e.g., the ribosomal genes of *Drosophila* (99). The manner in which fertilizer treatments can produce genotrophs is not known.

Paramutation

In 1956, R.A. Brink (100) made an observation that ran counter to one of the main tenets of Mendelian genetics: the integrity of the genes in heterozygotes. So far, there had been no proved exceptions to the rule that, when a heterozygote forms gametes, the alleles emerge uncontaminated by their temporary association with each other. Now it was found that this rule was broken regularly and predictably by two alleles of a colour gene in maize: R (dark seed colour) and R^{st} (stippled seeds). Heterozygotes for these two alleles regularly yielded R-bearing gametes in which the colour-forming potential of the R-allele had been weakened. Although this change had a strong tendency to persist through the generations, it was not fully stable and, in the absence of the R^{st} gene, there was a gradual restoration of the colour-forming ability of R. This phenomenon, called 'paramutation', has been studied extensively by

Brink and his students, with highly interesting results. Paramutation differs from mutation proper in several important features: it occurs regularly in certain heterozygotes; its direction can be predicted from the particular allele combination; and the genetic change is metastable, with a tendency to partial reversion. We shall have no space in this book to deal more fully with this fascinating subject and its relation with position effects and with the controlling elements to be discussed in the next chapter. Excellent reviews are available (e.g. 101, 102).

References

1. Novick, A. and Szilard, L. (1950), 'Experiments with the chemostat on spontaneous mutations of bacteria', *Proc. Nat. Acad. Sci. U.S.A.* **36**, 708-719.
2. Kubitschek, H.E. and Bendigkeit, H.E. (1964), 'Mutation in continuous cultures. I. Dependence of mutational response upon growth-limiting factors', *Mutation Res.* **1**, 113-120.
3. Rinehart, R.R. (1969), 'Spontaneous sex-linked recessive lethal frequencies from aged and non-aged spermatozoa of *Drosophila melanogaster*', *Mutation Res.* 7, 417-423.
4. Auerbach, C. (1959), 'Spontaneous mutations in dry spores of *Neurospora crassa*', *Zeitschr. Vererb Lehre* **90**, 335-346.
5. Ryan, F.J., Nakada, D. and Schneider, M.J. (1961), 'Is DNA replication a necessary condition for spontaneous mutation?' *Zeitschr. Vererb Lehre* **92**, 38-41.
6. Zamenhof, S. (1960), 'Effects of heating dry bacteria and spores on their phenotype and genotype', *Proc. Nat. Acad. Sci. U.S.A.* **46**, 101-105.
7. Webb, S.J. (1967), 'Mutation of bacterial cells by controlled desiccation', *Nature* **213**, 1137-1139.
8. Zamenhof, S., Eichhorn, H.H. and Rosenbaum-Oliver, D. (1968), 'Mutability of stored spores of *Bacillus subtilis*', *Nature* **220**, 818-819.
9. Drake, J.W. (1966), 'Spontaneous mutations accumulating in bacteriophage T4 in the complete absence of DNA replication', *Proc. Nat. Acad. Sci. U.S.A.* **55**, 738-743.
10. Lindgren, D. (1972), 'The temperature influence on the spontaneous mutation rate. I. Literature review', *Hereditas* **70**, 165-178.

11. Greer, S. and Zamenhof, S. (1962), 'Studies on depurination of DNA by heat', *J. Mol. Biol.* **4**, 123-141.
12. Tamm, C., Hodes, M.E. and Chargaff, E. (1952), 'The formation of apurinic acid from the desoxyribonucleic acid of calf thymus', *J. Biol. Chem.* **195**, 49-63.
13. Bautz-Freese, E. (1961), 'Transitions and transversions induced by depurinating agents', *Proc. Nat. Acad. Sci. U.S.A.* **47**, 540-545.
14. Drake, J.W. and McGuire, J. (1967), 'Characteristics of mutations appearing spontaneously in extracellular particles of bacteriophage T4', *Genetics* **55**, 387-398.
15. Benzer, S. (1961), 'On the topography of the genetic fine structure', *Proc. Nat. Acad. Sci. U.S.A.* **47**, 403-415.
16. Speyer, J.F., Karam, J.D. and Lenny, A.B. (1966), 'On the role of DNA polymerase in base selection', *Cold Spring Harbor Symp. Quant. Biol.* **31**, 693-697.
17. Speyer, J.F. and Rosenberg, D. (1968), 'The function of T4 DNA-polymerase', *Cold Spring Harbor Symp. Quant. Biol.* **33**, 345-350.
18. Drake, J.W. (1973), 'The genetic control of spontaneous and induced mutation rates in bacteriophage T4', *Genetics (Suppl)* **73**, 45-64.
19. Freese, E.B. and Freese, E. (1967), 'On the specificity of DNA polymerase', *Proc. Nat. Acad. Sci. U.S.A.* **57**, 650-657.
20. De Vries, F.A.J., Swart-Idenburg, J.H. and De Waard, A. (1972), 'An analysis of replication errors made by a defective T4 DNA polymerase', *Molec. Gen. Genetics* **117**, 60-71.
21. Bernstein, H. (1971), 'Reversion of frameshift mutations stimulated by lesions in early function genes of bacteriophage T4', *J. Virology* **7**, 460-466.
22. Hershfield, M.S. and Nossal, N.G. (1973), '*In vitro* characterization of a mutator T4 DNA polymerase', *Genetics Suppl.* **73**, 131-136.
23. Bessman, M.J., Muzyczka, N., Goodman, M.F. and Schnaar, R.L. (1974), 'Studies on the biochemical basis of spontaneous mutation. II. The incorporation of a base and its analogue into DNA by wild-type, mutator and antimutator DNA polymerase', *J. Mol. Biology,* **88**, 409-423.
24. Bernstein, C., Bernstein, H., Mufti, S. and Strom, B. (1972), 'Stimulation of mutation in phage T4 by lesions in gene 32 and by thymidine imbalance', *Mutation Res.* **16**, 113-119.

25. Liberfarb, R.M. and Bryson, V. (1970), 'Isolation, characterization and genetic analysis of mutator genes in *Escherichia coli* B and K-12', *J. Bacter.* **104**, 363-375.
26. Treffers, H.P., Spinelli, V. and Belser, N.O. (1954), 'A factor (or mutator gene) influencing mutation rates in *Escherichia coli', Proc. Nat. Acad. Sci. U.S.A.* **40**, 1064-1071.
27. Cox, E.C. (1973), 'Mutator gene studies in *Escherichia coli*: the mutT gene', *Genetics* **73**, (Suppl.) 67-80.
28. Siegel, E.C. and Kamel, F. (1974), 'Reversion of frameshift mutations by mutator genes in *E.coli', J. Bacter.* **117**, 994-1001.
29. Coukell, M.B. and Yanofsky, C. (1970), 'Increased frequency of deletions in DNA polymerase mutants of *Escherichia coli', Nature* **228**, 633-635.
30. Ishii, Y. and Kondo, S. (1972), 'Spontaneous and radiation-induced deletion mutations in *Escherichia coli* strains with different repair capacities', *Mutation Res.* **16**, 13-25.
31. Gefter, M.L., Hirota, Y., Kornberg, T., Wechsler, T., Bernoux, J.A. and Bernoux, C. (1971), 'Analysis of DNA polymerases II and III in mutants of *Escherichia coli* thermosensitive for DNA synthesis', *Proc. Nat. Acad. Sci. U.S.A.* **68**, 3150-3153.
32. Tait, R.C. and Smith, D.W. (1974), 'Role for *Escherichia coli* polymerases I, II and III in DNA replication', *Nature* **249**, 116.
33. Hall, R.M. and Brammar, W.J. (1973), 'Increased spontaneous mutation rates in mutants of *E.coli* with altered DNA polymerase III', *Molec. Gen. Genetics* **121**, 271-276.
34. Bazill, G.W. and Gross, J.D. (1973), 'Mutagenic DNA polymerase in *Bacillus subtilis', Nature New Biology* **243**, 241-243.
35. Cox, E.C., Degnen, G.E. and Scheppel, M.L. (1972), 'Mutator gene studies in *Escherichia coli*: the mutS gene', *Genetics* **72**, 551-567.
36. Fowler, R.G., Degnen, G.E. and Cox, E.C. (1974), 'Mutational specificity of a conditional *Escherichia coli* mutator, mut D5', *Molec. Gen. Genetics* **133**, 179-191.
37. De Lucia, P. and Cairns, J. (1969), 'Isolation of an *E.coli* strain with a mutation affecting DNA polymerase', *Nature* **224**, 1164-1166.
38. Kondo, S. (1973), 'Evidence that mutations are induced by errors in repair and replication', *Genetics* **73**, (Suppl) 109-122.

39. Smirnov, B.G., Filkova, E.V. and Skavronskaya, A.G. (1973), 'Base repair substitution caused by the *uvr502* mutation affecting mutation rates and UV-sensitivity of *Escherichia coli*', *Molec. Gen. Genetics* **126**, 255-266.
40. Smirnov, G.B., Filkova, E.V., Skavronskaya, A.G., Saenko, A.S. and Sinzinis, B.I. (1973), 'Loss and restoration of viability of *E.coli* due to combinations of mutations affecting DNA-polymerase I and repair activities', *Molec. Gen. Genetics* **121**, 139-150.
41. Okazaki, R., Arisawa, M. and Sugino, A. (1971), 'Slow joining of newly replicated DNA chains in DNA polymerase-I deficient *Escherichia coli* mutants', *Proc. Nat. Acad. Sci. U.S.A.* **68**, 2954-2957.
42. Böhme, H. (1967), 'Genetic instability of an ultraviolet-sensitive mutant of *Proteus mirabile*', *Biochem. Biophys. Res. Comm.* **28**, 191-196.

43a. Bridges, B.A., Mottershead, R.P., Rothwell, M.A. and Green, M.H.L. (1972), 'Repair-deficient bacterial strains suitable for mutagenicity screening: tests with the fungicide Captan' *Chem. Biol. Interactions* **5**, 77-84.

43. Mohn, G. (1968), 'Korrelation zwischen verminderter Reparaturfähigkeit für UV-Läsionen und hoher Spontanmutabilität eines Mutatorstammes von *E.coli* K-12', *Molec. Gen. Genetics* **101**, 43-50.
44. Hill, R.F. (1968), 'Do dark repair mechanisms for UV-induced primary damage affect spontaneous mutation?' *Mutation Res.* **6**, 472-475.
45. Zamenhof, S., Heldenmuth, L.H. and Zamenhof, P.J. (1966), 'Studies on mechanisms for the maintenance of constant mutability: mutability and the resistance to mutagens', *Proc. Nat. Acad. Sci. U.S.A.* **55**, 50-58.
46. Novick, A. (1955), 'Mutagens and antimutagens', *Brookhaven Symp. Biol.* **8**, 201-216.
47. Johnson, H.G. and Bach, M.K. (1966), 'The antimutagenic action of polyamines: suppression of the mutagenic action of an *E.coli* mutator gene and of 2-aminopurine', *Proc. Nat. Acad. Sci. U.S.A.* **55**, 1453-1456.

47a. Flink, I. and Pettijohn, D.E. (1975), 'Polyamines stabilize DNA folds', *Nature*, **253**, 62-63.

48. Friedman, S.M. and Weinstein, I.B. (1964), 'Lack of fidelity in the translation of symthetic polyribonucleotides', *Proc. Nat. Sci. U.S.A.* **52**, 988-996.

49. Bresler, S.E., Mosevitsky, M.I. and Vyacheslavov, L.G. (1973). 'Mutations as possible replication errors in bacteria growing under conditions of thymine deficiency', *Mutation Res.* **19**, 281-293.
50. Coughlin, C.A. and Adelberg, E.A. (1956) 'Bacterial mutations induced by thymine starvation', *Nature* **178**, 531-532.
51. Holmes, A.J. and Eisenstark, A. (1968), 'The mutagenic effect of thymine starvation on *Salmonella typhimurium*', *Mutation Res.* **5**, 15-21.
52. Bridges, B.A., Law, J. and Munson, R.J. (1968), 'Mutagenesis in *E.coli.* II Evidence for a common pathway for mutagenesis by ultraviolet light, ionizing radiation and thymine deprivation', *Molec. Gen. Genetics* **103**, 266-273.
53. Von Borstel, R.C., Quah, S.K., Steinberg, C.M., Flury, F. and Gottlieb, D.J.C. (1973), 'Mutants of yeast with enhanced spontaneous mutation rates', *Genetics* **73** (Suppl.), 141-151.
54. Loprieno, N. (1973), 'A mutator gene in the yeast *Schizosaccharomyces pombe*', *Genetics Suppl.* **73**, 161-164.
55. Demerec, M. (1963), 'Selfer mutants of *Salmonella typhimurium*', *Genetics* **48**, 1519-1531.
56. Magni, G.E. (1963), 'The origin of spontaneous mutations during meiosis', *Proc. Nat. Acad. Sci. U.S.A.* **50**, 975-980.
57. Magni, G.E. (1969), 'Spontaneous mutations', *Proc. 12th Internat. Congr. Genetics* **3**, 247-259.
58. Friis, J., Flury, F. and Leupold, U. (1971), 'Characterization of spontaneous mutations of mitotic and meiotic origin in the *ad*-1 locus of *Schizosaccharomyces pombe*', *Mutation Res.* **11**, 373-390.
59. Loprieno, N., Bonatti, S., Abbondandolo, A. and Guglielminetti, R. (1969), 'The nature of spontaneous mutations during vegetative growth in *Schizosaccharomyces pombe*', *Molec. Gen. Genetics* **104**, 40-50.
60. Paszewska, A. and Surzycki, S. (1964), 'Selfers' and high mutation rate during meiosis in *Ascobolus immersus*', *Nature* **204**, 809.
61. Stadler, D.R. and Kariya, B. (1969), 'Intragenic recombination at the mtr locus of *Neurospora* with segregation at an unselected site', *Genetics* **63**, 291-316.
61a. Guerdoux, J.L. (1969), 'Réversion à haute fréquence d'un gène nicotinique moins chez le *Coprinus radiatus*', *Molec. Gen. Genetics* **105**, 334-343.

62. Demerec, M. (1937), 'Frequency of spontaneous mutations in certain stocks of *Drosophila melanogaster', Genetics* **22**, 469-478.

63. Ives, P.T. (1950), 'The importance of mutation rate genes in evolution', *Evolution* **3**, 236-252.

64. Green, M.M. and Lefevre, G. (1972), 'The cytogenetics of mutator gene-induced X-linked lethals in *Drosophila melanogaster', Mutation Res.* **16**, 59-64.

65. Slatko, B.E. and Hiraizumi, Y. (1973), 'Mutation induction in the male recombination strains of *Drosophila melanogaster', Genetics* **75**, 643-649.

65a. Hiraizumi, Y., Slatko, B., Langley, C. and Nill, A. (1973), Recombination in *Drosophila melanogaster* males. *Genetics* **73**, 439-444.

66. Kimura, M. (1967), 'On the evolutionary adjustment of spontaneous mutation rates', *Genet. Res., Camb.* **9**, 23-34.

67. Drake, J.W. (1969), 'Spontaneous mutation', *Nature* **221**, 1128-1132.

68. Gibson, T.C., Scheppe, M.L. and Cox, E.C. (1970), 'Fitness of an *Escherichia coli* mutator gene', *Science* **169**, 686-688.

69. Atwood, K.C., Schneider, L.K. and Ryan, F.J. (1951), 'Selective mechanisms in bacteria', *Cold Spring Harbor Symp. Quant. Biol.* **16**, 345-355.

69a. Springgate, C.F. and Loeb, L.A. (1973), 'Mutagenic DNA polymerase in human leukemic cells'. *Proc. Nat. Acad. Sci. U.S.A.* **70**, 245-248.

70. Goldschmidt, R. (1952), 'Evolution, as viewed by one geneticist', *Amer. Sci.,* **40**, 84-98.

71. Wilson, A.C., Sarich, V.M. and Maxon, L.R. (1974), 'The importance of gene rearrangements in evolution: evidence from studies of chromosomal, protein and anatomical evolution', *Proc. Nat. Acad. Sci. U.S.A.* **71**, 3028-3030.

72. D'Amato, F. and Hoffman-Ostenhof, O. (1956), 'Metabolism and spontaneous mutations in plants', *Adv. Genetics* **8**, 1-28.

73. Floris, C. and Meletti, P. (1972), 'Survival and chlorophyll mutation in *Triticum durum* plants raised from aged seeds', *Mutation Res.* **14**, 118-122.

74. Lewis, C.M. and Holliday, R. (1970), 'Mistranslation and ageing in *Neurospora', Nature* **228**, 877-880.

75. Ann. Report of the Director of Genetics. *Carnegie Inst. Wash. Year Book* **57**, 390-406.
76. Yoshikawa, H. (1966), 'Mutations resulting from transformation of *Bacillus subtilis', Genetics* **54**, 1201-1214.
77. Herdman, M. (1973), 'Mutations arising during transformation in the blue-green alga *Anacystis nidulans', Molec. Gen. Genetics* **120**, 369-378.
78. Taylor, A.L. (1963), 'Bacteriophage-induced mutation in *Escherichia coli', Proc. Nat. Acad. Sci. U.S.A.* **50**, 1043-1051.
79. Daniell, E., Roberts, R. and Abelson, J. (1972), 'Mutation in the lactose operon caused by bacteriophage Mu', *J. Mol. Biol.* **69**, 1-8.
80. Shimada, K. and Weissberg, R. (1973), '*E.coli* mutants produced by the insertion of bacteriophage λ DNA', *Genetics* **73**, (Suppl) 81-83.
81. Kelly, B.L. and Sunshine, M.G. (1967), 'Association of temperate phage P2 with the production of histidine negative segregants by *Escherichia coli', Biochem. Biophys. Res. Comm.* **26**, 237-243.
82. Jordan, E., Saedler, H. and Starlinger, P. (1968), '0^0 and strong-polar mutations in the *gal* operon are insertions', *Molec. Gen. Genetics* **102**, 353-363.
83. Shapiro, J.A. (1969), 'Mutations caused by the insertion of genetic material into the galactose operon of *Escherichia coli', J. Mol. Biol.* **40**, 93-106.
84. Fiandt, M., Szybalski, W. and Malamy, M.H. (1972), 'Polar mutations in *lac, gal* and λ phage consist of a few IS-DNA sequences inserted with either orientation', *Molec. Gen. Genetics* **119**, 223-231.
85. Nomura, M. and Engbaek, F. (1972), 'Expression of ribosomal protein genes as analyzed by bacteriophage Mu-induced mutations', *Proc. Nat. Acad. Sci. U.S.A.* **69**, 1526-1530.
86. De Crombrugghe, B., Adhya, S., Gotterman, M. and Paston, I. (1973), 'Effect of *Rho* on transcription of bacterial operons', *Nature New Biology,* **241**, 260-264.
87. Minamori, S. and Ito, K. (1971), 'Extrachromosomal element δ in *Drosophila melanogaster.* Induction of recurrent lethal mutations in definite regions of second chromosomes', *Mutation Res.* **13**, 361-369.
88. Ito, K. (1974), 'Mutagenicity of an extrachromosomal element δ for X-chromosomes in *Drosophila melanogaster', Jap. J. Genetics* **49**, 25-32.

89. Levitan, M. (1963), 'A maternal factor which breaks paternal chromosomes', *Nature* **200**, 437-438.
90. Nichols, W.W. (1966), 'Studies on the role of viruses in somatic mutation', *Hereditas* **55**, 1-27.
91. Prokofyeva-Belgovskaya, A.A. (1969), 'Action of viruses on chromosomes' in *Principles of Human Cytogenetics,* Meditsina, Moscow 1969.
92. Sprague, G.F., McKinney, H.H. and Greely, L. (1963), 'Virus as mutagenic agent in maize', *Science* **141**, 1052-1053.
93. Burdette, W.J. and Yoon, J.S. (1967), 'Mutations, chromosomal aberrations and tumors in insects treated with oncogenic virus', *Science* **155**, 340-341.
94. Gershenson, S., Alexandrov, Y.N. and Malinta, S.S. (1970), 'Production of recessive lethals in *Drosophila* by viruses non-infectious for the host', *Mutation Res.* **11**, 163-173.
95. Aula, P. and Nichols, W.W. (1968), 'Lysosome and virus induced chromosome breakage', *Exp. Cell Res.* **51**, 595-601.
96. Durrant, A. (1962), 'The environmental induction of heritable changes in *Linum', Heredity* **17**, 27-61.
97. Evans, G.M. (1968), 'Induced chromosomal changes in *Linum', Heredity* **23**, 301-310.
98. Cullis, C.A. (1973), 'DNA differences between flax genotrophs', *Nature* **243**, 515-516.
99. Tartof, K.D. (1973), 'Regulation of ribosomal RNA gene multiplication in *Drosophila melanogaster', Genetics* **73**, 57-71.
100. Brink, R.A. (1956), 'A genetic change associated with the R locus in maize which is directed and potentially reversible', *Genetics* **41**, 872-889.
101. Brink, R.A., Styles, E.D. and Axtell, J.D. (1968), Paramutation: directed genetic change', *Science* **159**, 161-170.
102. Brink, R.A. (1974), 'Paramutation', *Ann. Rev. Genetics* **7**, 129-152.

CHAPTER TWENTY-TWO

Instabilities

Genes with unusually high mutation frequencies, often of the order of per mille or percent, are called unstable. They are found in untreated organisms and may also be produced by mutagenic treatments. We shall start with the discussion of instabilities that had arisen spontaneously or through genetic manipulation but without mutagenic treatment and shall follow this by the discussion of induced instabilities. As pointed out before, there is no hard and fast distinction between spontaneous and induced mutations, and some examples might have found their place equally well in either section.

Instabilities in untreated organisms

Controlling elements

The first to describe and analyse unstable genes in flowering plants and *Drosophila* was Demerec (1,2). He made the interesting observation that some of these genes are stable at certain stages of the life cycle of the organism, unstable at others. In *Delphinium,* for example, one unstable gene produces either tiny coloured spots, which must have arisen very late in development, or very large coloured sectors, which must have arisen early in embryonic development. Another gene appeared to be equally mutable at all developmental stages, producing sectors of all intergrading sizes. Similar observations have subsequently been made on other flowering plants, in particular maize and *Antirrhinum*; these have recently been reviewed by Fincham (3,4). In *Drosophila*, mutability was found to be under the control of age, sex, germinal *versus* somatic tissue, and genetic background (2).

By far the most thorough analysis of mutable genes was carried out by Barbara McClintock in maize; its results have revolutionized our ideas about the nature of the eukaryotic genome. McClintock was the first to postulate that the genome is a dual structure, consisting of structural genes and of controlling elements. When this conclusion was published over 20 years ago (5), it was heterodox in the extreme. Subsequently, controlling elements were recognized in bacteria, where their action could be analysed in great detail. The duality of the genome and the distinction between structural and regulatory elements forms the firm basis of all modern theories of development. In 1971, McClintock was given the National Medal of Science, the highest award the USA can confer on a scientist. It is impossible even to attempt a brief review of her monumental work in this book. I can only indicate those findings that are relevant for the problem of spontaneous instabilities. Their important implications for a study of gene action and development will not be considered.

Characteristic of the controlling elements in maize is their arrangement into two-element systems. One of these elements modifies the activity and mutability of a structural gene, sometimes of several neighbouring ones, with which it is closely associated in the *cis*-position. It is subordinate to a second element which may be anywhere in the genome. In parallel with the controlling elements of bacteria, McClintock has called the first element 'operator', the second, 'regulator' (6). While the terminology is convenient, it should be realized that there are great differences between the regulatory systems in bacteria and maize.

The most startling property of the maize controlling elements is their ability to transpose spontaneously from one chromosomal location to another on the same or a different chromosome. These transpositions can be recognized by their effects on the newly acquired neighbouring genes; they are not due to cytologically detectable translocations. It is the mobility of the controlling elements within the genome that leads to instabilities. It is also this mobility that makes the action of the elements in development amenable to analysis. It should be realized, however, that mobility may not be the normal behaviour of controlling elements. Those studied by McClintock occurred in plants with a history of breakage-fusion-bridge cycles (Chapter 2), and this may have resulted in disruption of the normal functioning of the control system concerned.

In the present context, it is the transpositions themselves that concern us because they are the origin of what used to be called 'mutable genes'. The hierarchy between the two elements of a system is shown by the fact that the regulator element transposes autonomously, while the operator element does so only on a signal from its specific regulator. In cells lacking the appropriate regulator, the operator element remains locked at its site close to or within (7) the structural gene that it controls. Under these conditions, it completely or partially inhibits activity of this gene. The inhibited gene behaves like a stable amorph or hypomorph (Chapter 3) until the missing regulator element of the system is introduced by crossing. This results in mobility of the operator element away from its original site to a new one somewhere else in the genome, and this in turn entails three consequences. (1) The activity of the originally inhibited gene becomes released. (2) Another gene becomes inhibited through insertion of the operator near to it or inside it; this can be observed only in the rare instances when the newly inhibited gene happens to be monitored in the cross, or in the case of operator elements that produce chromosome breaks in their neighbourhood. (3) The progeny contains reversions to the original phenotype or mutants to new phenotypes at the locus from which the operator element has disappeared.

As an example, let us consider one of the genes that are concerned with anthocyanine formation. In cooperation with other genes, it produces purple kernels. Strains that have an operator gene inserted close to this structural element but lack the regulator element of the same system have colourless kernels. This property is stably inherited as a Mendelian trait as long as the appropriate regulator element is not introduced into the cell. When this happens, both phenotype and hereditary transmission change in a remarkable manner. The plant now forms kernels with a pattern of coloured spots, and the progeny contains plants with different degrees of anthocyanine forming ability. While it is doubtful whether the coloured spots on the kernels should be attributed to a release of gene activity or to mutation, the colour-forming plants have all the attributes of mutants. Their genetic determinants map in the same gene; their phenotypes are those typically found in allelic series of colour genes; and in the absence of the regulator element, each 'mutant' breeds as a stable Mendelian trait.

Clearly, neither these mutations nor the amorphs and hypomorphs from which they arose are likely to be due to base substitutions or to the minute deletions or duplications that result in frameshifts. Their closest resemblance is, on the one hand, to insertion mutations and, on the other hand, to mutations produced by 'eduction' of bacteriophage (Chapter 21). This similarity has been stressed recently in a paper that deals with the turning-off and turning-on of the *gal* operon in *E.coli* under the influence of an insertional element, whose effect varies with its direction of insertion (8). A related interpretation compares the controlling elements in maize with bacterial episomes (9), and this finds support from bacterial instabilities that appear to be under episomal control (10,11).

Meanwhile, evidence for the existence of similar controlling elements in other eukaryotes is slowly accumulating, although in no case has the analysis gone even approximately as far as in maize. No two-element controlling systems have been reported in other eukaryotes; but this may not constitute an essential difference from maize. Some maize genes, too, are autonomously unstable, probably because they carry the controlling elements 'packaged' together with the structural one. On the other side, there are remarkable similarities between the effects of the controlling elements in maize and the behaviour of some unstable genes in other organisms. In maize, each controlling system produces its characteristic pattern of somatic mutations shown, e.g., as the anthocyanine spotting pattern of the kernels. The same is true for *Delphinium* (1), *Antirrhinum* (3), and *Drosophila* (2). In maize, certain controlling elements cause not only gene inactivation and mutation, but also chromosome breakage and deficiencies; the same applies to some mutable genes in *Drosophila* (12). In maize, controlling elements may change their 'state' abruptly, possibly through some loss or gain of active material, and this results in alterations of the degrees to which the various effects – inhibition of gene action, mutations, deficiencies etc – are expressed. In *Drosophila*, a mutable duplication was shown to exist in two different 'states' distinguished by their effect on the phenotypic expression of a neighbouring gene (13,14). In mice, the X-chromosome element that controls X-inactivation in females and inhibits gene action in an inserted autosomal fragment, exists in two or three different states with different degrees of inhibitory action (15). In maize, the operator elements of several systems have been mapped close to or within (7) the affected structural gene; in

Drosophila, evidence has been presented that a mutable gene may be due to the insertion of foreign DNA (13). Finally, the most characteristic property of controlling elements in maize, their ability for spontaneous transposition, has been found to be possessed by several chromosomal elements of *Drosophila* (12,16).

Exosomes

In Chapter 21 we have seen that transformation in prokaryotes may induce mutations in genes close to the transformed one. An essentially different situation has been described for transformation in eukaryotes. Here it is the transformed gene itself that may remain unstable, alternating in its expression between the transformed and the original state. It has been suggested that this is due to the formation by the donor DNA of an 'exosome', i.e. a particle that remains in close association with its homologous host element but never becomes linearly integrated into the chromosome. This model was derived from extensive studies on *Drosophila* (17,18). It has also been tentatively applied to similar phenomena in *Bombyx* (19), flowering plants (20,21), and *Neurospora* (22). Since exosomes are not presumed to produce mutations in the proper sense, they are of marginal relevance to this book, and only the findings on *Drosophila* will be briefly discussed.

When *Drosophila* females are induced to lay their eggs very rapidly by means of a special collecting device, the 'ovitron', these eggs lack the outer chorion and are permeable to DNA. Eggs from strains that were homozygous for recessive marker genes were exposed to DNA from their own strain or to DNA from a wild-type strain carrying the normal alleles of the marker genes. After the first kind of treatment, all emerging flies were of the host type. After the second type of treatment, a fraction of them showed 'transformed' spots, i.e. small areas in which the host phenotype had been replaced by that of the donor, such as red facets in otherwise white eyes, or dark bristles in an otherwise yellow environment. Some of these mosaically 'transformed' flies yielded similar 'mosaic transformants' among their progeny, indicating involvement of their gonads in transformation. Lines could be established in which the same kind of mosaicism was transmitted through the generations, although occasionally it was lost in a subline. Yet complete transformations, e.g. flies with wholly red eyes or wholly wild-type hypodermis, were not encountered in any generation. Thus, the peculiar features to be explained were (1) the restriction of transformation to small islands of cells and (2) the absence of complete transformants in spite of regular transmission of mosaic transformation. The exosome model explains these findings in the following way. The donor DNA forms a close association with its homologous segment on the host chromosome. It replicates in step with it and usually is transmitted with it in mitosis and meiosis; occasionally it may be lost, yielding lines that no longer show mosaic transformation. As long as the exosome is attached to a gene, this gene offers two alternative templates for transcription: the original mutant one, and the transformed wild-type one. In support of this model it was found that (1) mosaically transformed flies still carry the original mutant information and (2) the exosome maps at the mutant locus that it suppresses. Curiously enough, in the one case where this was not so, the exosome mapped very close to or at the site of a suppressor locus for the mutant gene.

Tandem duplications

One of the earliest discovered mutant genes in *Drosophila* was Bar. It is a sex-linked gene with semi-dominant action, which reduces eye size drastically in males and homozygous females, while heterozygous females have kidney-shaped eyes (Fig. 22.1). The gene has been mentioned before in Chapter 3 as example for a neomorph, and in Chapter 5 as a marker gene in tests for sex-linked lethals. It was early noticed that stocks homozygous for Bar occasionally produce revertants to wild-type, and that these are matched in frequency by flies with a more extreme type of Bar, called double-Bar. This suggested an event that yields complementary products, and it was postulated (23,24) that Bar is a tandem duplication which, through 'unequal' or 'oblique' crossing-over (Fig. 22.2), gives rise to the normal single condition and the exaggerated Bar effect of the triplication. Cytological examination confirmed the hypothesis (Fig. 22.3).

A similar case has been described more recently for a regulatory gene in *Drosophila*. The effect of the Hairless gene is enhanced by a closely neighbouring gene, called Enhancer-Hairless; reversion to unmodified Hairless occurs with unusually high frequency. The suspicion that Enhancer-Hairless was a tandem duplication was confirmed in experiments in which two different duplications were introduced into the vicinity of the non-modified Hairless gene: both had the same effect as the original enhancer (25).

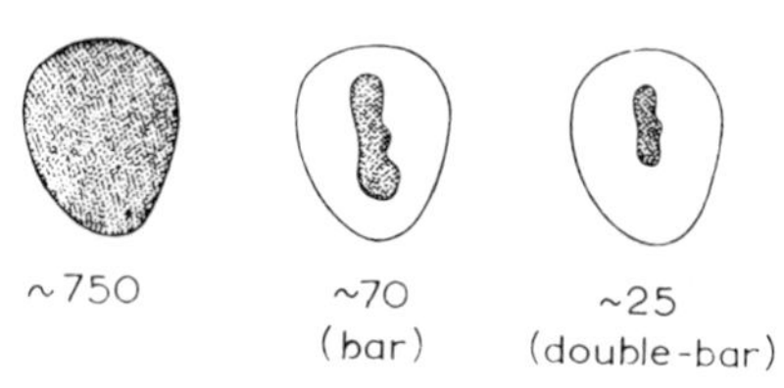

Fig. 22.1. Mutant eye shapes in *Drosophila*. (*a*) wild-type, (*b*) Bar, (*c*) double-Bar. The figures indicate the approximate number of facets in the eye

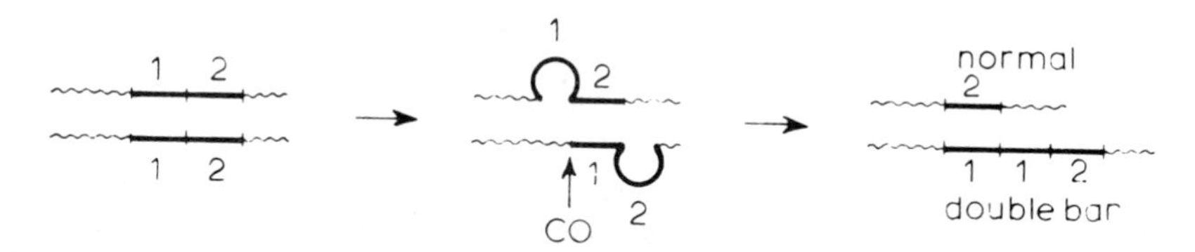

Fig. 22.2. The origin of Bar-reversions and double-Bar through oblique crossing-over in a tandem duplication.

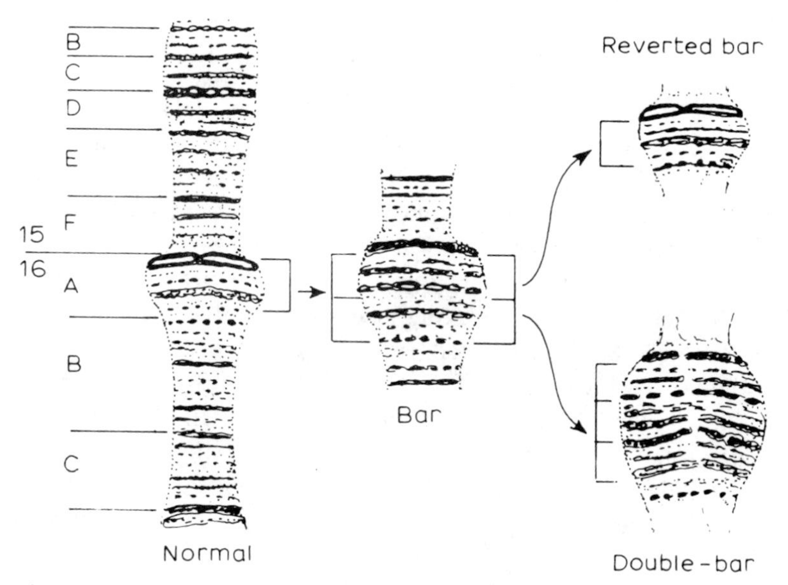

Fig. 22.3. The Bar region of the X-chromosome in the salivary glands of *Drosophila melanogaster*. (Fig. 1, in Bridges (24) courtesy Am. Ass. for the Advancement of Science.

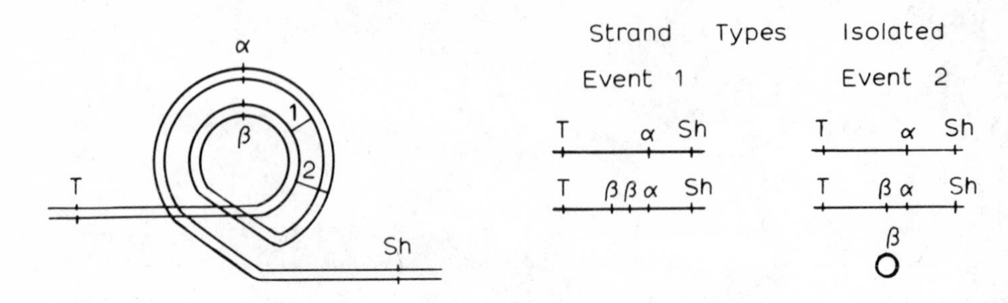

Fig. 22.4. Proposed mechanism for reversions of a tandem duplication without recombination. *a* and *β* are the two elements of a duplication in maize, similar to the Bar duplication. T and Sh are flanking markers. Only one chromosome is shown, separated into sister chromatids. At the left, the homologous *a* and *β* loci have formed a loop at meiosis. Two types of intrachromosomal exchange can occur. Exchange 1, involving sister chromatids, yields a strand that has lost the duplication and a complementary strand with a triplication. Event 2, which involves a single chromatid, yields a parental strand with the duplication and a strand that has lost the duplication; the *β* segment forms an acentric ring and will be lost at mitosis. (Fig. 4, in Laughnan, Mutation and Plant Breeding (Bibliography). Courtesy Nat. Acad. Sciences, Nat. Research Council U.S.A.)

Other cases of instability in *Drosophila* (26,27) and maize (J.R. Laughnan in 'Mutation and Plant Breeding', Bibliography) have been shown to be due to tandem duplications. Reversions may or may not be accompanied by recombination between flanking markers; they may even occur under conditions, such as the presence of an inversion or deletion, when crossing-over is prevented. In these cases, reversions have been attributed to sister-strand crossing-over or to intrastrand crossing-over within the double loop formed by the duplication (Fig. 22.4). In *E.coli* crossing-over within a duplication loop accounts for the instability of a mis-sense suppressor in a duplicated *t*-RNA gene (28). In *Drosophila*, unequal sister-strand crossing-over has been suggested as the mechanism by which bobbed mutants, which carry reduced numbers of tandemly repeated units in the structural gene for ribosomal RNA, revert to normal in certain genotypes (29).

Translocations

In Chapter 2 (Fig. 2.3), we have seen that a translocation heterozygote

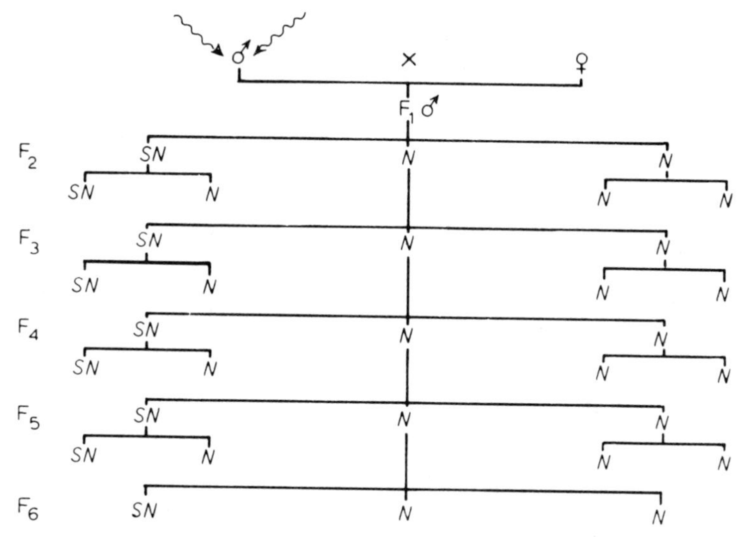

Fig. 22.5. The inheritance of Snaker (*SN*) in the mouse. *N* denotes normal phenotype. (The P_1 male had been treated with triethylenemelamine) (Fig. 1, in Cattenach (30), Courtesy Berlin-Heidelberg-New York: Springer)

produces six types of gamete, only two of which are viable. The other four carry a duplication for one of the translocated segments, and a deficiency for the other. Occasionally it may happen that one of these unbalanced types yields viable individuals, usually of abnormal phenotype. Where this applies, phenotypically normal heterozygotes for the full translocation segregate 'mutant' offspring in a proportion that depends on the segregation pattern of the quadrivalent and on the viability of the unbalanced zygotes. Seemingly, a new dominant mutation has arisen in the parental gonad and, since the same 'mutant' continues to occur among the progeny of certain normal individuals of the same line, the line seems to carry a highly mutable gene. Cytological or genetical analysis is necessary to show that the mutant phenotype is the expression of chromosome imbalance rather than of a mutant gene. Because of the importance of this kind of instability for the genetics of human abnormalities, I shall briefly discuss an example from mammalian genetics.

Fig. 22.5 shows a pedigree of 'Snaker', a behavioural mutant in the mouse (30). The fact that the translocation responsible for this particular pedigree had been produced by a chemical mutagen is irrelevant; if it had arisen spontaneously, it would have yielded the same pattern of inheritance. The phenotypically normal P_1 male gave rise to three types of progeny:

(a) normal mice that produced only normal progeny;
(b) Snakers that bred as heterozygotes for a dominant abnormality;
(c) phenotypically normal mice that gave the same segregation pattern as the P_1 male. An interpretation that fitted this pedigree and was borne out by further genetical tests was the following.

$$\male P_1 \ \frac{Dp}{+}\ \frac{Def}{+} \times \frac{+}{+}\ \frac{+}{+}$$

$$F_1 \ \underbrace{\underset{a_1}{\frac{+}{+}\ \frac{+}{+}}\ ;\ \underset{a_2}{\frac{Dp}{+}\ \frac{+}{+}}}\ ;\ \underset{b}{\frac{+}{+}\ \frac{Def}{+}}\ ;\ \underset{c\,(=P_1)}{\frac{Dp}{+}\ \frac{Def}{+}}$$

$$\frac{Dp}{+}\quad \frac{Def}{+}$$

reconstituted *c*

The P_1 male and his type *c* progeny carried a balanced translocation, probably of the insertional type, a small piece ofone chromosome being inserted into a non-homologous one. The Snakers (*b*) were heterozygous for the deficiency, but did not carry the duplication; apparently a haplo-insufficient gene (Chapter 3) had been uncovered by the deficiency. The normal mice (*a*) were of two types: those with only normal chromosomes (a_1) and those carrying a small duplication, complementary to the deficiency in the Snakers and without effect on the phenotype (a_2). The fact that animals of type *c* could be 'reconstituted' by crossing Snakers with type a_2 normal mice proved the correctness of this interpretation. The cytological analysis was inconclusive, probably because of the small size of the chromosome segment or segments involved; but in a similar case in mice it could be shown that the insertion of a piece of autosome into the X was responsible for the recurrent segregation of the same abnormal phenotype in successive generations (31).

Similar pedigrees have repeatedly been reported for human abnormalities caused by chromosomal imbalance. Thus Down's syndrome ('mongolism'), which is due to trisomy of the small chromosome 21, may arise either through non-disjunction or through segregation in a translocation heterozygote that carries one Chromosome 21 attached to another autosome. In the first case, the condition is not inherited, although a mother may have several non-disjunctional children. In the second case, there is a familial incidence through the generations, similar to that of Snaker in the mouse. Fig. 22.6 shows a pedigree of such a family.

Transposed duplications

The duplications that arise from translocations or transpositions are not arranged in tandem. There is therefore no reason to expect that they should be unstable through oblique crossing-over. Yet in several cases a pronounced instability has been found to characterize translocation-duplications. In *E.coli*, transposed duplicated segments of the tryptophan operon reverted at frequencies of about

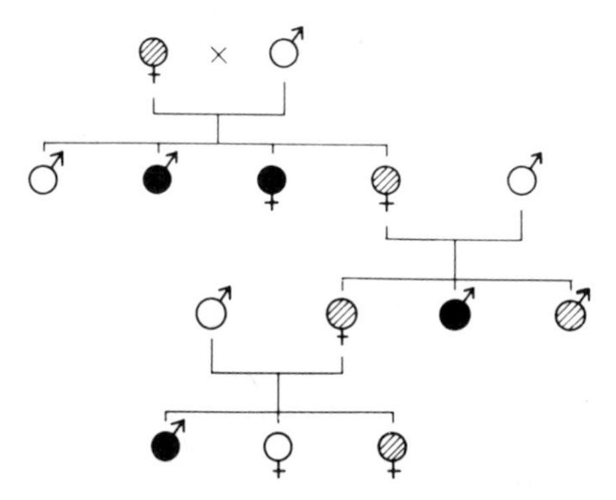

Fig. 22.6. Pedigree of a family in which Down's syndrome is due to a translocation. The solid symbols stand for persons with Down's syndrome. The shaded symbols stand for phenotypically normal translocation heterozygotes. Modified from Fig. 70 in Petit et Prévost, Génétique et Evolution. Courtesy Hermann, Paris.

1%, and this could not be accounted for by a simple crossover model (32). A similar case in *Aspergillus* has been analysed in great detail and with interesting results (33,34).

In 1966, Bainbridge and Roper reported the repeated occurrence of an unstable mutation in *Aspergillus nidulans*. The mutation was called *crinkled* because of its abnormal colony growth; it also had a markedly reduced linear growth rate. It segregated out in a number of crosses between phenotypically normal parents, but in each case one of the parents carried a translocation between Chromosomes III and VIII. Since in haploid organisms all but very small deletions are lethal, the authors explained their findings by the hypothesis that the translocations yielding crinkled segregants were between chromosome segments of very unequal lengths, so that one of the unbalanced combinations would carry a large duplication and a very small deletion, while in the other the situation was reversed. The latter was assumed to be responsible for the dead ascospores found in all these crosses; the former was assumed to result in crinkled mutants. Genetical analysis fully confirmed the hypothesis. In particular, the use of marker genes made it possible to show that crinkled colonies carried a piece of Chromosome III attached to Chromosome VIII (Fig. 22.7). The effect is, however, not specific

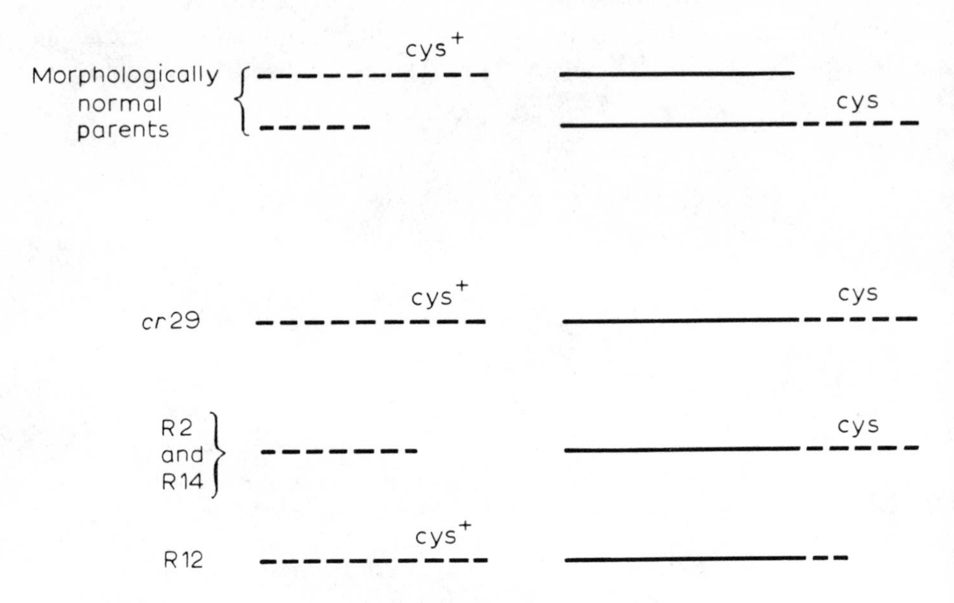

Fig. 22.7. The origin of crinkled and its revertants in *Aspergillus nidulans.* Broken lines represent the relevant part of Chromosome III, solid lines that of Chromosome VIII. *cys* (cystine-less) is a marker gene. *cr29* is a crinkled mutant; R2, R12, and R14 are reversions. Note that in R2 and R14 the loss of duplicate material has occurred in the original position, while in R12 it has occurred in the transposed position. (Fig. 1, in Bainbridge (33). Courtesy Cambridge University Press.

for a certain chromosome region or even for a certain pair of chromosomes. Crinkled colonies were also obtained from duplicated segments of other chromosome regions and chromosomes. Common to all crinkled strains was a high frequency of mitotic reversion to improved growth rate and morphology, resulting in sectors of more normal appearance that outgrew the rest of the colony. This instability has been called 'mitotic non-conformity'. Genetic analysis showed that the revertants had lost part or all of the duplication

(Fig. 22.7). The fact that it made no difference whether the loss affected the duplicate segment in its original or in its transposed position showed that the crinkled phenotype is not due to a position effect but to chromosomal imbalance.

The origin of mitotic non-conformity raised a problem that has not yet been solved. As I said at the beginning of this section, unequal crossing-over can hardly explain the loss of material from transposed duplications; in the case of *Aspergillus,* it was also excluded by experiment. So was the possibility that, in *Aspergillus,* spontaneous deletions are always frequent but that only those covered by a duplication can survive. If this were true, diploids should yield many deletions but this was not the case. When, however, a diploid carried a transposed duplication, i.e. when it carried a chromosome segment in triplicate, then loss of superfluous genetic material was again restricted to the triplicate segment. Thus, the duplication itself is responsible for its instability. At the moment, the most promising hypothesis, which is also applicable to the bacterial case (p.443), assumes that instability is connected with replication of the chromosomes at membrane sites. If these sites should be specific for chromosome segments rather than for whole chromosomes, then competition between duplicate segments for the same site might lead to under-replication of one of the segments. Tests for this hypothesis are of interest not only for the problem of mitotic non-conformity but also for an understanding of chromosome replication.

Induced instabilities

UV and most, probably all, chemical mutagens can induce instabilities; whether X-rays can do so is doubtful. Tests for the induction of instabilities by alkylating agents and for their pattern of inheritance in *Drosophila* and yeast have been described in Chapter 15. In yeast, similar results have been obtained with nitrous acid and UV. In *Drosophila,* calf-thymus DNA produced instabilities for small deletions (Chapter 18).

When tested in the same system, mutagens vary greatly in the frequencies with which they yield instabilities; on the other hand, the same mutagen may produce different proportions of instabilities in different systems. Thus, in the red-to-white system of fission yeast (Chapter 11), replicating instabilities were frequent after treatment with UV, EMS and NG, but rare after treatment with nitrous acid or hydroxylamine (35). Also in bacteriophage, nitrous acid appears to produce few unstable changes; otherwise even one-stranded phages should yield a proportion of mottled plaques, contrary to what has been observed (Chapter 18). In *E.coli*, however, nitrous acid yielded high frequencies of unstable mutations from lac^+ to lac^-. Heat had a similar effect; this may form a link with spontaneous instabilities (36).

In considering possible causes of induced instabilities two main points have to be kept in mind.

(1) The instabilities are able to replicate as such, i.e. an unstable organism or cell produces three types of progeny: stable non-mutants, stable mutants, and instabilities that again segregate these three types, often in proportions that remain more or less the same over many generations (37) (Fig. 15.7). Any parent mosaic that yields more than one mosaic in its progeny must carry an instability that can replicate in the unstable state. This property distinguishes replicating instabilities from delayed mutations, i.e. mutations that appear one or more cell generations after treatment. Explanations that can account for delayed mutations are not necessarily applicable to the origin of replicating instabilities. In particular, this is true for an explanation that is often considered to account for the induction of replicating instabilities by alkylating agents. It is based on the results of experiments with phage T4 (38). Phage particles were treated with EMS *in vitro* and applied to bacteria at low multiplicity. Single bursts were analysed for the proportion of mutants at the rII locus; the number of phage replications between infection and burst was estimated from the burst sizes. Clones of different sizes yielded different proportions of mutants; the distribution could be well fitted by the assumption that the mutational lesion, presumably an alkylated guanine, retained the same probability of producing a mutation over the 8 or more phage generations scored. Other cases of delayed mutations may have the same origin but, since alkylated guanine cannot replicate as such, replicating instabilities must arise in different ways.

(2) Induced replicating instabilities are locus-or site-specific; this distinguishes them from instabilities that are due to transposable controlling elements. The locus-specificities of instabilities in *Drosophila,* and the site-specificities of instabilities in yeast have been described in Chapter 15. Replicating instabilities that had been

induced by the addition of calf-thymus DNA to the food of *Drosophila* larvae (39) yielded overlapping deficiencies in successive generations (Fig. 22.8). In all these cases, the site, locus or chromosomal segment affected varied between different unstable lines; this makes it impossible to attribute the specificities to more general specific effects of the mutagen used, such as the preference of calf thymus DNA for a certain part of the second chromosome.

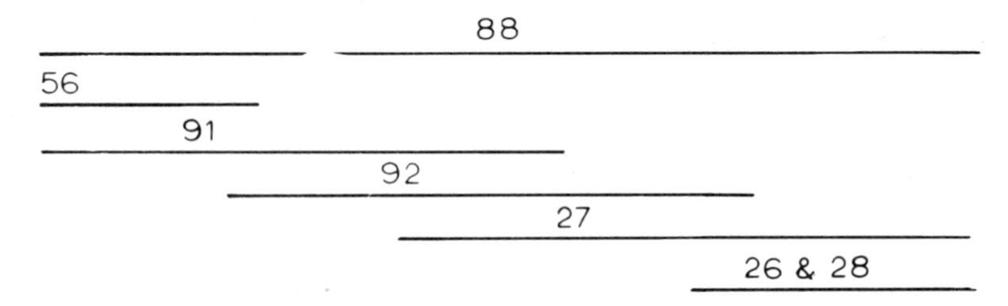

Fig. 22.8. Complementation map of *de novo* arisen lethals in the F_1, F_2, and F_3 of a male that had been raised on food containing calf thymus DNA. Lethals 26, 27 and 28 had arisen in F_1, lethal 56 in F_2, and lethals 88, 91 and 92 in F_3. (Fig. 1, in Mathew (39). Courtesy Cambridge University Press.)

At our present state of knowledge, it is difficult to find a satifactory testable model that fulfills these requirements. Obviously, if a mutagen produces a translocation that, like Snaker in mice, segregates the same abnormal progeny through the generations, this is formally an induced replicating instability. Indeed Snaker, as mentioned above, had been chemically induced; the similarity of the Snaker pedigree to that of a replicating instability in *Drosophila* (Fig. 15.6) is striking. However, only a very small proportion of replicating instabilities are due to translocations. Tandem duplications are a more likely source for, once produced, they usually undergo some amount of oblique crossing-over. The hypothesis that replicating instabilities arise from induced tandem duplications is especially attractive for *Drosophila* where such duplications form a high proportion of chemically induced changes (Chapter 15). An attempt to test the hypothesis was unsuccessful (40): no replicating instabilities were found among the progeny of flies that had been raised on formaldehyde-treated food, a treatment that is known to yield many tandem duplications (Chapter 18). In any case, duplications of the sizes observed in *Drosophila* cannot be responsible for the site-specific instabilities in yeast. Loss or insertion

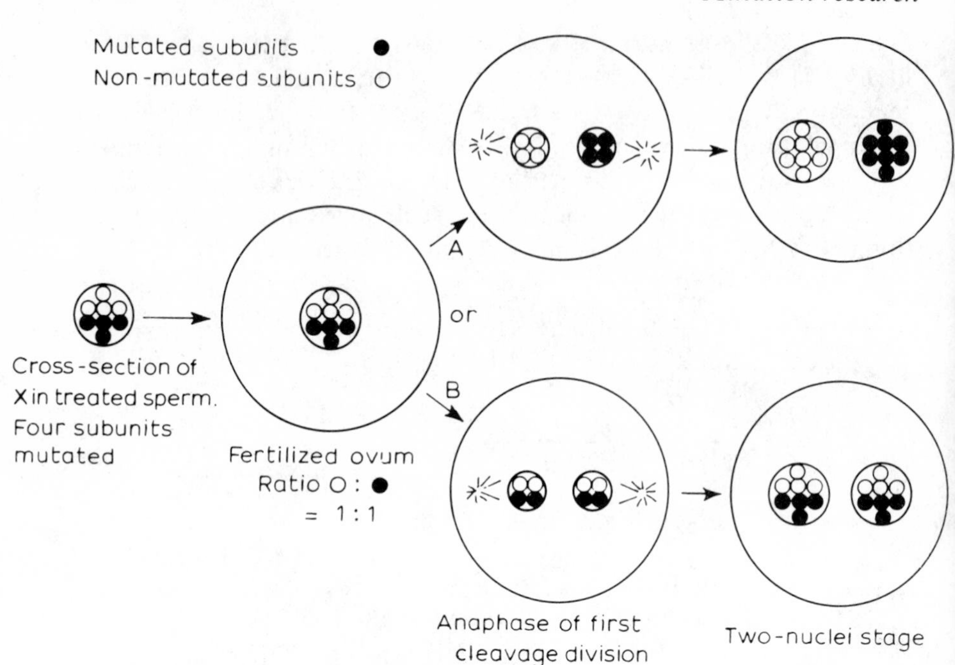

Fig. 22.9. The sorting-out of mutated and non-mutated strands in a multistranded chromosome. (Fig. 4, Auerbach (41). Courtesy Long Island Biological Association.)

of controlling elements or episomes under the influence of the treatment are possibilities to be considered, but there is no proof for any of these.

A possibility that, for eukaryotes, has been considered from the start (41) and is still being considered is the assumption that the chromosome is multistranded and that mutagens may affect only one or a few of these strands, leaving the remainder in their original state. In the course of successive cell divisions, chromosomes with different proportions of mutant strands segregate out and continue to do so until lines with only mutant or only non-mutant strands have been obtained (Fig. 22.9). The model may well apply to some cases, but it is difficult to apply it to the long-continued lines of instabilities in fission yeast (37). Recently, a new and attractive hypothesis has been brought forward; in its formal consequences, it is very similar to the multistrand model (42). Instead of assuming that the eukaryote chromosome consists of a multiplicity of strands, it assumes that the eukaryote gene consists of a multiplicity of tandemly repeated units. A 'rolling helix' transmits information between units of the same gene, and a mutant change in one of the units may eventually appear in all of them.

References

1. Demerec, M. (1931), 'Behaviour of two mutable genes of *Delphinium ajacis*', *J. Genetics* **24**, 179-193.
2. Demerec, M. (1941), 'Unstable genes in *Drosophila*', *Cold Spring Harbor Symp. Quant. Biol.* **9**, 122-126.
3. Fincham, J.R.S. (1970), 'The regulation of gene mutation in plants', *Proc. Roy. Soc. Lond. B* **176**, 295-300.
4. Fincham, J.R.S. (1973), 'Localized instabilities in plants – a review and some speculations', *Genetics* **73**,(Suppl.) 195-205.
5. McClintock, B. (1951), 'Chromosome organization and genic expression', *Cold Spring Harbor Symp. Quant. Biol.* **16**, 13-47.
6. McClintock, B. (1965), 'The control of gene action in maize', *Brookhaven Symp. Biol.* **18**, 162-184.
7. Nelson, O.E. (1968), 'The waxy locus in maize. II. The location of the controlling element alleles', *Genetics* **60**, 507-524.
8. Saedler, H., Reif, H.J., Hu, S. and Davidson, N. (1974), 'IS2, a genetic element for turn-off and turn-on of gene activity in *E.coli*', *Molec. Gen. Genetics* **132**, 265-290.
9. Peterson, P.A. (1970), 'Controlling elements and mutable loci in maize: their relationship to bacterial episomes', *Genetica* **41**, 35-56.
10. Dawson, G.W.P. and Smith-Keary, P.F. (1963), 'Episomic control of mutation in *Salmonella typhimurium*', *Heredity* **18**, 1-20.
11. Schwartz, N.M. (1965), 'Genetic instability in *Escherichia coli*', *J. Bacter.* **89**, 712-717.
12. Green, M.M. (1973), 'Some observations and comments on mutable and mutator genes in *Drosophila*', *Genetics* **73** (Suppl), 187-194.
13. Rasmuson, B., Green, M.M. and Karisson, B.M. (1974), 'Genetic instability in *Drosophila melanogaster.* Evidence for insertion mutations', *Molec. Gen. Genetics* **133**, 237-248.
14. Rasmuson, B. and Green, M.M. (1974), 'Genetic instability in *Drosophila melanogaster* A mutable tandem duplication', *Molec. Gen. Genetics* **133**, 249-260.
15. Cattanach, B.M. and Isaacson, J.H. (1967), 'Controlling elements in the mouse X chromosome', *Genetics* **57**, 331-346.

16. Ising, G. and Ramel, C. (in press), '*The behaviour of a transposing element in Drosophila melanogaster*' in: The Genetics and Biology of *Drosophila*. Ashburner, M. and Novitski, E. (eds).
17. Fox, A.S. and Yoon, S.B. (1970), 'DNA-induced transformation in *Drosophila*: locus specificity and the establishment of transformed stocks', *Proc. Nat. Acad. Sci. U.S.A.* **67**, 1608-1615.
18. Fox, A.S., Yoon, S.B. and Gelbart, W.M. (1971), 'DNA-induced transformation in *Drosophila*: genetic analysis of transformed stocks', *Proc. Nat. Acad. Sci. U.S.A.* **68**, 342-346.
19. Nawa, S., Sakaguchi, B., Yamada, M. and Tsugita, M. (1971), 'Hereditary changes in *Bombyx* after treatment with DNA', *Genetics* **67**, 221-234.
20. Ledoux, L. and Huart, R. (1974), 'DNA-mediated genetic correction of thiamineless *Arabidopsis thaliana*', *Nature* **249**, 17-21.
21. Hess, D. (1973), 'Transformationsversuche an höheren Pflanzen: Untersuchungen zur Realisation des Exosomen-Models der Transformation bei *Petunia hybrida*', *Zeitschr. Pflanzenphysiol.* **68**, 432-440.
22. Mishra, N.C. and Tatum, E.L. (1973), 'Non-Mendelian inheritance of DNA-induced inositol-independence in *Neurospora*', *Proc. Nat. Acad. Sci. U.S.A.* **70**, 3875-3879.
23. Muller, H.J., Prokofyeva-Belgovskaya, A.A. and Kossikov, K.V. (1936), 'Unequal crossing-over in the *Bar* mutant as a result of duplication of a minute chromosome section', *C.R. (Dokl) Acad. Sci. U.S.S.R. N.S.*, **1**, 83-88. (Russian and English).
24. Bridges, C.B. (1936), '*Bar* as duplication', *Science* **83**, 528-530.
25. Nash, D. (1970), 'The mutational basis for the 'allelic' modifier mutants *enhancer* and *suppressor* of *hairless* of *Drosophila melanogaster*', *Genetics* **64**, 471-479.
26. Peterson, H.M. and Laughnan, J.R. (1963), 'Interchromosomal exchange at the *Bar* locus in *Drosophila*', *Proc. Nat. Acad. Sci. U.S.A.* **50**, 126-133.
27. Kalisch, W.E. and Becker, H.J. (1970), 'Über eine Reihe mutabler Allele des white-locus bei *Drosophila melanogaster*', *Molec. Gen. Genetics* **107**, 321-335.
28. Hill, C.W., Foulds, J., Soll, L. and Berg, P. (1969), 'Instability

of a missense suppressor resulting from a duplication of genetic material', *J. Mol. Biol.* **39**, 563-581.

29. Tartof, K.D. (1974), 'Unequal mitotic sister chromatid exchange as the mechanism of ribosomal RNA gene magnification', *Proc. Nat. Acad. Sci. U.S.A.* **71**, 1272-1276.
30. Cattanach, B.M. (1965), 'Snaker: a dominant abnormality caused by chromosomal imbalance', *Zeitschr. Vererb. Lehre* **96**, 275-284.
31. Ohno, S. and Cattanach, B.M. (1962), 'Cytological study of an X-autosome translocation in *Mus musculus', Cytogenetics* **1**, 129-140.
32. Jackson, E.N. and Yanofsky, C. (1973), 'Duplication-translocation of tryptophan operon genes in *Escherichia coli', J. Bacter.* **116**, 33-40.
33. Bainbridge, B.W. and Roper, J.A. (1966), 'Observations on the effects of a chromosome duplication in *Aspergillus nidulans', J. Gen. Microbiol.* **42**, 417-424.
34. Nga, B.H. and Roper, J.A. (1969), 'A system generating spontaneous intrachromosomal changes at mitosis in *Aspergillus nidulans', Genet. Res., Camb.,* **14**, 63-70.
35. Nasim, A. (1967), 'The induction of replicating instabilities by mutagens in *Schizosaccharomyces pombe', Mutation Res.* **4**, 753-763.
36. Zamenhof, S. (1961), 'Gene unstabilization induced by heat and by nitrous acid', *J. Bacter.* **81**, 111-117.
37. Nasim, A. (1974), 'UV-induced replicating instability in *Schizosaccharomyces pombe', Mutation Res.* **22**, 25-31.
38. Green, M.M., MacDonald, D. and Krieg, D.R. (1961), 'The delayed origin of mutants induced by exposure of extracellular phage T4 to ethylmethane-sulfonate', *Proc. Nat. Acad. Sci. U.S.A.* **47**, 64-72.
39. Mathew, C. (1965), 'The production of recessive lethals by calf-thymus DNA in *Drosophila', Genet. Res., Camb.,* **6**, 165-174.
40. Ratnayake, W.E. (1970), 'Lack of evidence for the hypothesis that chromosomal duplications are the causes of replicating instabilities', *Ceylon J. Sci. Biol. Sciences* **8**, 49-52.
41. Auerbach, C. (1951), 'Problems in chemical mutagenesis', *Cold Spring Harbor Symp. Quant. Biol.* **16**, 199-213.
42. Thomas, C.A. Jr. (1973), 'The rolling helix: a model for the eukaryotic gene?' *Cold Spring Harbor Symp. Quant. Biol.* **38**, 347-352.

CHAPTER TWENTY-THREE

Applied mutation research

The application of mutation research to concerns of the human society poses its own problems and requires its own techniques. It is a wide field. In this chapter, I can do no more than indicate the nature of the problems, very briefly mention some of the techniques, and provide references for further reading.

Broadly speaking, applied mutation research tries to satisfy two opposite demands: (1), to increase the number of useful mutations; (2), to minimize the number of harmful mutations.

I. INCREASING THE NUMBER OF USEFUL MUTATIONS

All existing organisms are the result of long periods of evolution during which their genotype has become adapted to their peculiar ecological niche. An increased overall mutation frequency will therefore always create a preponderance of harmful mutations, with beneficial or otherwise desirable ones forming only a small minority. The same applies to the domesticated animals and plants that are the result of artificial evolution by breeding. Here, too, most new mutations are likely to disrupt the genotypical adaptation of the organism to its agricultural environment. At present we have no means for specifically inducing desired mutations (Chapter 20); probably, we shall never have more than very inadequate means for doing this. It follows that attempts to create useful variations by mutations have to be restricted to species with a high reproductive rate allowing a great amount of wastage, i.e. to micro-organisms and cultivated plants. Whether, in future, transformation or related techniques of information transfer into DNA will make it possible to produce directed variations in animals and man is a problem that is outside the scope of this book.

In certain situations, mutations that are harmful to the cells or organisms treated are beneficial to man; in particular, this applies to cancer therapy and sterilization of noxious insects. The genetical effects involved are usually chromosome breaks or rearrangements, and no specificity beyond good chromosome breaking ability is required of the mutagen used.

Mutation breeding

From the end of the 1930's, ionizing radiation has been used for the production of mutations in agricultural plants. Later on, chemical mutagens became available, and a number of them, e.g. ethylene imine and ethylmethanesulphonate, find widespread application. The problems, techniques and results of mutation breeding have been discussed in two symposia held in 1961 in the USA (Mutation and Plant Breeding, Bibliography) and in 1972 in Vienna (Induced Mutations and Plant Improvement, Bibliography). In a lecture delivered in 1972, Nilan (1) lists the main directions in which mutation breeding can be improved. Mutagens, he states, 'must be made more effective (more numerous mutations per dose of mutagen), more efficient (higher frequency of such desirable events as point mutations in relation to such undesirable events as sterility and chromosome aberrations), and more specific (higher frequencies of mutations at specific loci)'. Along the first two lines, marked success has been obtained, mainly because the ratio between point mutations and chromosome breaks differs between mutagens and, for a given mutagen, can be modified by the experimental conditions. Thus, the full potentiality of sodium azide (Chapter 18) as a mutagen for barley was realized only when it was applied at pH3 where it produces high frequencies of morphological mutations not accompanied by appreciable frequencies of chromosome aberrations. The high efficiency of alkylating agents as compared with ionizing radiation (Chapter 16) makes them very suitable for mutation breeding. Nilan quotes a striking example for the production of mildew resistant mutants in barley by either X-rays or EMS: while the number of partially or completely resistant mutants per 100 000 F_2 plants from treated seeds was 5.7 after X-radiation, it was 12.4 after treatment with EMS.

As would be expected, the greatest difficulty is offered by the third requirement: locus specificity of the induced mutations. In fact, what the mutation breeder needs is not so much locus specificity as phenotypic specificity, i.e. the specific, or at least preferential, induction of certain *types* of mutation, such as straw stiffness, disease resistance etc, no matter what their genetical basis. Several examples of locus or phenotypic specificity have been described, although their causes are not yet understood. In some cases, this kind of specificity, too, might be determined by the ratios between intergenic and intragenic changes that different treatments produce. This applies probably to the finding that 'albinas', which may be due to deletions as well as to point mutations, are much more frequent among chlorophyll mutations produced by irradiation than among chemically induced ones. It may also explain a moderate degree of locus specificity shown by neutrons for one of several loci that determine strawstiffness in barley (erectoides mutants); for mutations at the preferred locus are often connected with chromosome aberrations.

For other locus or phenotypic specificities, this explanation is unlikely. Khvostova and Tarasenko (2) have discussed this in relation to work on higher plants. A particularly striking case of phenotypic specificity has been described for the tomato (3). Treatment of the seeds with three chemical mutagens – hydrazine, hydroxylamine or EMS – produced very different mutational spectra. Thus, anthocyanine mutations occurred exclusively after treatment with hydrazine, while certain morphological mutations that were quite frequent after treatment with hydroxylamine or EMS were rare or absent in the hydrazine series. Although the three mutagens probably react with different nucleotide bases (Chapter 17), it is most unlikely that this should cause such striking phenotypic specificities. A more plausible explanation may be derived from the analysis of a similar type of specificity found after treatment of *Drosophila* larvae with hydrazine or hydroxylamine (Chapter 20). There it was found that specificity arose at the last step in the mutation process: the growth of mutant cells into clones. The fact that treated tomato seeds – which already contain a multicelled embryo – yielded wholly mutant plants, supports the assumption that the environment of the treated seed exercised heavy selection for or against mutant cells.

Manipulation of the 'sieves' in the mutation process (Chapter 20) would seem one means for obtaining at least a certain degree of phenotypic specificity. Unfortunately, the correct choice of sieves can only be a hit-and-miss process at present and is likely to remain so in the near future. A more rational approach to the creation of locus specificity has been described by Swaminathan (4). It is based on the finding that certain chemical mutagens act preferentially on derepressed genes (Chapter 20). Preliminary results with treatment of seeds at various stages of germination gave promising results.

The recessivity of most mutations creates difficulties in the mutation breeding of diploid plants. Melchers has stressed the advantage of using haploid plants for the purpose (5). The possibilities of growing plants from mutagenically treated cell or tissue cultures have been discussed by Nickell and Heinz (6).

While chromosome rearrangements are usually undesirable in projects for improved plant varieties, they may be important in the creation of new karyotypes; this has been discussed by Gaul (7). The ability of most mutagens to increase cross-over frequencies may be used to obtain rare recombinants.

Mutation breeding in micro-organisms is favoured by the large numbers that can be raised, the short generation times, and the relatively small outlay in equipment and labour. Even without means for producing specific mutations, practically all kinds of desired mutant can be obtained by treatment with UV or chemicals.

Genetic control of insect pests

In an article bearing this title, Smith and von Borstel (8) have reviewed and critically discussed genetical techniques for reducing the size of a noxious insect population and for eventually eliminating it from a circumscribed area. The most obvious and earliest used method was the release of large numbers of mutagenically treated males that had remained sexually competitive with normal males but, because of induced dominant lethals in their sperm, produced no viable progeny. This 'male-sterile' technique had its first spectacular success when applied to a cattle-pest, the screw-worm fly, in the southern regions of the United States. Where the method is not applicable, e.g. because the effort to release repeatedly large numbers of sterile males is too great, or because the margin between the sterilizing and sexually incapacitating doses is too small, other techniques are available or are being developed. One, already suggested in 1940 by Serebrovsky in a Russian journal (9) and subsequently proposed independently in Britain by Curtis (10), makes use of the inherited semi-sterility of translocation heterozygotes (Chapter 2). Translocations are induced in the laboratory, and homozygotes for one or several of them are released into the wild population. Mated to normal insects, they produce offspring that is more or less sterile, depending on the number of translocations for which it is heterozygous. The advantage of this method is that the partial sterility is inherited as a dominant character and that, therefore, laboratory insects need to be released only occasionally in order to maintain the gradual decline of the population. Additional possibilities, in particular the use of conditional lethals (Chapter 3), are discussed in reference 8. More recently, Denell (11) has suggested the use of mutations for male sterility in insects with the XY type of sex determination.

Cancer therapy

We have seen (Chapter 2) that chromosome breaks and rearrangements are lethal to dividing cells, but can be tolerated by non-dividing ones. This is the basis for cancer treatment by means of chromosome-breaking agents such as ionizing radiation or 'radiomimetic' chemicals like nitrogen mustard or TEM. Although the correlation between the chromosome-breaking and carcinostatic abilities of chemicals is good, it is far from perfect. There can be many reasons for the exceptions; in Chapter 18 we have discussed the special case of monofunctional alkylating agents. These are not carcinostatic in spite of their ability to cause chromosome breakage, possibly because of the very long latency period of most of the induced breaks.

The main problem of cancer therapy by radiation or radiomimetic chemicals consists of finding conditions under which the effect on the tumour cells is maximal while still remaining tolerable for the surrounding normal cells. It is here that studies on cell killing and chromosomal damage in model systems can be of help. Any conditions that enhance the sensitivity of the malignant cells while leaving that of the normal ones unchanged or only moderately changed, offer possibilities for improved treatment techniques. So do conditions that make the normal cells more resistant while leaving the malignant ones sensitive. Some of these possibilities have been discussed by Alper (12). Special importance attaches to the oxygen effect on radiation damage (Chapter 7). Solid tumours often contain an anoxic core that remains undamaged by the treatment and from which the tumour can regenerate. Attempts to reach these recalcitrant cells are made by fractionating the dose. This gradually reduces the size of the tumour and this, in turn, improves the blood supply to its core. In some hospitals, the patient is made to breathe oxygen during treatment. The normally oxygenated tissue that surrounds the tumour will not be further sensitized. Radiation of high LET, as we have seen in Chapter 7, does not depend on oxygen for its chromosome-breaking ability; this suggests the suitability of neutrons for cancer therapy. It is also possible that suitable fractionation intervals might increase the sensitivity of the chromosomes; we may recall here the very drastic effect of certain fractionation regimes on mutation frequencies in mice (Chapter 9). Finally, it may become possible to apply radiation-sensitizing substances, e.g. bromouracil, specifically to tumour cells, or radioprotective substances specifically to the healthy environment (Chapter 7).

II. REDUCING THE NUMBER OF HARMFUL MUTATIONS

Radiation hazards

Very soon after his discovery of the mutagenic action of X-rays, Muller started to alert the medical profession to the genetical consequences of carelessly and avoidably exposing the human gonads to radiation. In spite of a sustained campaign by himself and other geneticists, not much attention was paid by either the medical profession or the general public. This changed after the second world war, when fall-out from nuclear explosions became an additional source of radiation, and one to which not only individuals but whole populations were exposed. This danger has now receded, but other sources of radiation hazards to individuals and populations have been opened up through the increasing uses of nuclear power for energy supply.

In 1956, the British Medical Research Council and the National Academy of Sciences of the USA independently prepared reports (13,14) in which the biological hazards from environmental ionizing radiation were assessed. Guidelines were provided for limiting the overall dose to the population as well as the individual doses to specially exposed persons such as workers in atomic energy establishments. Because of the linear dose-effect curve for X-ray induced mutations (Chapter 5), no dose was considered to be wholly without genetical dangers, and calculations of permissible doses had to weigh the risks against the benefits to be derived from the uses of radiation in medicine, defence, and industry.

In the intervening years, much new material has come to light, which necessitates re-thinking of the previous conclusions. In 1972, this task was carried out by an advisory committee of the National Academy of Sciences (15). Their report provides a very clear discussion of the issues involved, of the uncertainties that beset attempts to assess the expected damage, and of the means by which, in spite of these uncertainties, conclusions have been drawn and practical measures have been recommended. The major source of information for this task were the reports of the United Nations Scientific Committee on the Effects of Atomic Radiation (UNSCEAR). This committee is meeting regularly every few years in order to survey and evaluate the data obtained in the intervening period. The latest meeting took place in 1972(16). The reports go into great detail; for the public, they are accompanied by a short report summarizing the main conclusions (17). The most recent review of the field has been provided by Sankaranarayanan (18).Searle (19) has reviewed the evidence from radiation experiments on mice, which form the most important basis for the estimates of radiation hazards to man.

Although there can be no doubt that ionizing radiation causes genetic damage to man, many questions are still under debate. They concern the manner in which the damage will affect individuals and populations, the way in which it will be spread over successive generations, and the quantitative evaluation of the damage done by defined types of exposure. Crow (20) has discussed the basic questions lucidly in an article in which he also deals with the important concepts of 'genetic load' and 'genetic death', which were introduced by Muller as measurements of population damage. The former is the frequency of deleterious mutations in the population; the latter, the extinction of a gene lineage through death or infertility of an individual carrying the gene. Crow concludes that the main danger to mankind is the production of recessive lethal or harmful mutations which, under the cover of diploidy, may persist over many generations until they are lost by death or infertility of the homozygotes. In *Drosophila,* many lethal or harmful recessives are not fully recessive and reduce viability of the heterozygotes to a slight but noticeable degree. If this should be true also for man, then the greatest amount of genetic radiation damage would be due to such heterozygous effects, which might manifest themselves in many ways, such as lowered disease resistance, reduced stamina, an increased tendency to break down under physical or mental stress etc. Not only will these disabilities affect already the first generation; they may also be transmitted through many generations, leaving a trail of more or less drastic human misery until the responsible gene is finally eliminated. Doubtless,, there also exist recessive genes that, through heterosis, increase the fitness of the heterozygotes. Their role in radiation damage has been much discussed, but there is general agreement that they can in no way compensate for the ill effects of deleterious genes.

Some of the statements in Crow's article have been superseded by more recent findings. The most important of these refers to the effects of dose rate. In 1957, Crow could still write that 'the number of mutations induced is strictly proportional to the amount of radiation, irrespective of the intensity or spacing of the dose'. We now know (Chapter 9) that this is true only for spermatozoa, while in spermatogonia and oocytes repair processes remove a larger proportion of potential mutations after chronic than after acute exposures. Since spermatogonia and oocytes, because of their long life span, are much more at risk than the short-lived spermatozoa, this finding has to be taken account of in calculations of radiation damage. It is true that there is a lower threshold to the decrease of mutation frequency with dose rate; it is even possible that, at very low dose rates, mutation frequency may again increase slightly (Chapter 9). All the same, corrections for the effect of dose rate must reduce the estimate of radiation damage and increase the value of the permissible dose. So – if the preliminary results on mice and *Drosophila* should be confirmed (Chapter 9) – would the finding that very low doses, even when delivered acutely, are less effective than predicted from the linear dose-effect curve. Nonetheless, it is highly improbable that continuous exposure to even very low doses of man-made ionizing radiation at low dose rates should not produce *some* genetic damage and, although the *percentage* of affected individuals may be very small, their actual *number* will still be large.

In any case, when one is dealing with the life and happiness of future generations, one should always err on the side of caution and exposures should be reduced to an extent that is compatible with the welfare of the present one. As Crow states: 'Every preventable instance of disease or death is one too many'.

A genetic hazard that had not yet been realized in 1957 is the creation of chromosomal imbalance by non-disjunction, translocation, or chromosome loss. Both trisomy and monosomy (Chapter 2) give rise to serious abnormalities, ranging from abortions (many of which are due to chromosome imbalance) over grossly abnormal and short-lived infants to viable but severely handicapped individuals. We have already mentioned Down's syndrome (Mongolism), which is due to trisomy for the small chromosome 21 (Fig. 22.6). Monosomy or trisomy for one of the larger autosomes is lethal, but sex-chromosome imbalance does not usually interfere with survival. In contrast to *Drosophila*, in which the Y-chromosome plays no role in sex-determination, so that XXY flies are females and XO flies are males, mammals carry male-determining factors on the Y-chromosome. Zygotes that, through non-disjunction in either the father or the mother, carry a Y in addition to two X-chromosomes, develop into sterile XXY males (Klinefelter syndrome). Zygotes that, through non-disjunction or chromosome loss, carry an X but no Y, develop into sterile females (Turner syndrome). The frequencies with which these and other types of chromosomal imbalance occur spontaneously makes them a major genetical hazard. How much their incidence is likely to be increased by additional background radiation is under debate. It is true that X-rays produce translocations and chromosome breaks (resulting in chromosome loss) in mammals as they do in *Drosophila*; but translocations are very rare after low radiation doses. Non-disjunction was the first genetical X-ray effect to be discovered in *Drosophila* (Chapter 5), but experiments on mice gave no evidence that irradiation yields increased frequencies of imbalance for the sex-chromosomes. Nor was there an increased incidence of children with Down's syndrome among the offspring of Japanese mothers that had survived the atom bomb explosions. Recently, however, there have been disturbing data both on humans (21) and mice (22), which show an interaction between low radiation doses and maternal age in the production of non-disjunction.

Environmental chemical mutagens

Having been at last successfully alerted to the genetic dangers of ionizing radiation, civilized man did not need much persuasion to accept the geneticists' warning against the dangers of potential mutagens among the countless new substances that are used in medicine, industry, pest control, food preservation etc. In fact, there is a risk that geneticists, by stressing these dangers, may make the public over-anxious. Yet, it is quite possible and even probable that the overall effect of all these chemicals, acting separately or in combination, may be a far greater genetic hazard than that posed by ionizing radiation. It is the inescapable duty of geneticists to assess this hazard as accurately as possible, of governments to lay down guidelines for the clearance of substances for use, and of industry to follow these guidelines and collaborate in the testing. All this is, indeed, under way. A proposal for a 'three-tier' screening procedure of potential mutagens has been outlined by Bridges (23).

An 'Environmental Mutagen Society (EMS)' was founded in USA in 1970, and in Europe in 1971. Such societies exist now in many countries. The American EMS publishes regular surveys of the already enormous literature on the subject (Bibliography). The periodical 'Mutation Research' has a separate section on 'Environmental Mutagenesis and Related Subjects', which includes reports from relevant scientific meetings and chapters from a forthcoming handbook of testing procedures. The three-volume handbook 'Chemical Mutagens' (Bibliography) deals with the principles and methods of testing. The book by Fishbein *et al.* (Bibliography) is a compilation of what was known about chemical mutagens in 1970. A special book is devoted to reviews on 'Chemical Mutagenesis in Mammals and Man' (Bibliography). In all these areas, new information is accumulating rapidly.

It should be clearly realized that the genetical damage caused by chemicals is of the same elusive but far reaching and widespread type as that anticipated from radiation. A chemical that increases the frequency of non-disjunction would constitute a particularly ominous hazard. So far, no good techniques of testing routinely for non-disjunction in animals are available. In the soybean, Vig (24) has developed a system for scoring non-disjunction as well as mutation, rearrangement formation and recombination.

Mutagenic effects should not be confused with teratogenic ones on the developing embryo. The thalidomide tragedy, e.g., was not caused by a mutagenic but by a teratogenic action of the drug.

Those among its victims that reproduce have no higher probability of putting malformed children into the world than has the bulk of the population. It is true that the cytotoxic action of many mutagens may lead to teratogenic effects on embryos; but the opposite is not true, and many teratogens are not mutagenic (see article by Kalter in Chemical Mutagens, vol. 1, Bibliography).

Endeavours to assess genetic damage from environmental chemicals are confronted with very great difficulties, much greater than those encountered in the assessment of radiation damage. While X-rays and other ionizing radiations can penetrate to the chromosomes of every cell and, almost certainly, produce chromosome breaks and mutations in all of them, this cannot be presumed for chemicals. We have seen that chemicals may show very marked specificities for organisms or, even, for different cells within the same organism. Thus, urethane is mutagenic for *Drosophila* but not for *Neurospora*; sodium azide for barley but not for *Drosophila* (Chapter 18). The addition of formaldehyde to the food of *Drosophila* larvae produces many mutations in spermatogonia but none in female germ cells (Chapter 18). A substance that is mutagenic in a test system is therefore not necessarily mutagenic for man. Conversely, chemicals that fail to produce mutations in test organisms may do so after metabolic transformation in a mammal. This has been shown to be the case for a number of chemicals, and various tests have been developed for treating lower organisms with mammalian metabolites of suspect substances.

It is also not possible to assume for chemicals, as may be assumed for X-rays, that they will always produce the various types of genetic effect in roughly the same proportions. Indeed, we know of many instances where they fail to do so. Thus, in *Drosophila*, the ratio of rearrangements to point mutations varies greatly between chemicals and, after treatment with alkylating agents, can be increased tenfold or more by storage (Chapter 15). In the *Drosophila* X-chromosome, the ratio of small deletions to visible, presumably intragenic, mutations is much higher for alkylating agents than for hydroxylamine (Chapter 17). Under these circumstances, it becomes very important to establish correlations between those effects that are most likely to cause genetic damage and those that are most suitable for routine tests, which include not only tests for mutations and rearrangements but also for somatic recombination.

Dose effect curves create another problem. For X-rays, they may be assumed to be roughly linear under almost all circumstances. Extrapolation from higher to lower doses is therefore legitimate although, as we have just seen, even here this will entail inaccuracies. Dose-effect curves for chemicals acting inside cells or organisms cannot be presumed to be linear; they may be steeper or flatter than this, or they may be bi- or polyphasic (Chapter 16). Some chemicals might even have a threshold for their mutagenic action and fail to produce mutations below a given concentration. *Dose-rate effects* have so far been little considered and, indeed, they may be difficult to obtain for short-lived chemicals, often tested on short-lived cells. Yet, as we have seen in Chapter 16, dose-rate may be an important factor in determining mutagenic effects.

As an example for the complexity of the problems, let me briefly discuss the results obtained with caffeine. Recent reviews can be found in 'Chemical Mutagenesis in Mammals and Man' (Bibliography) and in the introductory article to vol 26, No. 2, of Mutation Research. Only a few key references will be given here; many additional ones are found in the review articles. Caffeine, because of the great amount consumed in coffee and tea, has been one of the first substances for which mutagenicity testing was considered necessary. Indeed, mutagenic effects of caffeine on fungi (25) and bacteria (26) were reported already in 1948 and 1951. Tests on *Drosophila* gave, at best, ambiguous results; in mice, even very prolonged treatments failed to produce visible mutations, dominant lethals or translocations (27,28); for dominant lethals, this was subsequently confirmed in several laboratories.

The ability of caffeine to break the chromosomes of plants had likewise been discovered very early (29). At the time, this did not attract much general attention; but the attitude of the public had changed when Ostertag (30) in 1965 reported that, in human leukocytes and HeLa cells, high doses (500-10,000 μg/ml) of caffeine produce chromosome breaks, very few of which rejoin to form rearrangements. His conclusion that caffeine is 'one of the most dangerous mutagens for man' is not shared by others. Adler (see 'Chemical Mutagenesis in Mammals and Man'; Bibliography) points to a number of unwarranted assumptions that underly this conclusion, such as that (1) the linear dose-effect curve found for high doses can be extrapolated to concentrations below 500 μg/ml; (2) there is no effect of dose rate; (3) germ cells *in vivo* respond to caffeine as do cultured cells *in vitro*; and others. Even if all these assumptions were justified, it would remain very doubtful whether the observed effect – chromosome breakage without reunion – would result in any damage other than dominant lethals leading to very early and, possibly, unnoticed abortions. Meanwhile, Kihlman has found that caffeine produces chromosome breaks by two quite different mechanisms. One of these – the one described by Ostertag and possibly restricted to animal cells cultured at 37° C – is independent of oxidative phosphorylation but dependent on DNA synthesis and results only in chromosome fragmentation. The other, found by Kihlman and his collaborators in plants and in Chinese hamster cells cultured at 30°, requires oxidative phosphorylation, is not restricted to the S-phase and results in chromosome aberrations as well as in fragmentation. Apart from the unphysiologically low temperature at which this potentially more dangerous effect occurs, it too requires very high concentrations of caffeine, and Kihlman concludes that a possible genetic risk from the consumption of caffeine is, at worst, exceedingly low. This conclusion is shared by Adler.

One might, thus, dismiss caffeine altogether from the danger list, were it not for its potentiating effect on mutagenesis by UV and certain chemicals. This aspect is considered in several articles in *Mutation Research* 26, No. 2. According to Kihlman, the concentration required for a potentiating effect on chromosome damage is still well above the range to which coffee and tea drinkers are exposed; but this need not be true for the effect on mutation frequencies. The origin of this interaction effect is still under study; it probably arises at the repair step and, under certain conditions, may lead to *reduced* rather than *increased* frequencies of UV-induced mutations (31).

Fortunately, it is not to be expected that many chemicals will have such complex effects on mutagenesis as do purine-derivatives. It is obvious that, as a first approach, the testing of suspected mutagens in the environment will have to be restricted to assessing the reality and degree of their mutagenic action. It is, however, well to keep in mind that the overall effects of all these chemicals and of radiation may well be synergistic or, in some cases, antagonistic.

Mutation and the origin of cancer. Mutagens and carcinogens

At the beginning of this century, the German biologist Boveri put forward the theory that cancer cells may arise from normal ones by somatic mutation; this theory, which appeared first in a German medical journal, was later published in bookform and translated into English (32). The somatic mutation theory of cancer has never quite disappeared from the scientific consciousness, although it has had to compete with others (for a discussion see the article by Miller and Miller in Chemical Mutagens, vol. 1, Bibliography). The somatic mutation theory gained strong support from the finding that most potent mutagens – X-rays, UV, alkylating agents – are also carcinogens. The opposite, however, did not seem to be true: many powerful carcinogens, e.g. 2-aminofluorene, methylcholanthrene and dibenzanthracene, did not appear to be mutagenic. Recently, however, it has been found that many of them are converted into mutagens by mammalian metabolism (33-36), and this has led to a revival of the somatic mutation theory. The fact that certain cancers are now known to be induced by virus action conflicts less with this theory than would have been the case ten years ago; for the borderline between the induction of mutations by chemical changes within a gene or by the insertion of foreign nucleotides into a gene has become tenuous (Chapter 21). If, indeed, it can be assumed that most mutagens produce the type of somatic change that results in cancer, then somatic mutations or related processes (such as prophage induction) would form a much more urgent and immediate source of biological danger than that due to germinal mutations. At present, the tasks of estimating mutagenic and carcinogenic hazards from environmental chemicals are entrusted to different agencies; this may well change in future. Meanwhile, it has been suggested that the laborious and time-consuming tests for carcinogenesis in animals may be replaced by the rapid and cheap tests for mutagenicity in bacteria. This question plays a prominent role in present-day discussions on biological dangers from environmental agents.

References

1. Nilan, R.A. (1973), 'Increasing the effectiveness, efficiency and specificity of mutation induction in flowering plants', In: 'Genes, Enzymes and Populations', pp. 205-222, ed. A. Srb, Plenum Publ. Corporat.
2. Khvostova, V.V. and Tarasenko, N.D. (1970), 'Problems of the specificity of experimental mutagenesis in higher plants', *Uspekhi sovremmennio biologii* **69**, 409-423.

3. Jain, H.K., Raut, R.N. and Khamankar, Y.G. (1968), 'Base specific chemicals and mutation analysis in *Lycopersicon'*, *Heredity* **23**, 247-256.
4. Swaminathan, M.S. (1969), 'Mutation Breeding', *Proc. 12th Int. Congr. Genetics,* **3**, 327-347 (Science Council of Japan).
5. Melchers, G. (1972), 'Haploid higher plants for plant breeding', *Z. Pflanzenzüchtg.* **67**, 19-32.
6. Nickell, L.G. and Heinz, D.J. (1973), 'Potential of cell and tissue culture techniques as aids in economic plant improvement', In: 'Genes, enzymes and population', pp. 109-128, ed. A. Srb, Plenum Publ. Corporat.
7. Gaul, H. (1965), 'Induced mutations in plant breeding', In 'Genetics Today' vol. 3, Pergamon Press, New York, 689-709.
8. Smith, R.H. and Borstel, von R.C. (1972), 'Genetic control of insect populations', *Science* **178**, 1164-1174.
9. Serebrovsky, A.S. (1940), 'On the possibility of a new method for the control of insect pests', *Zool. Zh.* **19**, 618-630.
10. Curtis, C.F. (1968), 'A possible genetic method for the control of insect pests, with special reference to tsetse flies *(Glossina spp)', Bull. Ent. Res.* **57**, 509-523.
11. Denell, R.E. (1973), 'Use of male sterilization mutations for insect control programmes', *Nature* **242**, 274-275.
12. Alper, T. (1967), 'Application of radiobiology in radiotherapeutic developments', In: 'Modern Trends in Radiotherapy I., pp. 1-33, ed., Deeley and Wood, Butterworth.
13. 'The hazards to man of nuclear and allied radiations', Medical Research Council, London, Her Majesty's Stationery Office, 1956.
14. 'The biological effects of atomic radiation'. Summary reports from a study by the National Acad. Sciences. Nat. Res. Council, Washington, D.C. 1956.
15. 'The effects on populations of exposure to low levels of ionizing radiation'. Report of the advisory committee on the biological effects of ionizing radiation. Division of medical sciences. Nat. Acad. Sci. USA, Nat. Res. Council, Washington, D.C. 1972.
16. 'Ionizing radiation: levels and effects'. A report of the United Nations Scientific Committee on the effects of atomic radiation. 2 volumes. United Nations New York. 1972.

17. 'Report of the United Nations Scientific Committee on the Effects of Atomic Radiation'. General Assembly, official records: 27th session. Suppl. No. 25 (A/8725) 1972.
18. Sankaranarayanan, K. (1974), 'Recent advances in the assessment of genetic hazards of ionizing radiation', *Atomic Energy Review* **12**, 47-74.
19. Searle, A.G. (1974), 'Mutation induction in mice' *Adv. Rad. Biol.* **4**, 131-207.
20. Crow, F.J. (1957), 'Possible consequences of an increased mutation rate', *Eugenics Quarterly* **4**, 67-80.
21. Uchida, I.A., Holunga, R. and Lawler, C. (1968), 'Maternal radiation and chromosomal aberration'. *Lancet* II, 1045-1049.
22. Yamomoto, M., Shimada, T., Endo, A. and Watanabe, G. (1973), 'Effects of low-dose X-radiation on the chromosomal non-disjunction in aged mice'. *Nature New Biology* **244**, 206-207.
23. Bridges, B.A. (1974), 'The three-tier approach to mutagenicity screening and the concept of radiation-equivalent dose'. *Mutation Res.* **26**, 335-340.
24. Vig, B.K. (1973), 'Somatic crossing over in *Glycine max* (L) Merrill: effect of some inhibitors of DNA synthesis on the induction of somatic crossing over and point mutations', *Genetics* **73**, 583-596.
25. Fries, N. and Kihlman, B. (1948), 'Fungal mutations obtained with methylxanthines', *Nature* **162**, 573.
26. Novick, A. and Szilard, L. (1951), 'Experiments on spontaneous and chemically induced mutations growing in the chemostat', *Cold Spring Harbor Symp. Quant. Biol.* **16**, 337-343.
27. Cattanach, B.M. (1962), 'Genetical effects of caffeine in mice', *Zeitschr. Vererb. Lehre.* **93**, 215-219.
28. Lyon, M.F., Phillips, J.S.R. and Searle, A.G. (1962), 'A test for mutagenicity of caffeine in mice', *Zeitschr. Vererb. Lehre.* **93**, 7-13.
29. Kihlman, B.A. and Levan, A. (1949), 'The cytological effect of caffeine', *Hereditas* **35**, 109-111.
30. Ostertag, W., Duisberg, E. and Stürmann, M. (1965), 'The mutagenic activity of caffeine in man', *Mutation Res.* **2**, 293-296.

31. Loprieno, N. and Schüpbach, M. (1971), 'On the effect of caffeine on mutation and recombination in *Schizosaccharomyces pombe*', *Molec. Gen. Genetics* **110**, 348-354.
32. Boveri, T. (1929), 'The origin of malignant tumors', Williams and Wilkins, Baltimore.
33. Ames, B.N., Durston, W.E., Yamasaki, E. and Lee, F.D. (1974), 'Carcinogens are mutagens: a simple test system combining liver homogenates for activation and bacteria for detection', *Proc. Nat. Acad. Sci. U.S.A.* **70**, 2281-2285.
34. Maher, V.M., Douville, D., Tomura, T. and Lancker, van J.L. (1974), 'Mutagenicity of reactive derivatives of carcinogenic hydrocarbons: evidence of DNA repair', *Mutation Res.* **23**, 113-128.
35. Isono, K. and Yourno, J. (1974), 'Chemical carcinogens as frameshift mutagens: *Salmonella* DNA sequence sensitive to mutagenesis by polycyclic carcinogens', *Proc. Nat. Acad. Sci. U.S.A.* **71**, 1612-1617.
36. Commoner, B., Vithayathil, A.J. and Henry, J.I. (1974), 'Detection of metabolic carcinogen intermediates in urine of carcinogen-fed rats by means of bacterial mutagenesis', *Nature* **249**, 850-852.

Bibliography

Biophysik: Das Trefferprinzip in der Biologie Timoféeff-Ressovsky, N.W. and Zimmer, K.G.S. Hirzel, Verlag, Leipzig.	*page* 1, 4, 80, 83
Actions of Radiations on Living Cells (1946), Lea, D.E. Cambridge University Press.	1, 16, 97
What is Life? (1948), Schrodinger, E., Cambridge University Press.	1
'Chromosome Aberrations Induced by Ionizing Radiations' (1962), Evans, H.J., *Int. Rev. Cytol.*, **13**, 221-321.	5
Mutation as a Cellular Process (1969), Wolstenholme, G.E.W. and O'Connor, M., Ciba Foundation Symposium, Churchill; London.	11, 240, 396 417
General Genetics (2nd Edition) (1965), Srb, A.M., Owen, R.D. and Edgar, R.S., Freeman; San Francisco.	18, 161
Mutation: An Introduction to Research on Mutagenesis (1962), Auerbach, C., Oliver and Boyd; Edinburgh.*	23, 73, 75, 76 78, 134, 160, 170, 173, 176, 182
The Molecular Basis of Mutation (1970), Drake, J.W., Holden-Day; San Francisco.	39, 311, 315, 321, 412
Genetic Complementation (1966), Fincham, J.R.S., Benjamin, New York, Amsterdam	40

* This book is out of print but some copies are still available from the author.

Genetic Variations of Drosphila melanogaster (1968), Lindsley, D.L. and Grell, E.H., Carnegie Inst. Washington Publication No. 627. — *page* 67

'Mutation in Eukaryotes' (1971), Auerbach, C. and Kilbey, B.J., *Ann. Rev. Genetics* **5**, 163-218. — 127, 378

'Mechanisms of Radiation Mutagenesis in Cellular and Subcellular Systems' (1969), Bridges, B.A., *Ann. Rev. Nuclear Science,* **19**, 139-178. — 148

Radiation Biology Vols I, II (1954, 55) Hollaender, A., (ed.), McGraw-Hill; New York. — 159, 203, 204, 350

Chemical Mutagens: Principles and Methods for their Detection (1973), Hollaender, A., (ed.), Plenum Press, New York, London. — 160, 256, 308, 309, 336, 461, 465

Molecular Biology of Bacterial Viruses (1963), Stent, G., Freeman; San Francisco. — 173

The Genetics of Bacteria and their Viruses (2nd Edition), (1968), Hayes, W., Blackwell Scientific Publications, Oxford, Edinburgh. — 173, 189

Fungal Genetics (1963), Fincham, J.R.S. and Day, P.R., Blackwell Scientific Publications, Oxford. — 173

Approaches to the Genetic Analysis of Mammalian Cells (1962), Merchant, D.J. and Neel, J.V., (ed), University of Michigan Press; Ann Arbor. — 173

Molecular Genetics (1971), Stent, G.S., Freeman; San Francisco. — 188

Problems of Radiation Genetics (1964), Dubinin, N.P., (translated by Beale, G.H.), Oliver and Boyd, Edinburgh and London. — 203

'Structural Defects in DNA and their Repair in Micro-organisms' (1966), *Rad. Res. Suppl.,* **6**, Academic Press; New York and London. — 234, 244

Cold Spring Harbor Symposium of Quantitative Biology (1968), **33**, Cold Spring Harbor Laboratory, Long Island, New York. — 234

'Ultraviolet-induced Mutation and DNA Repair' (1969), Witkin, E.M., *Ann. Rev. Genetics,* **3**, 525-552. — 234, 386

Genetic and Allied Effects of Alkylating Agents (1966), Loveless, A., Butterworths; London. *Page* 257

Action of Chemicals on Dividing Cells (1966), Kihlman, B.A., Prentice-Hall, Englewood Cliffs, USA. 264

'The Genetic Control of Mutation' (1973), Drake, J.W., (ed.), *Genetics* supplement, **73.** 419

Mutation and Plant Breeding (1961), Publ. 891, Committee on Plant Breeding and Genetics, Nat. Acad. Sciences, Nat. Res. Council, Washington. 440, 453

Induced Mutations and Plant Improvement (1972), ST1/PUB/297 International Atomic Energy Agency, Vienna. 453

Chemical Mutagenesis: A Survey of Literature. (1973) Oak Ridge Nat. Lab., U.S. Atomic Energy Commission. 461

Chemical Mutagens: Environmental Effects on Biological Systems (1970), Fishbein, L., Flamm, W.G. and Falk, H.L., Academic Press; New York and London. 461

Chemical Mutagenesis in Mammals and Man (1970), Vogel, F. and Röhrborn, (ed.), Springer-Verlag, Berlin, Heidelberg, New York. 461, 463

Author Index

A page number in **bold** type designates the page on which the full reference to an author's paper is given.

Subject Index

Page numbers in **bold** type designate sections of text in which the subject is referred to in the prefatory heading or sub-heading.